Marketing Recorded Music

This fourth edition of *Marketing Recorded Music* is the essential resource to help you understand how recorded music is professionally marketed. Updated to reflect the digital era, with new chapters on emerging media, streaming, and branding, this fourth edition also includes strategies for independent and unsigned artists. Fully revised to reflect international marketing issues, *Marketing Recorded Music* is accompanied by a companion website with additional online resources, including PowerPoints, quizzes, and lesson plans, making it the go-to manual for students, as well as aspiring and experienced professionals.

Tammy Donham is Associate Professor of Recording Industry Studies at Middle Tennessee State University. She frequently teaches online courses and was the recipient of the 2016 Outstanding Achievement in Instructional Technology Award. Donham is a graduate of Leadership Music (class of 2012), as well as a member of the Country Music Association.

Amy Sue Macy, Professor in the Department of Recording Industry at Middle Tennessee State University, received both her undergraduate degree in music education and her master's degree in business administration from Belmont University. For 15 years, she worked for various labels, including MTM, MCA, Sparrow Records, and the RCA Label Group.

Clyde Philip Rolston is Professor of Music Business in the Mike Curb College of Entertainment and Music Business at Belmont University. Prior to joining the faculty at Belmont University, he was the Vice President of Marketing at Centaur Records, Inc., where he also engineered and produced many projects, including recordings by the Philadelphia Trio and the London Symphony Orchestra.

"A definitive resource for teaching marketing and promotion in the music industry, professors should consider adding this as required text for their course."

Terrance Tompkins, MBA, Program Coordinator/Assistant Professor, Hofstra University

Marketing Recorded Music

How Music Companies Brand and Market Artists

4th Edition

Tammy Donham, Amy Sue Macy and Clyde Philip Rolston

Focal Press
Taylor & Francis Group

NEW YORK AND LONDON

Cover image: Shutterstock

Fourth edition published 2022
by Routledge
4 Park Square, Milton Park, Abingdon, Oxon, OX14 4RN

and by Routledge
605 Third Avenue, New York, NY 10158

Routledge is an imprint of the Taylor & Francis Group, an informa business

© 2022 Tammy Donham, Amy Sue Macy and Clyde Philip Rolston

First edition published by Routledge 2006
Third edition published by Routledge 2015

British Library Cataloguing-in-Publication Data
A catalogue record for this book is available from the British Library

Library of Congress Cataloging-in-Publication Data
A catalog record for this book has been requested

ISBN: 978-0-367-72117-6 (hbk)
ISBN: 978-0-367-69394-7 (pbk)
ISBN: 978-1-003-15351-1 (ebk)

DOI: 10.4324/9781003153511

Typeset in Kuenstler
by Apex CoVantage, LLC

Access the companion website: www.routledge.com/cw/donham

Contents

Figures

Tables

About the Authors

Tammy Donham received her undergraduate degree in marketing from Western Kentucky University and her master's degree in business administration from Middle Tennessee State University. Donham worked for Fruit of the Loom, Inc., in several marketing capacities before moving to Nashville in 1996, where she worked for nearly 17 years for the Country Music Association (CMA).

She held various marketing-related positions within CMA, ultimately rising to vice-president of marketing. While in her post as VP, she oversaw all marketing, creative services and research efforts for the CMA Awards, CMA Music Festival, and CMA Country Christmas events and television specials, including broadcast, digital, radio, out-of-home, and print initiatives. She was CMA's lead liaison with ABC Television Marketing, Synergy, and Affiliate teams and worked closely with these and other event partners to maximize promotional and brand-building opportunities for CMA properties across all platforms. Donham is a graduate of Leadership Music, as well as a member of the Academy of Television Arts & Sciences and the Country Music Association.

Donham began teaching Marketing of Recordings and Digital Strategies for the Music Business in the Recording Industry Department at Middle Tennessee State University in the fall of 2013.

LinkedIn: linkedin.com/in/tammydonham

Amy Sue Macy, Professor in the Department of Recording Industry at Middle Tennessee State University, received both her undergraduate degree in music education and her master's degree in business administration from Belmont University. For 15 years, she worked for various labels, including MTM, MCA, Sparrow Records, and the RCA Label Group. Amy created strategic marketing plans for launching new releases into the marketplace, working closely with artists and their managers, including Martina McBride, Kenny Chesney, Clint Black, Alabama, and Lonestar. While maintaining a marketing focus, Amy was responsible for all sales at the national retail level with clients including Wal-Mart, Kmart, Target, Best Buy, and Musicland, to name a few, all the while communicating key marketing strategies coast to coast with RCA Label Group's national distributor Bertelsmann Music Group (BMG).

Since securing her teaching gig at MTSU, she has served as the music business internship coordinator for eight years and has taught Marketing of Recordings, Record Retail Operations, Survey of the Recording Industry, the Lecture Series, an Old Time String Band Music Ensemble, and the student-run record label MATCH Records. She also teaches online utilizing the Desire2Learn Online System integrating Web conferencing software, collaborative study through wikis, virtual tests, and links to various websites and live "spreadsheet" activities. She has been the recipient of various teaching awards, including the 2009 Distinguished Educator in Distance Learning and the 2009–2010 Outstanding Achievement in Instructional Technology.

Amy's love of music extends beyond the business world. She initially moved to Nashville to be a musician and has toured professionally with several artists internationally. She is a vocalist and is accomplished on guitar, fiddle, and banjo. Amy has performed in musicals at Nashville's famous Ryman Auditorium as well as the Tennessee Performing Arts Center, has been an artist in residence at the Country Music Hall of Fame, and has recently created a concert series focusing on the "story behind the song" that highlights Civil War and Irish music that has migrated to America.

LinkedIn: linkedin.com/in/amymacy

Clyde Philip Rolston is Professor of Music Business in the Mike Curb College of Entertainment and Music Business at Belmont University. Prior to joining the faculty at Belmont University, he was a vice president of marketing at Centaur Records, Inc. While with Centaur Records, Dr. Rolston engineered and produced many projects, including recordings by the Philadelphia Trio and the London Symphony Orchestra. He is an active member of the Music and Entertainment Industry Educators Association. Dr. Rolston received his bachelor's and master's degrees from Louisiana State University and a PhD in marketing from Temple University. He has taught merchandising and marketing to music business students for over 20 years. His research interests include music marketing, music acquisition and consumption, international music business, and consumer behavior.

For the past three years, he has co-produced with Dan Keen a show for Music City Roots featuring alumni and current students from Belmont University. He has published articles on the music industry in the *MEIEA Journal*, the *Journal of Arts Management, Law and Society*, the *NARAS Journal*, and numerous conference proceedings.

Contributors

Charles Alexander

ORCID #: 0000-0003-0996-0406

Charles Alexander is an independent singer/songwriter, digital strategist, and music technology educator. He is founder of Outside the Box Music, a digital media development, strategy, and artist services company. He is a streaming marketing and strategy specialist. He was responsible for Keeley Valentino generating over 12 million streams on Spotify in one year for her song "Nashville," the first truly independent artist on the platform to do so. Alexander has worked with artists as diverse as Ryan Cabrera, Keb' Mo', Gabe Dixon, and Jackie Venson. He helps artists, musicians and songwriters create and extend their online presence, especially on digital music services and TikTok. He also helps implement strategies and campaigns to increase fan acquisition, fan engagement and brand awareness. He established "Rock The Net," a first of its kind digital music education initiative geared towards educating music industry professionals on how to leverage the internet.

He frequently writes about the future of music marketing technology and trends. He is also an in-demand speaker at events such as South by Southwest, Music Biz, Americana Music Association Conference, and Folk Alliance International.

He is a NewSong Music Competition Regional Finalist and performs at venues such as the Bluebird Cafe in Nashville, Tennessee, on a regular basis.

Michelle Conceison

ORCID #: 0000-0002-2922-4855

Michelle Conceison is an artist manager, entrepreneur, and assistant professor in the recording industry program at Middle Tennessee State University (MTSU). She founded Mmgt in 2004, an artist management and music marketing firm in Nashville. Prior to music, she worked in digital media, leading teams at multiple global advertising agencies. Conceison has served on the boards of Folk Alliance International and International Bluegrass Music Association and leads mentorship programs for Women In Music and East Coast Music Association. Her fields of research include audience and industry stakeholder analysis, marketing, media, merchandising, and leadership in music organizations.

Acknowledgments

Tammy Donham

I would like to thank my co-authors, Amy Sue Macy and Clyde Philip Rolston, for asking me to be a part of this journey and allowing me to experience the joys and challenges of writing a textbook. Your reassurance and guidance were truly comforting.

I have always felt both proud and lucky to have spent the bulk of my marketing career in the industry that I love, but to now be able to teach at the collegiate level is yet another dream come true. Thank you to my colleagues at Middle Tennessee State University for giving me this opportunity and enabling this life I live to unfold.

Thank you to my many industry and Leadership Music friends who contributed to this work. I truly appreciate your willingness to share your thoughts and expertise with me. A huge note of gratitude to our friend Haley Jones at MRC for making magic happen. *You* are a true rock star!

And last, to my husband Chris, who provided both technical and emotional support, as well as my son Keevan, who kindly accepted my missed movie nights. You both supported and encouraged me through the intense research and writing periods, and you make me realize every day how blessed I am.

Amy Sue Macy

I would like to express deep appreciation to both Tammy Donham and Clyde Philip Rolston for going along with this adventure of writing *another* textbook. Thank you to MTSU's Department of Recording Industry Department Chair John Merchant and my many colleagues whose academic integrity is so encouraging. To the students of the RI Department – you keep me motivated by allowing me to be a part of your dreams. Many thanks to my enduring music industry pals who continually pick up the phone and allow me to pick their brains about the current state of the business: Nancy Quinn of Thirty Tigers; David Macias of Thirty Tigers; Rob Simbeck, writer; Danny

Bess, CFO Warner Music Group; Katie McCartney, general manager of Monument Records; Troy Vest of A3 Merchandise; Tim Roberts of CBS Radio; Haley Jones of MRC; Shane Finch of Enco Systems; and the many more friends who are constant encouragers to the effort. A special shout-out to Charles Alexander and Michelle Conceison for their unique contributions to this edition of RMM.

I'm grateful to my sister, Robin, for being so "there" on the westward front. Thanks to my Mom and Dad for giving me the love of learning. Thank you to my family, who has endured more pizza and cereal than one household should. You are so special: Doug, Emma Jo, and Millie Mae. I so appreciate you allowing me to take over the kitchen table to write this edition. Dinner is now served!

Clyde Philip Rolston

I would like to thank my coauthors for their persistence and willingness to always pitch in when I needed them, for taking on the most difficult chapters and delivering such fine work. Many thanks to my students, past and present, and my colleagues who gave feedback and input that make this edition better than the last. A special thanks to Belmont University for allowing me the time off to write this book, specifically Dean Doug Howard and Rush Hicks. Mostly I would like to thank my wife, Allison, whose faith, love, and support keep me focused on what is truly important.

The editors and publishers have made every effort to trace copyrights to their proper holders, including U.S. and world holders for original materials, and would like to gratefully acknowledge permission to reproduce this work.

After this textbook went into production, the data analytics company MRC, who has been the source of data for the *Billboard* charts, has rebranded to the name Luminate. The data sourcing and interface does not appear to be changing, but please note that any mentions in the book to MRC are now in reference to Luminate.

www.luminatedata.xyz

CHAPTER 1

Introduction

Do you need music? Based solely on the fact that you are reading this, it is a safe bet that you answered that question, "Yes!" Do you *really need* music? If you still said yes, then you might just have the right attitude to succeed in the music industry, but the truth is none of us need music in the Maslovian sense. It is not necessary to survive, like food and water. There are other ways to express our emotions or our artistic side, make us feel better or earn a living. But we love music and fill our days with it. Music definitely makes life better, and it is everywhere – movies, television, radio, grocery stores and even our telephones. We listen to music from the time we wake up in the morning until we go to bed at night. We can't imagine doing anything else to make a living.

WHY ARTISTS MIGHT WANT TO SIGN WITH A RECORD LABEL

So, you've made up your mind. You want to be a working musician, writing, recording, and performing your songs for the fans. How many fans? Dozens? Hundreds? Thousands? What does success in the music business mean to you? Will you be content performing locally on the weekends and paying the bills working full-time as an accountant? How about living in a van and cheap hotels, driving from city to city to play in bars and small venues? Or do you have to have the rock 'n' roll lifestyle – the one that allows you to own a thirty thousand–square foot mansion with a pool and a five-car garage with a different exotic car in every bay? To do that, you are going to need a record label.

CONTENTS

But if I do that, I'll have to give up creative control of my music, won't I? I don't want to do that. Besides, I can record my songs on my laptop and distribute them world-wide on Apple Music and Spotify. I don't *need* a record label. Chance the Rapper did it without a record label, didn't he? And he won a Grammy!

Yes, it is true that Chance the Rapper was not signed to a record label, but he did have an exclusive deal with Apple's iTunes. According to iSpot.TV and Phillips (2017), that two-week exclusive on iTunes was in exchange for $500,000 and appearing in an ad that Apple paid for and aired for 10 months on media such as Facebook, Twitter, and YouTube, generating over 4.6 million impressions with music fans. That would kickstart anyone's career.

Of course, Chance is the exception and not the rule. In 2018, Spotify claimed over three million creatives on their site, but at the end of 2020, fewer than 900 were making more than $1 million per year (from publishing and recordings) and about 7,500 were making more than $100,000 per year (Ingham, 2020). That means about 0.28% of the creatives on Spotify were making a living there.

And the competition is getting worse. New tracks are being added to Spotify at the rate of 60,000 songs a day (Ingham, 2020). In 2017, zmonline.com reported that 20% of all the tracks on the site had never been streamed ("Spotify releases playlist"). Still want to get into the music business without a label?

Record labels have the money, the personnel, and the experience to help an artist navigate the complexities of the industry. Being signed to a legitimate label gives the artist instant credibility. It is a vetting process that few get through. With a recording advance, the artist can afford to hire a top-notch producer, and the experienced marketing team will make the artist stand out from the crowd (5 Reasons, 2020).

REVENUE STREAMS

Because record labels provide most of the musical recordings, they play a role in every revenue stream in the music business. If the artist is a songwriter, then they sit atop all three revenue streams, increasing their chances of more income. In this simplified representation of the industry, the revenue streams may appear to be siloed, but in reality, they are very much interconnected. For example, the songwriter in the publishing stream provides the song used by the artist in both the recording stream and the live stream. Recorded music is used by broadcasters and others, generating income for the writer and the publisher, and so on.

The Revenue Streams

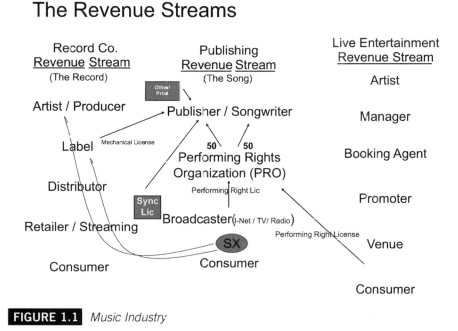

FIGURE 1.1 *Music Industry*

Without the song, there isn't a sound recording. Without the sound recording, there isn't the radio exposure, the download, or the stream. Without that exposure, there isn't the demand for the tour tickets, the venues, or the merchandise. The talent heads each revenue stream, and the consumer anchors them all.

MUSIC EVERYWHERE

Music permeates our lives. It seems to be literally everywhere. We listen to music on its own, but it is also part of almost every other form or entertainment – sporting events, gambling halls – except books and magazines, and they often write about the entertainment industry. Every leisure activity seems to have its own soundtrack. The global market for recorded music was basically flat from 2010 to 2015 and has been growing ever since. Even in midst of the 2020 pandemic, global recorded music grew by over 7% (IFPI, 2021). Media Fuse predicts that by 2026, total recorded music revenue will exceed $1 trillion.

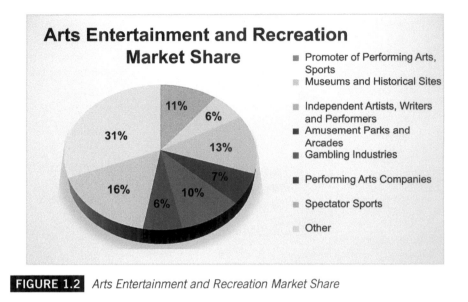

FIGURE 1.2 *Arts Entertainment and Recreation Market Share*

THE NEW DEAL

Most artist contracts now include not just the artist's revenue from selling recordings but from multiple revenue sources, including touring, merchandise, and publishing. The artist agrees to this "360 deal" in exchange for a higher royalty rate, and the label can use income from sources other than record sales to recoup the expenses of making and marketing the recording.

The 360 deal works best if the artist is also the songwriter and understands the value of exploiting other forms of entertainment to generate additional income.

Justin Timberlake is a perfect example. He is a musical artist, songwriter, actor, and movie producer. He wrote and recorded the song "Can't Stop the Feeling" for the movie *Trolls*, in which he voiced the part of Branch and served as the executive music producer. The song was the top-selling song of 2016 in the United States and won numerous awards, including a Grammy and a People's Choice Award. The official video has received nearly 1.4 billion views as of March 2021. Additional music videos were made in support of the song and the *Trolls* movie as well as dozens of fan-produced videos emulating the dance moves of the original song. "Can't Stop the Feeling" garnered more attention when it was used in a flash mob dance video in support of Hillary Clinton's 2016 presidential campaign. The "#Pantsuitpower Flashmob for Hillary" video was viewed over 2 million times on Facebook. The song was licensed for use in several television shows around the world and has been covered in live performances by country artists Hunter Hayes and Lady A. As of March 2021, the song has sold 3.7 million copies, been streamed 1.4 billion times, and been played on radio 7.5 billion times, according to Music Connect. Like Pharrell Williams' "Happy," the song demonstrates how a single recording can lead to a multitude of opportunities for an artist.

PUBLISHING

Artists may earn publishing income if they write or publish the songs they record. In 2018, the U.S. music publishing business generated $3.3 billion in revenue, a near 12% increase over the previous year and a 55% increase since 2014 (Ingham, 2019). Publishers and songwriters earn royalties for every use of their song. This includes performance royalties for every public performance, mechanical royalties for every recording or stream, and synch royalties for use in television, film, and commercials.

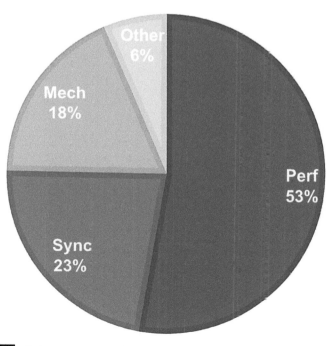

 **FIGURE 1.3** *Publishing Income Sources*

TOURING AND MERCHANDISE

With the decline in physical sales and downloading, artist income comes primarily from streaming and touring. Before the 2020 pandemic hit, the concert business was expecting another banner year. Both IBIS and PricewaterhouseCoopers predicted steady 2% annual growth through 2030 (Young, 2020). Before the industry was shut down by the pandemic, Queen grossed $67.45 million with 16 shows, and Elton John earned $61.85 million on 25 shows. Slipknot managed to squeeze in 28 European shows and grossed $19.82 million before the theaters went dark (2020's Worldwide Year End). According to *Forbes'* list of highest paid celebrities, Elton John has earned the most of any musician in 2020 at $81 million (Forbes, 2020). This, in contrast to his 2019 $194-million tour, serves as a stark reminder that these musicians are the source of income for a lot of other people, and those

Table 1.1	Top-Grossing Tours of 2019			
Rank	**Artist**	**Gross**	**Tix Sold**	**# of Shows**
1	Ed Sheeran	$223,653,796	2,587,445.00	54
2	Pink	$223,653,796	1,816,917.00	68
3	BTS	$196,414,822	1,611,963.00	42
4	Elton John	$193,969,305	1,461,089.00	105
5	The Rolling Stones	$177,806,770	784,652.00	16
6	Metallica	$177,434,939	1,734,159.00	46
7	Ariana Grande	$118,279,032	1,086,969.00	77
8	Paul McCartney	$113,705,245	830,035.00	32
9	Backstreet Boys	$93,310,105	999,242.00	95
10	Kiss	$88,707,212	855,752.00	72
11	Cher	$83,115,897	684,370.00	66
12	Shawn Mendes	$82,678,309	1,120,741.00	90
13	Michael Bublé	$80,055,915	667,927.00	58
14	Jonas Brothers	$78,698,351	727,544.00	51
15	Spice Girls	$78,203,580	697,357.00	13
16	Billy Joel	$76,804,820	598,963.00	22
17	Post Malone	$75,906,144	736,683.00	55
18	Justin Timberlake	$75,650,311	567,965.00	38
19	Eagles	$73,646,789	410,404.00	29
20	Hugh Jackman	$70,710,437	792,537.00	69

(Source: *Billboard* Boxscore)

concert grosses are before all those people get paid. Still, they bring home a pretty good paycheck.

The tour gross numbers do not include another important source of income for the artist (and record labels with 360 deals), and that is music festivals. Many of these festivals are smaller and don't carry the status of a Coachella or Bonnaroo but are outgrossing those high-profile events. The top-grossing festival in 2019 was the Outside Lands Music & Arts Festival, headlined by Paul Simon, Childish Gambino, the Lumineers, and others. It sold 205,500 tickets and grossed $29.6 million. The top 20 festivals combined grossed over $205 million. According to Vivid Seats, the average ticket price for a festival in 2018 was $659 (Vividseats.com), while the average price for concert was about $95 (Cohen, 2019).

Music merchandise was a $3.5-billion industry in 2018 (Millman, 2020), and though it was severely impacted by the 2020 pandemic, it is expected to bounce back along with the touring business. Merchandise company atVenu saw an 11% increase in per head sales in 2019 and a 29% increase in per show gross. Music made up only 10% of merch sales. Venues with a capacity of 500 or less had the third-highest per-person sales average at $6.43 each. The average for a venue of 20,000 plus was $11.40 (atVenu). One of the allures of the music business is the amount of money that can be made if you reach the top. Whether you are the artist or the label, it takes an experienced team of professionals to achieve those goals. Being a part of the team

FIGURE 1.4 *Top-Grossing Tours of 2019*

that fosters a creative community, the satisfaction of sharing in the creation of the music and getting it to the fans is rewarding both financially and personally. The chapters ahead will show you how it is done. It is not hard work if you love what you are doing, and the rewards are incalculable.

BIBLIOGRAPHY

"5 Reasons Record Labels Still Matter." October 19, 2020. From https://mn2s.com/news/label-services/5-reasons-record-labels-still-matter/. Accessed March 27, 2021.

"2019 Music Merch Industry Trends." n.d. *atVenu*. From www.atvenu.com/post/2019-merch-trends Access March 29, 2021.

"2020's Worldwide Year End." n.d. *Touring Data*. From https://touringdata.wordpress.com/2021/02/01/2020s-worldwide-year-end/.

"Apple Music TV Commercial, 'Chance the Rapper: Coloring Book". *iSpot.TV*. From www.ispot.tv/ad/AgLW/apple-music-chance-the-rapper-coloring-book?conv=1.

Cohen, Arianne (2019). "Bad News: You're Going to Shell Out $300+ for Concert Tickets Next Year." December 27, 2019. From www.fastcompany.com/90448082/bad-news-youre-going-to-shell-out-300-for-concert-tickets-next-year. Accessed March 29, 2021.

Forbes (2020). "The World's Highest-Paid Celebrities." From www.forbes.com/celebrities/.

IFPI (2021). Global Music Market Overview, 2021. International Federation of Phonographic Industries. http://www.gmr2021.ifpi.org/report. Retrieved March 12, 2021.

Ingham, Tim (2019). "How Much Money Is the US Music Publishing Industry Making? A Billion Dollars More Than It Was a Year Ago." June 16, 2019. From www.musicbusinessworldwide.com/how-much-money-is-the-us-music-publishing-industry-making-a-billion-dollars-more-than-it-was-four-years-ago/. Accessed March 29, 2021.

Ingham, Tim (2020). "Spotify Dreams of Artists Making a Living. It Probably Won't Come True." August 3, 2020. From www.rollingstone.com/pro/features/spotify-million-artists-royalties-1038408/. Accessed March 28, 2021.

Millman, Ethan. "How Do You Sell Concert Tees without Concerts? Merch Companies Are Up Against a Wall." *Rolling Stone*, April 28, 2020. https://www.rollingstone.com/pro/features/music-merchandisers-fighting-coronavirus-pandemic-990652/.

Phillips, Amy (2017). "Chance the Rapper Explains How He's Still Independent, Despite Apple Music Deal." March 17, 2017. From https://pitchfork.com/news/71701-chance-the-rapper-explains-how-hes-still-independent-despite-apple-music-deal/. Accessed March 28, 2021.

"Spotify Releases Playlist of All the Songs That No One Has EVER Played Before." June 16, 2017. From www.zmonline.com/the-latest/spotify-releases-playlist-of-all-the-songs-that-no-one-has-ever-played-before/. Accessed March 28, 2021.

Young, Clive (2020). "Concert Industry Continues Growth in 2020." *ProSound News, January 2, 2020*. From www.prosoundnetwork.com/live/concert-industry-continues-growth-in-2020. Accessed March 29, 2021.

CHAPTER 2

Marketing Concepts and Definitions

In this chapter, we look at some marketing theories and practices commonly taught in introductory marketing courses and how they apply to the music industry. We also look at how the Internet has impacted some of those practices.

SELLING RECORDED MUSIC

The concept of selling recorded music has been around for more than a century. While the actual storage medium for music has evolved, from cylinders to vinyl discs, to magnetic tape, digital discs, and now downloads and streaming services, the basic notion has remained the same: a musical performance is captured to be played back later, at the convenience of the consumer. Music fans continue to enjoy the ability to develop music collections, whether in the physical format or collections of digital files stored in the cloud. Consumers also enjoy the portability afforded by contemporary music listening devices that make their playlists available virtually everywhere. The ways in which consumers access music have undergone significant changes as CD and download sales have declined and vinyl and cassettes have made a comeback. According to the RIAA's 2020 Year-End Music Industry Revenue Report, 90% of the $12.1 billion in retail recorded music revenues came from streaming (Friedland, 2021).

Regardless of the medium, the importance of the recording cannot be understated. According to BMI's 2018–2019 Annual Report, 98% of performance royalties were generated by digital performances. That means that

CONTENTS

DOI: 10.4324/9781003153511-2

9

of the 2.19 trillion performances processed by BMI during the period, over 2 trillion were of **recorded** music. So, despite all the emphasis placed on live performances, they generate only 2% of performance royalties.

Music Purchase vs. Music Consumption

Music purchases should not be confused with music consumption. The consumption of music in the United States increased 11.6% by the end of 2020, thanks primarily to a 17% increase in audio streaming, but sales revenues fell for digital songs and albums (22.3% and 12.5%, respectively) and physical albums (7.4%) despite a 46.2% increase in vinyl sales (MRC, 2021).

Thus, recorded music is finding ways to make money, much the same as television programming has done for over 70 years. For much of this time, the television programming industry relied solely upon advertising revenue to fund some of the most popular television shows in history. Other, more recent forms of revenue have come from premium (fee-based) programs, physical sales of shows (DVDs), and network-owned streaming services. Fee-based programming did not occur until the premium channels such as HBO and Showtime started developing their own proprietary shows. Even with these other forms of income, the bulk of revenue for producing television content still comes from advertising. Streaming services, however, use a different business model. Netflix relies entirely on revenue from subscriptions, while Hulu offers both ad-supported and ad-free plans.

As the paradigm shifts from physical sale of recordings to a more complex model of generating revenue, marketing efforts must also evolve to respond to the plethora of income possibilities.

WHAT IS MARKETING?

In today's marketplace, the consumer is showered with an array of entertainment products from which to choose, making the process of marketing more important than ever. Competition is fierce for the consumers' entertainment budget. But before explaining how recorded music is marketed to consumers, it is first necessary to gain a basic understanding of marketing.

Kotler and Armstrong define marketing as "the process by which companies create value for customers and build strong customer relationships in order to capture value from customers in return." Marketing involves satisfying customer needs or desires. To study marketing, one must first understand the notions of *product* and consumer (or *market*). The first questions a marketer should answer are, "What markets are we trying to serve?" and, "What are their needs?" Here we are using the term "need" not in the sense that Maslow defined needs (see Chapter 3) but in the generic sense that includes wants and desires, not just needs. Marketers must understand these consumer needs and develop products to satisfy those needs. Then, they must price the products effectively; make the products available in the marketplace; and inform, motivate, and remind the customer of their availability. In the music business, this involves supplying consumers with the recorded music they desire where they want it and at a price they are willing to pay.

Markets used to be physical places where buyers and sellers met, but today they are just as likely to be a website on the Internet or an app on your phone. A **market** is defined as a group of consumers who want or need your product and who have the willingness and ability to buy. This definition emphasizes that the consumer wants or needs something in contrast to the market being a physical place or an economic system. That want or need is satisfied by a product.

A **product** is defined as "something that is produced and offered for sale in the market" (Luten, 2020). The product is designed to satisfy a want or a need of the market. You may want a candy bar but not necessarily need one. You may need surgery but not necessarily want it. As we will see, that product may not be a physical good.

THE MARKETING MIX

The marketing mix, often called the Four Ps (or the five, six, or even seven Ps if you want to add emphasis on people, process, and/or presence to the traditional model), refers to a blend of product, distribution (place), promotion, and pricing strategies designed to produce mutually satisfying exchanges with the target market (Luten, 2020). Let's take a closer look at each of the components of the marketing mix.

PRODUCT

The marketing mix begins with the product. It would be difficult to create a detailed strategy for the other components without a clear understanding of the product to be marketed. Record label marketers are, however, often asked to create a basic marketing mix and marketing plan knowing very little about the final project because the music has not yet been recorded or mastered. Adjustments to the marketing mix can be made as the product develops and in response to the market.

An array of products may be considered to supply a particular market. Then the field of potential products is narrowed to those most likely to perform well in the marketplace. In most industries, this function is performed by the Research and Development (R&D) arm of the company, but in the music business, the artist and repertoire (A&R) department performs this task by searching for new talent and helping decide which songs will have the broadest consumer appeal. An additional challenge for the music industry is that many musicians resist the idea of change or creating for the market. Their position is closer to "This is my music. Take it or leave it." So, we typically start with a product (the artist and their music) and then try to match them with an existing market or markets.

PRODUCT POSITIONING

New products introduced into the marketplace must somehow identify themselves as different from those that currently exist. Marketers go to great lengths to **position** their products to ensure that their customers perceive their product as more suitable for them than the competitor's product. **Product positioning** is defined as the customer's *perception* of a product in comparison with the competition. In other words, product positioning takes place in the consumer's mind and it is relative to the other product offerings. From the label's perspective, this is achieved primarily through song selection, advertising, and publicity.

Consumer tastes change over time, and, as a result, new products must constantly be introduced into the marketplace. New technologies render old products obsolete and encourage growth in the marketplace. For example, the introduction of the compact disc (CD) in 1983 created opportunities for the record industry to sell older catalog product to customers who were converting their music collections from vinyl LPs to CDs. Similarly, when a recording artist releases a new recording, marketing efforts are geared toward selling the new release rather than older recordings, although the new release usually creates some consumer interest in earlier releases, and they may be featured alongside the newer release at retail.

THE PRODUCT LIFE CYCLE

The **product life cycle (PLC)** is a concept used to describe the course that a product's sales and profits take over what is referred to as the *lifetime of the product*, from its inception to its removal from the market.

It is characterized by four distinct stages: introduction, growth, maturity, and decline. Preceding this, before the product is introduced into the marketplace, is the *product development* stage. Though most marketing texts ignore the development stage, this is a period of important activity for entertainment marketers. Even before the product, a single or EP, is released to the market, promotion activities are underway at both the consumer and industry level, creating a buzz and increasing the chances of success for the product.

For most products, the *introduction* stage is typically a period of slow growth as the product is introduced into the marketplace. Profits are nonexistent because of heavy marketing expenses, and production expenses are yet to be recouped. The *growth* stage is a period of rapid acceptance into the marketplace and profits increase. *Maturity* is a period of leveling in sales, mainly because the market is saturated – most consumers have already purchased the product. Marketing is more expensive (to the point of diminishing returns) as efforts are made to reach resistant customers and to stave off competition. *Decline* is the period when sales fall off and profits are reduced. At this point, for most products, prices are cut to maintain market share, but the music industry is different. Recorded music is typically discounted early in the PLC to achieve a higher chart position and reward early adopters (Kotler and Armstrong, 1996). The PLC can be applied to individual products (a particular album), product forms (artists and music genres), and even product classes (CDs, vinyl, or downloads), which are referred to as formats in the music business. Product formats have the longest life cycle – the compact disc has been around since the early 1980s (and is currently well into the decline phase). However, the life cycle of a music album is now a matter of months (down from two years). It is at this point that the label will generally terminate actively marketing the album. When a product reaches the decline stage, the company may withdraw it from the market or, as in the case of vinyl records, efforts or circumstances may lead to a revival of the product that will sustain a smaller number of competitors for an extended maturity stage.

Not all products will have the nice bell-shaped curve. A hit song may have a very steep, short introduction and growth stage and just as steep and quick decline, looking more like an inverted V than a bell. A product class, like vinyl records, may never be withdrawn from the market but, instead, achieve a mature stage where sales level off and maintain a steady, albeit lower, level of sales for many years

DIFFUSION OF INNOVATIONS AND ADOPTER CATEGORIES

Diffusion of innovations is defined as the process by which the use of an innovation is spread within a market, over time, and over various categories of adopters (*Dictionary of Marketing Terms*, American Marketing Association, 2008). The concept of diffusion of innovations describes how a product typically is adopted by the marketplace and what factors can influence the rate (how fast) or level (how widespread) of adoption. The rate of adoption is dependent on *consumer traits*, the *product*, and the company's marketing efforts.

Consumers are considered **adopters** if they have purchased and used the product. Potential adopters go through distinct stages when deciding whether to adopt (purchase) or reject a new product. In one model of this process, the stages are referred to as *AIDA*, which under one scenario (purchases driven by affect or emotion) is represented as Attention, Interest, Desire, and Action. Another scenario (cognitive or driven by logic) uses Awareness, Information, Decision, and Action. These stages describe the psychological progress a buyer must go through to get to the actual purchase.

First, a consumer becomes *aware* that they need to make a purchase in this product category. Perhaps the music consumer has grown bored with his or her collection and directs their *attention* toward buying more music. The consumer then seeks out *information* on new releases and begins to gain an *interest* in something in particular, perhaps after hearing a song on the radio or attending a concert. The consumer then makes the *decision* (and *desires*) to purchase a particular recording. The *action* is the actual purchase. Some products (like a car or music to study by) have elements of both the emotional and the logical models present in the decision-making process because their purchase invokes both emotional and rational responses.

The AIDA Model

Awareness → Information → Decision → Action

Cognitive

Awareness → Interest → Desire → Action

Emotional

FIGURE 2.1   *The AIDA Model*

Cognitive vs. Emotional Decisions

Much research has been done in the field of advertising on the relative roles of affective (emotional) factors vs. cognitive factors in motivating consumers. Cognition-based attitudes are thoughts and beliefs about something – an "attitude object." Affect-based attitudes are emotional reactions to the attitude object. Attitudes can have both of these components and can be based more or less on either cognitions or affect. While there is probably an element of both in each purchase decision, the relative contribution of each, as well as the process involved, may differ depending upon the type of product and the level of involvement. It is possible to love a particular recording of a song but not like the recording artist. Consequently, you may listen to the song when it streams but refuse to buy concert tickets from that artist.

An alternative to the AIDA model, the hierarchy of effects model consists of three major stages: *cognition* (awareness or learning), *affect* (feeling, interest or desire), and *conation* (action or behavior) – CAB. The order in which these stages are processed is subject to the type of product and the level of involvement. According to Richard Vaughn (Feb/Mar, 1986), product category and level of involvement may determine the order of effects as well as the strength of each effect. His model uses involvement (high/low) and think/feel (cognitive or affective components) as the two dimensions for classifying product categories and ordering these three steps

The resulting model is a four-quadrant grid with quadrant one (informative) containing high involvement cognitive products, quadrant two (affective) for high involvement emotional purchases, quadrant three (habitual) for low involvement rational purchases, and quadrant four (satisfaction) for low-involvement emotional purchases. Music purchases generally fall into the satisfaction quadrant, although as prices increase, so does the level of involvement. Some music purchases can be considered impulse buys, if correctly positioned at retail. Other purchases may require creating the emotional (affective) response in the customer. Since music by its very nature, evokes emotional responses, exposure to the product is known to increase purchase rate.

Types of Involvement: Purchase and Product

The purchase process may take only seconds or may take weeks, depending on the importance of the decision and the risk involved in making the wrong decision. **Purchase involvement** refers to the amount of time and effort a buyer invests in the search, evaluation, and decision processes during the purchasing process. If the consumer is not discriminating or the consequences of a poor decision are not a major financial or social risk, then the process may take only moments and is thus a low-involvement decision. On the other hand, if the item is expensive, such as a new car, or highly visible, like clothing, and if the consequences of a wrong decision are severe, then the process may take much longer. Several factors influence the consumers' level of involvement in the purchase process, including previous experience, ease of purchase, product involvement, and perceived risk of negative consequences. As relative cost increases, so does the level of involvement.

Table 2.1	Product Involvement: A.k.a. the Collector

PRODUCT INVOLVEMENT: A.K.A. THE COLLECTOR

"Product involved" consumers tend to be early adopters of new products. This is the guy in your high school class who spent all his money from his part-time job at the record store on music and was the first to hear of new groups before they became famous. Product involvement is product specific, and product-involved consumers often become *opinion leaders* for their peer group. You might ask your friend who works in the record store his advice on music but not on fashion, since his wardrobe consists entirely of jeans and rock-related t-shirts. Product involvement is different from purchase involvement. Product involvement tends to be more central to one's life or self-image – he's the car guy or the jazz guy. Purchase involvement is temporary or short term and tends to be highest when there is social or economic risk to the decision. You may have done a lot of research and gathered a lot of information before you bought your car to be sure it was reliable and economical, but now that you have made your choice, your identity and leisure time don't revolve around cars. You don't read and talk cars all the time. It's just a tool, a way to get to where you need to go. If you are "the car guy" and subscribe to several car magazines; watch car shows; and basically eat, drink, and sleep cars, then you are product involved. In the entertainment world, product-involved consumers are the collectors: the person who has ten recordings of the same song because there are (minute) differences in every one.

One major advantage of being product involved is that it may decrease purchase involvement, the time spent making the decision, because you routinely spend the time and effort gathering all the information that is needed to make your selection.

Adopter Categories

Consumers who adopt a new product (make a purchase) are divided into five categories: innovators, early adopters, early majority, late majority, and laggards.

Innovators are the first 2.5% of the market (Rogers, 1995) and are eager to try new products. Innovators are above average in income, and thus the cost of the product is not of much concern (low financial risk). They are also less concerned with social and group norms and therefore perceive less social risk in trying new products or experiences. *Early adopters* are the next 13.5% of the market. Early adopters are more socially involved and are considered opinion leaders, but this makes them less willing to take social risks than the innovators. The enthusiasm of early adopters for a new product will do much to assist its diffusion to the majority; thus, marketers focus a lot of energy and effort on them.

The *early majority* is the next 34% and will weigh the merits of the product before deciding to adopt. They rely on the opinions of the early adopters

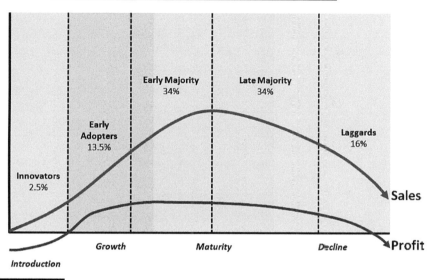

Classic PLC and Product Adoption Model Overlay

FIGURE 2.2 *Product Life Cycle*

in making their decision. The *late majority* represents the next 34%, and these consumers adopt after most of their friends have. The *laggards* are the last 16% of the market and generally adopt only when they feel they have no choice. Laggards adopt a product when it has reached the maturity stage and is being "deep discounted" or is widely available at discount stores. When introducing a new product, marketers target the innovators and early adopters. They will help promote the product through word of mouth. How the adopter categories drive the product life cycle is shown in Figure 2.2. Not shown are the non-adopters. Some people, no matter how much you try to convince them of what they are missing out on, will never like jazz, opera, or EDM. They are not in that target market and therefore will never buy that product and are classified as non-adopters.

INFLUENCES ON ADOPTION

Products (or innovations) also possess characteristics that influence the rate and level of adoption. Those include:

> "Relative Advantage: the degree to which an innovation is perceived as better than what it supersedes. Cassette tapes were perceived to be superior to vinyl because they could be played on portable devices.

> Compatibility: the degree to which an innovation is perceived as consistent with existing values and experiences. Using a 4K TV is no different than using an HD TV, it just looks better. No behavioral change was required of the consumer to use the new product.
>
> Complexity: the degree to which an innovation is perceived as difficult to understand and use. If your VCR is still flashing 12:00 then you are not an early adopter and complexity may be an issue that keeps you from adopting new technology.
>
> Trialability: the degree to which an innovation may be experienced without purchase or long-term commitment. This is why artists should give away free songs and allow sampling of every song on their new albums.
>
> Observability or Communicability: the degree to which the results of an innovation are visible to or can be communicated to others.
>
> The more easily you can see, hear, or describe the difference the faster consumers can make a decision to adopt, or not.
>
> Marketers can improve sales numbers by ensuring the product has a relative advantage, that it is compatible, that it is not complex, and that consumers can observe and try it before they purchase
>
> (Rogers, 1995)

One way marketers can increase the potential for success is by allowing customers to "try before they buy." First, listening stations in retail stores and online music samples increased the level of trialability of new music. Then streaming came along, and music fans could listen to entire songs or albums before making the decision to buy.

HEDONIC RESPONSES TO MUSIC

Researchers in the fields of psychology and marketing strive to understand the hedonic responses to music. **Hedonic** is defined as "of, relating to, or marked by pleasure." Hedonic products are those whose consumption is primarily characterized by an affective or emotional experience. It is a study of why people enjoy listening to music and what motivates them to seek out music for this emotion-altering experience. Recorded music is considered a tangible hedonic product, compared to viewing a movie or attending a concert, which is an intangible hedonic product. "The purchase of a tangible hedonic portfolio product, such as a CD, gives the consumer something to take home and experience at her convenience, possibly repeatedly" (Moe and Fader, 1999). MP3s, then streaming, fueled the increase in **consumption** of music, even though sales of recorded music have been declining steadily for over ten years. Still, Styvén (2010) found that high-involvement music consumers preferred physical product while embracing digital for its convenience.

Consumers like music for a variety of reasons, mostly connected to emotions or emotional responses to social situations involving music. A brief glance through articles and studies offers the following reasons.

- To evoke an emotional feeling or regulate a mood
- To evoke a memory or reminisce
- As a distraction or escape from reality
- To create a mood in an environmental setting or a cultural/sporting event
- To combat loneliness and provide companionship
- To foster social interaction with peers
- To calm and relax
- To stimulate (such as to stay awake while driving)
- For dancing or other aerobic activities
- To enhance/reinforce religious or cultural experiences
- For therapeutic purposes (such as pain reduction)
- To pass the time while working or waiting

By understanding the motivations that drive consumers to purchase or consume music, marketers can be more effective in providing the right music to the right customers at the right time. Even retail placement of music can benefit from understanding the context in which the music will be consumed. For example, music designed to inspire sports fans may be made available in locations that fans are likely to visit on their way to or while attending a sporting event.

PRICING STRATEGIES

Pricing is more complex than how much it cost to make the product and setting a selling price. Pricing structure must consider not only economic costs but market influences and business practices as well. Once the wholesale and retail pricing are determined, price-based incentives must then be considered. For example, should the product be put on sale and if so, when? Should coupons be issued? Should the retailers receive wholesale price breaks for quantity orders or other considerations?

There are generally three methods for deciding the retail price of a product: cost-based pricing, competition-based pricing, and consumer-based (value-based) pricing.

Cost-Based Pricing

Cost-based pricing is achieved by determining the cost of product development and manufacturing, marketing and distribution, and company overhead. Then an amount is added to cover the company's profit goal.

$$\text{Break-even point (units)} = \frac{\text{Total fixed costs}}{\text{Price} - \text{Variable costs per unit}}$$

The weakness with this method is that it does not consider competition and consumer demand.

When determining cost-based pricing, consideration must be given to the fixed costs of running a business, the variable costs of manufacturing products, and the costs related to marketing and product development. Fixed costs include items such as overhead, salaries, utilities, mortgages, and other costs that are not related to the quantity of goods produced and sold. Variable costs are usually associated with manufacturing and depend on the number of units produced. For recorded music, that would include discs, CD booklets, jewel cases, other packaging, and royalties. However, within the marketing budget of a record label, the costs of manufacturing and mechanical royalties are considered fixed once the number of units to be manufactured is determined and the recording costs have been computed. Then there are other (semi) variable costs, which can vary widely but are not directly related to the quantity of product manufactured. They would include recording costs (R&D) and marketing costs. For purposes of the formula that follows, the costs of marketing are considered variable costs, and the recording and manufacturing costs, having already been determined for the project, are considered fixed at this point.

$$\text{Break-even point (units)} = \frac{\text{Total fixed costs}}{\text{Price} - (\text{total variable costs/units})}$$

One of the major benefits of digital recording and distribution is the decreased cost of manufacturing. With no physical product to ship, store, break, package, or display, costs can be significantly reduced.

Competition-Based Pricing

Competition-based pricing attempts to set prices based on those charged by the company's competitors rather than demand or cost considerations. The company may charge more, the same, or less than its competitors, depending on its customers, product image, consumer loyalty, and other factors.

Table 2.2 Standard Business Model vs. Record Label Business Model			
	Variable Costs	**Fixed Costs**	**Semi-Variable**
Standard Business Model	Manufacturing cost per unit; royalties, licensing, or patent fees per unit; packaging per unit; shipping and handling per unit	Overhead, salaries, utilities, mortgages, insurance	Marketing, advertising, research and development
Record Label Business Model	All marketing aspects, including advertising, discounts, promotion, publicity expenses, sales promotions, and so on	Overhead, salaries, utilities, mortgages, insurance, manufacturing costs, production costs (A&R)	

Matching the competitor's price takes the cost of the product out of the decision-making process so that the consumer need only evaluate other product considerations, such as reliability or reputation. Pricing relative to the competition also impacts the brand image of the product. If the product is priced higher, it is likely to be perceived as higher quality, and if the product is priced lower than the competition, it is likely to be considered a bargain brand.

Companies are prohibited from coordinating pricing policies to maximize profits and reduce competition. Such **price fixing** is illegal under the Sherman Act (1890), which prevents businesses from conspiring to set prices (Finch, 1996). The Robinson-Patman Act (1936) prohibits any form of price discrimination, making it illegal to sell products to competing buyers at different prices unless those price differentials can be justified. But neither of these laws keeps competitors from shopping the competition and adjusting their prices accordingly, nor does it keep a major retailer like Wal-Mart or Apple from dictating prices.

Value-Based (Demand) Pricing

Demand-based pricing uses the buyer's perceptions of value to set prices beyond the cost plus a mark-up. The music business has, at times, sought to maintain a high perceived value for music by tweaking and repackaging the product. Many artists release special editions or boxed sets of albums in addition to the standard versions. These special editions offer features not found in the standard version at a higher price and profit. Even indie artists will create vinyl versions of their latest album because of the revenue it can generate at the merch table. It also gives hardcore fans a tangible connection with the artist. Electronic "digital booklets" may be made available to consumers who download albums. These digital booklets, usually in PDF file form, contain the artwork, liner notes, and lyrics normally found in a CD booklet and possibly other extras.

Consumers are willing to pay a premium price beyond the economic value of these special editions because they are limited releases or have exclusive content.

This same concept is applied in the concert side of the industry, which is selling deluxe packages that may include special merchandise, meals and a meet and greet with the artist, priced well beyond the actual costs and any normal profit, just because fans are willing to pay for the experience.

Crowdfunding

The use of crowdfunding to finance new projects continues to be popular with indie artists. Crowdfunding is the collecting of money from multiple sources to fund the production or venture rather than getting the funding from a single source. Crowdfunding replaces the label as the "investor" and gives the artist greater control over the project. Several companies offer crowdfunding services to would-be artists, including Patreon, Indiegogo, and ArtistShare. The most popular is Kickstarter. For their services, Kickstarter takes 5% of the money from successfully funded projects. Their financial partners who process the contributions, PayPal in the United States, take another 5%. Kickstarter only collects money from project supporters if the fundraising goal is met; however, competitors like Indiegogo have options that allow the artist to keep whatever is raised, less the fees, of course. To attract supporters, artists typically offer incentives ranging from a free download for a $1 donation to house party for a donation in the thousands.

Crowdfunding has not been without its controversies. Because the company gets a percentage of the money raised, there is no incentive for them to cut off the fundraising when the artist reaches their goal. In the case of former Roadrunner Records artist Amanda Palmer, more than 10 times the original goal was raised. This raises ethical issues for both the artist and for Kickstarter (Clover, 2012).

CROWDFUNDING: THE CASE OF AMANDA PALMER AND THE GRAND THEFT ORCHESTRA

On April 30, 2012, recording artist Amanda Palmer launched a Kickstarter campaign to raise $100,000 to release a new album. Palmer offered incentives to contributors based on the amounts of their gifts, from a download for $1; to a CD and art book for $125 to $5,000 for a show at your house; or $10,000 for dinner with her, and she would paint your portrait. Palmer raised over $379,000 from more than 6,600 supporters in just two days. When the campaign ended on May 31, 2012, Palmer had raised more than $1.19 million and had over 24,000 backers donating from $1 to $10,000. The album debuted at #10 on the *Billboard* charts. In response to a question on Twitter, Palmer posted the following on her blog on May 13, 2012:

> first i'll pay off the lovely debt – stacks of bills and loans and the like – associated with readying all of the stuff that had to happen BEFORE i brought this project to kickstarter. for the past 8 months or so, i wasn't touring – and therefore wasn't making much income – but every step of the way, there were expenses. so, during that time, i borrowed from various friends and family who i'd built up trust with over the years.

(Amanda Palmer, 2012)

Palmer goes on to lay out the estimated expenses for the entire project, including incentives that were part of the fundraising effort. In the end, she claimed she would do well if she had $100,000 left over after the entire project was completed.

Palmer then set out on a tour to promote the album, but not before putting out a request for "professional-ish" musicians to join the band and play for free. "We will feed you beer, hug/high-five you up and down, give you merch, and thank you mightily." She claimed that she did not have the resources to pay them. Her request was met with a huge backlash, including a petition posted at change.org and protests from musician unions (change.org, 2012).

Kickstarter also faced criticism for allowing Palmer to raise more funds than requested. Some critics argued that contributions should be limited to the amount requested to fund the project. Kickstarter posts the running total so that contributors can know how much money has already been raised before deciding to donate funds. Others argued that Kickstarter.com should not permit "established artists" to post funding requests, because the platform was created to allow aspiring artists to find support for their projects. Eventually the company responded with a press release defending its position. One industry analyst said,

> the Kickstarter team tried to shoot down one of the biggest complaints about celebrities using Kickstarter: that it takes away attention and funding from other

worthy projects from lesser-known people on website. . . . The Veronica Mars and Zach Braff projects have brought tens of thousands of new people to Kickstarter

the founders wrote. But the analyst adds,

Even if it's the case that famous people like Braff end up attracting new donors to the website, there's still a more fundamental question about whether the presence of these celebrities ends up encouraging or discouraging more people to take a chance on launching a campaign of their own.

(Fiegerman, 2013)

Of course, Kickstarter receives 5% of the money from funded projects, so they have a lot to gain from allowing larger sums to be raised. Only 43.94% of all project offerings have been successfully funded (www.kickstarter.com).

Amanda Palmer defended her actions as "the future of music." According to Palmer, "we're moving to a new era where the audience is taking more responsibility for supporting artists at whatever level" (Peoples, 2013). Palmer argues that the social media allows for an unprecedented connection with fans, that now the artist has opportunities to connect with fans on a personal basis, which is more important than the traditional music-industry based formula for artist control.

– From "Crowd Funding: A Case Study at the Intersection of Social Media and Business Ethics," Barry L. Padgett and Clyde Philip Rolston, Journal of the International Academy for Case Studies, Allied Academies, 2014).

Name Your Own Price

Radiohead was one of the first to use "name your own pricing" methods to sell their albums. After a small fee of about $.25, the fan could pay whatever price they wanted for a download of the entire *In Rainbows* album. The idea was great publicity, and it probably made the band more money than it would have made with the record label; still, **most people paid less than three dollars for the download.** Bandcamp, the online marketplace, offers artists the option of fixing the price for their music or allowing buyers to name their own price.

Both Palmer and Radiohead have the advantage of having once been on a major label and household name recognition. For a new or independent artist, these types of funding may be difficult to achieve.

Promotion

Promotion includes the activities of advertising, personal selling, sales promotion, and public relations. It involves informing, motivating, and reminding the consumer to purchase the product. In the recording industry, the traditional methods of promotion include radio promotion (getting airplay), advertising, sales promotion (working with retailers), publicity, tour support, and street teams. Positions dedicated to the Internet and social media marketing and co-branding, creating tie-ins with non-musical products, are relatively new.

Basic Promotion Strategies: Push vs. Pull

There are two basic promotion strategies: *push promotion* and *pull promotion* (Figure 2.3). A **push strategy** involves pushing the product through the distribution channel to consumers. Marketing activities are directed at motivating channel members (wholesalers, distributors, and retailers) to carry the product and promote it to consumers. In other words, wholesalers would be given incentives to persuade retailers to order and sell more product. This can be achieved through offering monetary incentives, discounts, free goods,

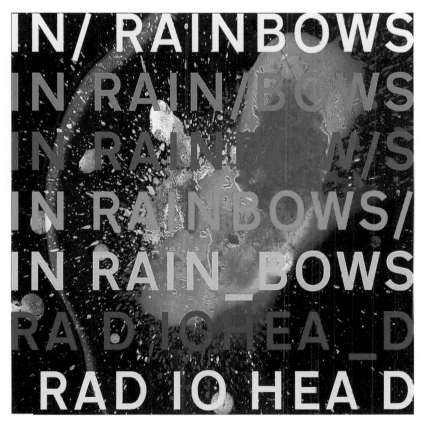

FIGURE 2.3   *Radiohead's* In Rainbows

advertising allowances, contests, and display allowances. All marketing activities are directed toward these channel members and are regarded as "trade" promotion and "trade" advertising. With the push strategy, channel members are motivated to "push" the product through the channel and ultimately on to the consumer.

Under the **pull strategy**, the company directs its marketing activities toward the final consumer, creating a demand for the product that will ultimately be fulfilled as requests for product are made from the consumer to the retailer and then from the retailer to the wholesaler. This is achieved by targeting consumers through advertising in consumer media and creating "consumer promotions." With a pull strategy, consumer demand pulls the product through the channels. Years ago, when MTV still played music videos and cable TV capacity was limited by an analog signal, they employed a pull strategy to get onto more cable systems, imploring teenagers to call their cable operators and tell them "I want my MTV." The campaign was so creative that it was immortalized in Dire Straits's song "Money for Nothing."

Different record companies have different philosophies on the emphasis of the two strategies. Some companies employ a balanced combination of consumer advertising and promotions (coupons and sale items) and trade promotion (incentives), while others focus primarily on consumers and a pull strategy, relying on the sales and promotions departments to win over retailers and radio, respectively.

Place

This aspect of marketing involves the process of distributing and delivering the products to the consumer. Distribution strategies entail making products available to consumers when and where they want them. The focus should be on consumer convenience. The various methods of delivery are referred to as **channels of distribution**. The process of distribution in the record business has undergone an evolution as digital downloads, then streaming, overtook the market share for physical recorded music products. Digital delivery had a dramatic effect on the physical marketplace, eroding market share, closing many retail music chains, and causing big box stores to drastically reduce or eliminate the floor space dedicated to selling recorded music.

Online retailers suffered a decline of their own as consumers moved from downloading to streaming. *Billboard* reported the first digital music sales decrease in 2013. In 2020, global digital download sales declined 18% over the previous year, while streaming revenue increased nearly 13.4% year over year. Overall, music revenue in the United States was up nearly 9.2% during the same time period, driven in part by a 28.7% increase in vinyl sales and a 13.4% increase in music streaming. Collectively, sales of physical product

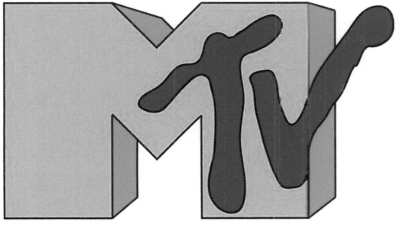

FIGURE 2.4 *MTV Logo*

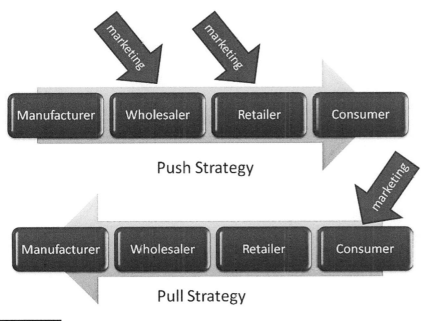

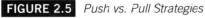

FIGURE 2.5 *Push vs. Pull Strategies*

declined 17.4% compared to the previous year (Friedland, 2021). Worldwide, the music industries earned and estimated $20.2 billion in total revenue in 2019 (Ingham, 2020). The U.S. industry revenue for 2020 was an estimated $12.2 billion (Friedland, 2021).

DISTRIBUTION SYSTEMS

Most distribution systems are made up of **channel intermediaries** such as wholesalers and retailers. These channel members are responsible for aggregating large quantities and assortments of merchandise and dispersing smaller quantities to the next level in the channel (such as from manufacturer to wholesaler to retailer). Manufacturers engage the services of distributors or wholesalers because of their superior efficiency in making goods widely accessible to target markets (Kotler, 1980).

The effectiveness of intermediaries can be demonstrated in the following charts. In the first example, no intermediary exists and each manufacturer must engage with each retailer directly. Thus, the number of contacts equals the number of manufacturers multiplied by the number of customers or retailers (M × C). In the second example, a distributor is included with each manufacturer, and each customer contacts only the distributor. The total

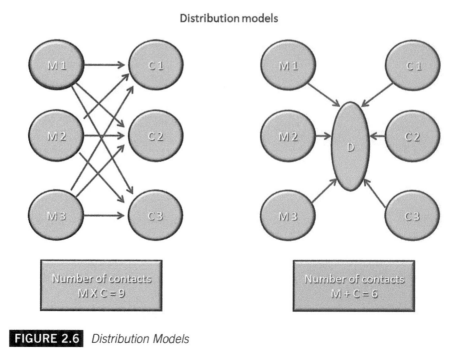

FIGURE 2.6 *Distribution Models*

number of contacts is the number of manufacturers plus the number of customers (M + C). The use of a distributor reduces the number of transactions by one third.

Types of Distribution Systems

Three basic types of physical distribution systems are corporate, contractual, and administered. All three are employed in the record business.

Corporate systems involve having one company own all distribution members at the other levels in the channel. A record label would use a distribution system owned by the parent company, as well as a company-owned retailer, in a fully integrated corporate system. In reality, the major record labels do not own record retailers, but they do own their own distribution. Such is the relationship between Mercury Records and Universal Music Group Distribution. Both companies are owned by Universal Music Group. This type of ownership is referred to as **vertical integration** or a **vertical marketing system** (VMS). When a manufacturer owns its distributors, it is called **forward vertical integration**. When a retailer such as Wal-Mart develops its own distribution or manufacturing firms, it is called **backward vertical integration**.

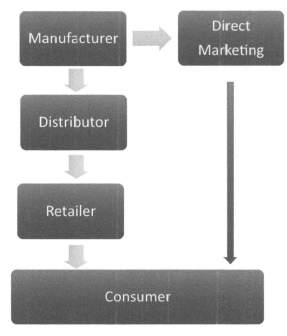

 *Distribution Chain*

Contractual distribution systems are formed by independent members who contract with each other, setting up an agreement for one company to distribute goods made by the other company. Independent record labels commonly set up such agreements with independent record distributors. Before the major record labels developed in-house distribution systems in the 1960s and 70s, nearly all recorded music was handled through this type of arrangement.

In **administered distribution systems**, arrangements are made for a dominant channel member to distribute products developed by an independent manufacturer. This type of arrangement is common for independent record labels that have agreements to be handled by the distribution branch of one of the major labels. However, branch distribution is on the decline as majors shift resources away from physical product and towards digital distribution. Digital distribution systems will be addressed in the chapter on distribution and retail.

Retail

Retailing consists of all the activities related to the sale of the product to the final consumer for personal use. Retailers are the final link in the channel of distribution. There are numerous types of retailers, each serving a special niche in the retail environment. Independent stores focus on specialty products and customer service, while mass merchandisers such as Wal-Mart concentrate on low pricing. Online retailing of physical and digital download product through Amazon.com, CD Baby, and others has been growing in the past decade, while brick-and-mortar retailers have declined. To put this in perspective, there were 11,645 record retailers in the United States in 2003. In 2021, there are only 2,544 (Thomas, 2021).

PRICE AND POSITIONING

Years ago, the retail industry developed marketing tools to measure the effectiveness of product location in their stores. Universal product codes (UPCs), or bar codes, and computerized scanners have helped retailers to determine the optimal arrangement of products within the store. They can move products around to various locations and note the effects on sales. As a result, retailers are charging manufacturers a rental fee for prime space in the store, including counter space, end caps, and special displays. Retail stores sometimes charge **slotting fees** – a flat fee charged to the manufacturer for placement of products on the shelves for a limited period of time. If the product fails to sell, the fee reduces the retail store's financial risk. A similar concept has developed in the record business, but it is also tied to promotion and favorable location for the product. It is referred to as *price and position* and will be discussed in the chapter on retailing.

Are the Four Ps Dead?

Seth Godin and others have argued that the Four Ps are no longer relevant in a modern world of digital marketing. We have moved from a mass-media, advertising-based model (interruption marketing) to a fragmented, interactive model (permission marketing) (Godin, 1999; Merzel, 2011). Like Goldhaber did with his work on the attention economy, Godin points out that we are inundated with advertising and cannot attend to even a small percentage of the messages we are exposed to. And we don't. But we are more likely to pay attention to messages we are interested in, particularly if we have given the sender permission to contact us. This emphasizes the importance of (1) being relevant and (2) maintaining an up-to-date database of contacts. Both issues will be discussed further in later chapters.

SUMMARY

The application of marketing theories and ideas is constantly changing. As new tools have become available, like the Internet or social media, for example, marketers adapt, embracing the new tools and sometimes abandoning old ones. But the underlying theories evolve more slowly than the tools that are used to implement them. This chapter has introduced the reader to some of the theories and practices from marketing that are most applicable to the music business and looked at some ways that technology and innovation have changed the way those practices are applied in the industry today.

GLOSSARY

Call to action—A statement usually found near the conclusion of a commercial message that summons the consumer to act, such as "call today" or "watch tonight at 11."

Channels of distribution—The various methods of distributing and delivering products ultimately to the consumer.

Competition-based pricing—Attempts to set prices based on those charged by the company's competitors.

Consumer-based pricing—Using the buyer's perceptions of value to determine the retail price, a reversal from the cost-plus approach.

Cost-based pricing—Determining retail price based on the cost of product development and manufacturing, marketing and distribution, and company overhead and then adding the desired profit.

Diffusion of innovations—The process by which the use of an innovation (or product) is spread within a market group, over time, and over various categories of adopters.

EAN—The European Article Numbering system. Foreign interest in UPC led to the adoption of the EAN code format, which is similar to UPC but allows extra digits for a country identification, in December 1976.

Hedonic—Of, relating to, or marked by pleasure.

Involvement—The amount of time and effort a buyer invests in the search, evaluation, and decision processes of consumer behavior.

Loss leader pricing—The featuring of items priced below cost or at relatively low prices to attract customers to the retail store.

Marketing—The performance of business activities that direct the flow of goods and services from the producer to the consumer (American Marketing Association, 1960). Marketing involves satisfying customer needs or desires.

Marketing mix—A blend of product, distribution, promotion, and pricing strategies designed to produce mutually satisfying exchanges with the target market. Often referred to as the four Ps.

Price fixing—The practice of two or more sellers agreeing on the price to charge for similar products.

Product class—A group of products that are homogeneous or generally considered substitutes for each other.

Product form—Products of the same form make up a group within a product class.

Product life cycle—The course that a product's sales and profits take over what is referred to as the lifetime of the product.

Product positioning—The customer's perception of a product in comparison with the competition.

Pull strategy—The company directs its marketing activities toward the final consumer, creating a demand for the product that will ultimately be fulfilled as requests for product are made from the consumer.

Push strategy—Pushing the product through the distribution channel to its final destination through incentives aimed at retail and distribution.

Universal Product Code (UPC)—An American and Canadian coordinated system of product identification by which a ten-digit number is assigned to products. The UPC is designed so that at the checkout counter, an electronic scanner will read the symbol on the product and automatically transmit the information to a computer that controls the sales register.

BIBLIOGRAPHY

American Marketing Association (1960). *Marketing Definitions: A Glossary of Marketing Terms, Committee on Definitions*, Chicago: American Marketing Association.

Clover, J. (2012, October 2). "Amanda Palmer's Accidental Experiment with Real Communism." Retrieved from www.newyorker.com/culture/culture-desk/amanda-palmers-accidental-experiment-with-real-communism.

Committee on Definitions (1960). *Marketing Definitions: A Glossary of Marketing Terms*, Chicago: American Marketing Association.

Dictionary of Marketing Terms: American Marketing Association (2004). Retrieved from www.marketingpower.com.

Fiegerman, S. (2013). Kickstart responds to critics of Zach Braff's campaign (May 10). Retrieved from http://mashable.com/2013/05/10/kickstarter-zach-braff-critics/

Finch, J. E. (1996). *The Essentials of Marketing Principles*, Piscataway, NJ: REA.

Friedland, J. (2021). *Mid-Year 2020 RIAA Revenue Statistics*, RIAA. Retrieved from www.riaa.com/wp-content/uploads/2020/09/Mid-Year-2020-RIAA-Revenue-Statistics.pdf.

Godin, S. (1999). *Permission Marketing: Turning Strangers into Friends and Friends into Customers*, New York: Simon & Schuster.

Goethem, Annelies van (date unknown). "Sad – in a Nice Way! The Royal Philharmonic Society Web Site." Retrieved from www.classicfm.co.uk/hearhere/Article.asp?id=856556.

Goldhaber, M. H. (1997, February 4). "The Attention Economy and the Net." *First Monday*.

Hefflinger, M. (2008). "Report: U.S. Music Consumption Up in 2007, Spending Down." *DMW Daily*. Retrieved from www.dmwmedia.com/news/2008/02/26/report:-u.s.-music-consumption-2007,-spending-down.

Imber, J. and Toffler, B.-A. (2008). *Dictionary of Marketing Terms*, Hauppauge, NY: Barron's Educational Series, Inc.

Ingham, T. (2020, May 4). "Global Recorded Music Industry Revenues Topped $20BN Last Year – But Streaming Growth Slowed." Retrieved from www.musicbusinessworldwide.com/the-global-recorded-music-industry-generated-over-20bn-last-year-but-streaming-growth-slowed/.

The International Federation of the Phonographic Industry (2014). Retrieved July 23, 2014, from www.ifpi.org/facts-and-stats.php.

Kotler, P. (1980). *Principles of Marketing*, Englewood Cliffs: Prentice-Hall.

Kotler, P., and Armstrong, G. (1996). *Marketing: An Introduction*, Upper Saddle River, NJ: Prentice Hall.

Kotler, P., and Armstrong, G. (2014). *Principles of Marketing*, Englewood Cliffs: Prentice-Hall.

Lathrop, T., and Pettigrew, J. (1999). *This Business of Music Marketing and Promotion*, New York: Billboard Books.

Luten, T. L. (2020). *Principles of Marketing for a Digital Age*, London: Sage.

Merzel, D. (2011, June 18). "The 4 P's Marketing are Dead! David Merzel's Blog. Ed. David Merzel. The International Federation of the Phonographic Industry." Retrieved from https://davidmerzel.wordpress.com/2011/06/18/the-4-ps-marketing-are-dead-via-davidmerzels-blog/

Moe, W., and Fader, P. (1999). "Tangible Hedonic Portfolio Products: A Joint Segmentation Model of Music CD Sales." Retrieved from www.atypon-link.com/AMA/doi/abs/10.1509/jmkr.38.3.376.18866.

MRC Year-end Report. (2021).

Nickels, W. G., and Wood, M. B. (1997). *Marketing: Relationships, Quality, Value*, New York: Worth Publishers.

Padgett, B. L., and Rolston, C. P. (2014). "Crowd Funding: A Case Study at the Intersection of Social Media and Business Ethics." *Journal of the International Academy for Case Studies*, Vol. 20, Iss. 3, pp. 61–66.

Palmer, A. (2012). "Blog post, (May 13)." Retrieved from http://amandapalmer.net/blog/

Peoples, G. (2013). "Amanda Palmer Q&A: Why pay-what-you-want is the way forward, and more (January 28)." Retrieved from http://www.billboard.com/biz/articles/news/indies/1533797/amanda-palmer-qa-why-pay-what-you-want-is-the-way-forward-and-more

Rogers, E. M. (1995). *Diffusion of Innovations* (4th edition), New York: The Free Press.

Styvén, M. E. (2010). "The Need to Touch: Exploring the Link Between Music Involvement and Tangibility Preference." *Journal of Business Research*, Vol. 63, pp. 1088–1094.

Thomas, B. (2021, February). "Record Stores in the U.S. Report 45122." Retrieved March 28, 2021, from https://my-ibisworld-com.bunchproxy.idm.oclc.org/us/en/industry/45122/industry-performance.

CHAPTER 3

Market Segmentation and Consumer Behavior

Before a marketing plan can be designed, it is necessary to fully understand the market – who your customers are. One should not make any marketing decisions until a thorough examination of the market is conducted. This chapter will explain how markets are identified and segmented and how marketers learn to understand groups of consumers and their shopping behavior. We will conclude by looking at the process and some influences on the consumer's decision-making process.

MARKETS AND MARKET SEGMENTATION

A market is defined as a set of actual and potential buyers of a product or service (Kotler and Armstrong, 2014). The market includes anyone who wants or needs your product and has the ability to buy it. Markets are identified by measurable characteristics of their members. Not all consumers of a product class (music) are potential buyers of your product offering (alternative rock). The basic goal of market segmentation, the subdividing of a market, is to determine the target market for your specific product.

Marketers segment markets for several reasons:

1. It enables marketers to identify groups of consumers with similar needs and interests and get to know the characteristics and buying behavior of the group members.

2. It provides marketers with information to help design custom marketing mixes that engage the particular market segment.

CONTENTS

DOI: 10.4324/9781003153511-3

CONTENTS

3. It is consistent with the marketing concept of satisfying customer wants and needs.

On the most basic level, music markets can be segmented into three sections: (1) current fans, (2) potential fans, and (3) those people who are not now, nor ever likely to be, fans. Perhaps this third group includes people who do not particularly care for the genre that your artist represents. It may include people who do not consume music, people who are unwilling or unable to pay for music, or those without access to become consumers. For example, if I don't have a computer or Internet access, I am not part of iTunes's target market. Businesses focus on the first two groups.

MARKET SEGMENTATION

Because most markets are so complex and composed of people with different needs, wants, and preferences, markets are typically subdivided so that promotional efforts can be tailored to fit the particular submarket. For most products, the total potential market is too diverse or heterogeneous to be treated as a single market. One need only walk the lot of the local car dealership or the aisles of the local grocery store to see that we have come a long way from the days of Henry Ford's Model T in any color you want "so long as it's black." To solve this problem, markets are divided into smaller, more homogeneous sub-markets. **Market segmentation** is defined as the "process of dividing a market according to similarities that exist among the various subgroups within the market. The similarities may be common characteristics or common needs and desires" (Dictionary of Marketing Terms, 2000). The members of the resulting segments are similar with respect to characteristics that are most vital to the marketing efforts. This segmentation may be made based on demographics, behavior, geography, psychographics, or some combination of two or more of these characteristics.

The process of segmenting markets is done in stages. In the first step, segmentation variables are selected, and the market is separated along those partitions. The most appropriate variables for segmentation will vary from product to product. The appropriateness of each segmentation factor is determined by its relevance to the situation. After this is determined and the market is segmented, each segment is then profiled to determine its distinctive demographic and behavioral characteristics. Then the segment is analyzed to determine its potential for sales. If the segment meets the criteria for successful segmentation (subsequently), the company's target markets are chosen from among the segments determined at this stage. If

FIGURE 3.1 *Target Market*

the segments do not meet the criteria, the process can be repeated, looking at different segmentation variables.

Take, for example, the market for radio listeners. Wikipedia lists over 60 music formats for radio. The 2010 Broadcast and Cable Yearbook listed over 90. No single radio station could possibly serve all the diverse musical tastes in the country. Instead, station owners, like Citadel Broadcast Corp. and Clear Channel Communications, segment the local market based on musical preference and offer multiple stations, each with a format corresponding to the musical tastes of a subset of the local market.

SEGMENTATION CRITERIA

In order to be successful, segmentation must meet these criteria:

Substantiality – the segments must be large enough to justify the costs of marketing to that particular segment. Costs are measured by how much is spent to reach each member of the market and the *conversion rate* – the percent of those you reach who follow through on the purchase.

Measurable – marketers must be able to conduct an analysis of the segment and develop an understanding of their characteristics. Historically, marketing decisions were made based on knowledge gained from analyzing the segment using surveys and other traditional research methods. Today, the Internet allows for data mining and behavioral targeting, creating market segments based on what Internet users purchase online and what types of sites they visit. For example, based on searches of weather reports and restaurant listings online, a search engine company can determine where someone lives. And based on searches they have conducted on the Web and what keywords they have used in those searches, the company can determine what products that person might be interested in receiving information about (Jesdanun, 2007).

Accessible – the segment must be reachable through existing channels of communication and distribution. The Internet has opened up accessibility to all marketers to reach members of their target market – as long as they are Internet users. It has also lowered the costs to reach target members, lowering the barriers to entry and allowing small, undercapitalized companies to compete with major players. Social norms and values, often codified into the law, also prevent or discourage business from accessing certain markets. The most obvious examples are selling alcohol or tobacco to minors, but they also include selling records labeled with Parental Advisory stickers to children.

Responsiveness – the segment must have the potential to respond to the marketing efforts in a positive way by purchasing the product. The use of Apple Pay, Android Pay, PayPal, Braintree, and gift cards sold through a multitude of outlets has opened up new payment methods, allowing businesses to expand their customer base to Internet shoppers who don't have access to credit cards (pre-teens) or may be reluctant to give out their credit card numbers online (the elderly).

Unique – the segments must be distinct enough to justify separate offerings, whether the distinctiveness calls for variety in product features or simply variety in marketing efforts.

Media fragmentation has allowed for more tightly defined market segments. The explosion of specialty magazines, TV channels, and websites, along with the ease of searching the Internet, has allowed marketers to more tightly target their messages to different audiences.

MARKET SEGMENTS

There is more than one way to segment markets, so marketers look for segmentation strategies that will maximize potential income. This is done by successfully targeting each market segment with a uniquely tailored plan – one that addresses the particular needs of the segment. Markets are most commonly segmented based upon a combination of geographic, demographic, and personality or psychographic variables and actual purchase behavior. The bases for determining geographic and demographic characteristics are quite standardized in the field of marketing because they are easily measurable. Psychographics, lifestyle, personality, behavioral, and purchase characteristics are not as standardized, and the categorization of these variables differs from textbook to textbook. Psychographics includes personality, beliefs, social class, and sometimes lifestyle. Behavioristic includes both attitudes toward the product and actual purchase behavior.

Traditionally, marketing has relied on demographic, geographic, or psychographic variables, either alone or in combination (using some demographics combined with some psychographics). Driven in large part by the ease of collecting online behavior (shopping) data, market segmentation has evolved to include more purchase behavior. Behavioral segmentation is a more effective way to segment markets because it is more closely aligned with the propensity to consume the product of interest.

GEOGRAPHIC SEGMENTATION

Geographic segmentation involves dividing the market into different geographical units such as cities, states, or regions. Markets may also be segmented based upon population density (e.g. urban, suburban, or rural) or even weather. Location may be used as a proxy for differences in income, culture, social values, and types of media outlets or other consumer factors (Evans and Berman, 1992). Companies like Claritas (owned by the Nielsen Co.) combine demographic, behavioral, and geographic data into behaviorally distinct segments within zip codes. These geo-demographic profiles, called PRIZM, include media consumption. While detailed descriptions are expensive, anyone can access brief profile descriptions on their website: https://claritas360. claritas.com/mybestsegments/. Media research companies such as the MRC and their subsidiary, Nielsen Audio, use geographic units called the *area of dominant influence (ADI)* or *designated market area (DMA)*. DMA is defined by Nielsen as an exclusive geographic area of counties in which the home market television stations hold a dominance of total hours viewed. The American Marketing Association describes both ADI and DMA as "the geographic area surrounding a city in which the broadcasting stations based in that city account for a greater share of the listening or viewing households than do broadcasting stations based in other nearby cities" (The AMA, 2018). Following is an index chart for contemporary hit radio (CHR) listening by geographic region.

This chart shows the American national listenership to CHR radio by geographic region. With the national index, or average, being 100, it is easy to see which regions of the country tend to prefer CHR and which ones do not listen to the format as much as the average.

Audience Share by Region
100 = National Average
Mon-Sun, 6AM-Mid, Persons 12+, AQH Shares

FIGURE 3.2   *Audience Share by Region*

(Source: Nielsen Audio)

When the marketer of recorded music and live performances uses geographic segmentation in this manner, it can help back the logic for the use of resources in support of the marketing plan. Marketing strategies are then easily tailored to particular geographic segments. For example, the label may use geographic segmentation to determine in which cities tour support money would be most effective.

DEMOGRAPHIC SEGMENTATION

Demographics are basic measurable characteristics of individual consumers and groups such as age, gender, ethnic background, education, income, occupation, and marital status. Demographics are the most popular method for segmenting markets because the information is easier and cheaper to measure than more complex segmentation variables, such as personality or consumer behavior. Fortunately, groups of people with similar demographics tend to have similar needs and interests that are distinct from other demographic segments.

Age is probably the demographic most associated with changing needs and interests. Consumers can be divided into age categories such as children, teens, young adults, adults, and older adults. The segment of "tweens" (preadolescents between the ages of 8 and 13) has been added to the mix because of their enormous spending power. Tweens are expected to number 23 million by 2020 (Jayson, 2009). This age group accounts for $43 billion in disposable income; over half have a Facebook account, and 78% have a cell phone. The typical tween spends 8 to 12 hours per day consuming media of one form or another, and more than half of the girls age 10 to 12 want to be famous (Younger, 2014).

Gender is also a popular variable for segmentation, as the preferences and needs of males are perceived as differing from those of females. Differential needs based upon gender are obvious for product categories such as clothing and personal hygiene, but even in the area of music preferences, differences in taste exist for males and females. Nielsen Audio reports that males are more likely to listen to alternative, rock, or news/talk radio, while women prefer top 40, country, adult contemporary, and contemporary Christian radio (Arbitron Radio Today: How America Listens to Radio, 2010).

Income segmentation is popular among certain product categories, such as automobiles, clothing, cosmetics, and travel, but is not as useful in the recording industry. *Educational* level is sometimes used to segment markets. Well-educated consumers are likely to spend more time researching purchases and are more willing to experiment with new brands and products (Kotler and Armstrong, 2008).

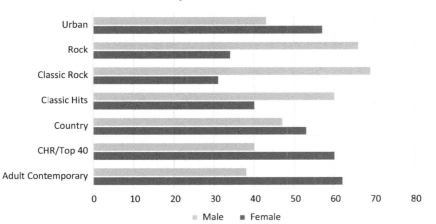

FIGURE 3.3 *Gender Comparison by Radio Format*

MULTIVARIABLE SEGMENTATION

The process of combining two or more demographic or other variables to further segment the market has proven effective in accurately targeting consumers. By considering age, gender, and income together, marketers can better tailor the marketing messages to reach each group. For example, older males may prefer news/talk radio, whereas younger males prefer alternative rock radio.

PSYCHOGRAPHIC SEGMENTATION

Psychographic segmentation involves dividing consumers into groups based upon opinions, motives, interests, lifestyles, or personality. Psychographic segmentation goes beyond demographics and provides information about what goes on in consumers' minds. While demographics may paint a broad picture of what consumers are like, psychographics adds the vivid detail, enabling marketers to shape very specific marketing messages to appeal to the target market.

By understanding the motive for making purchases, marketers can emphasize the product attributes that attract buyers. For example, if consumers are driven by price, pricing factors such as coupons can be emphasized. If another segment is driven by convenience, this issue can be addressed through widespread product distribution. Lifestyle segmentation divides consumers into groups according to the way they spend their time and the relative importance of things in their life. Imagine that your primary residence, your home, is on a piece of lakefront property. How would your life be different? You would likely have a boat tied to a dock, and all of this would be a short walk out your back door. You could be on the water in the middle of the lake in minutes, as opposed to having to hook the boat trailer up to the car and drive across town, wait for your turn at the boat launch, and then park the SUV before you could be in the boat on the lake. And if you lived on the waterfront, your house would be much more likely to have a nautical theme to it. If you played organized team sports as a child, your family probably built a lifestyle around traveling to practices, games, and tournaments that influenced much of your weekend activities and drove purchase and consumption behavior – ice chests, lawn chairs, suntan lotions, ballpark food, and hotels all become part of the lifestyle, at least during the season. Even the family's choice of cars was probably influenced by your participation in team sports!

Psychographics are more difficult to measure and require constant updating to stay abreast of changes in the marketplace. One system that uses

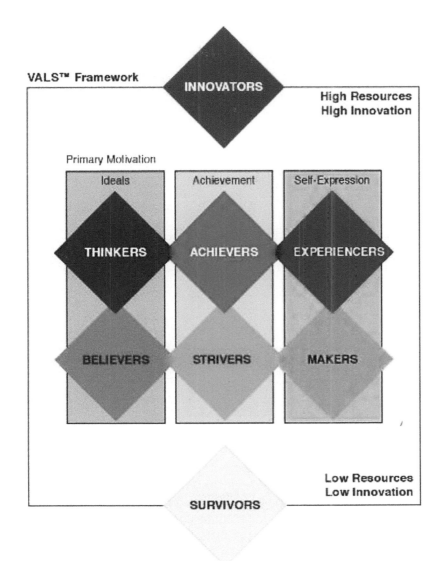

FIGURE 3.4   *The VALS Framework*

(Source: SRI Consulting Business Intelligence)

psychographics to identify market segments is the VALS segmentation framework, developed by the Stanford Research Institute. Originally, the segmentation was based on values (V) and lifestyles (LS), thus the acronym VALS, which is still used for branding purposes. But the version of VALS in current use segments consumer markets on the basis of selected psychological traits.

FIGURE 3.5   *Brand Camp Cartoon*

(Source: Tom Fishburne 2005)

The VALS system places consumers into three self-orientation categories and four levels of resources and innovations. The resulting eight segments provide insight into consumers based on motivations, beliefs, lifestyles, and resources. For example, *thinkers* are conservative and motivated by ideals; *experiencers* are young and enthusiastic and motivated by self-expression, excitement, and innovation. More information on the VALS segments is available at www.strategicbusinessinsights.com/vals/

PERSONALITY SEGMENTATION

Researchers at the University of Texas have found that personal music preferences can be linked to personality traits. Rentfrow and Gosling (2003) found that people's music preferences typically classify them into one of four basic dimensions: (1) reflective and complex, (2) intense and rebellious, (3) upbeat and conventional, or (4) energetic and rhythmic. Preference for each of the following music dimensions is differentially related to one of these basic personality traits.

Table 3.1 Personality/Genre Preference	
Personality	**Genre Preference**
Reflective and complex	Classical, jazz, blues, folk
Intense and rebellious	Alternative, rock, heavy metal
Upbeat and conventional	Country, pop, religious, soundtracks
Energetic and rhythmic	Rap/hip-hop, soul/funk, electronica/dance

BEHAVIORIAL SEGMENTATION

Behavioral segmentation is based on actual customer behavior toward products. Behavioral segmentation has the advantage of using variables that are closely related to the product and its attributes. Some of the more common behavioral segmentation variables are: product usage, benefits sought, user rate, brand loyalty, user status, readiness to buy, and purchase occasion. Let's look at a couple of behavioral segmentation variables particularly relevant to the music industry.

PRODUCT USAGE

Product usage involves dividing the market based on those who use a product, potential users of the product, and nonusers—those who have no usage or need for the product. How a product is used (and for what purposes) is also of importance. For example, Kodak found that disposable cameras were being placed on tables for wedding guests to help themselves to. As a result, Kodak modified the product by offering five-pack sets of cameras in festive packaging (Nickels and Wood, 1997). The recording industry adapted to usage situations by repackaging music that is customized for usage occasion, such as wedding music compilations, party mixes, romantic mood music (complete with recipes for romantic dinners), and so forth. (See section on hedonic responses to music in Chapter 2.)

BENEFITS SOUGHT

It is said that people do not buy products, they buy benefits. They buy aspirin to alleviate pain or reduce the chances of a heart attack, toothpaste to whiten their teeth or prevent cavities, and music to elevate mood or create atmosphere. **Benefit segmentation** divides the market according to benefits sought by consumers. In the music market, some consumers shop for classical music because it drowns out distracting noise while they work or study without distracting from the primary task. Others buy particular artists because they believe it will create the right atmosphere for a romantic evening.

BRAND LOYALTY

Brand loyalty is defined as the degree to which a consumer will repeatedly purchase a company's product or service. As a segmentation variable, brand loyalty can be used to divide the market into those who are and those who are not brand loyal. Artists' hardcore fans are brand loyal – they will buy most anything with the artist's name attached to it, often with little consideration of any other factors. The ad agency Saatchi & Saatchi calls extreme brand loyalty "lovemarks." Not surprisingly, the Beatles top their list of the 50 top music and radio brands. A demonstration of the power of brand loyalty can be seen in the Voyager Company's decision to make the Beatles *A Hard Day's Night* the first full-length commercial movie to appear on CD. When the Voyager Company chose to release *A Hard Day's Night* in May 1993, the number of U.S. households with computers had increased from 8.2% to just under 23%, but many of those computers still did not have CD-ROM drives (www.statista.com/). The Voyager Company chose *A Hard Day's Night* in large part because they knew loyal Beatles fans would buy the discs even if they could not play it on their computers yet, just because the Beatles' name was attached to it!

USER STATUS

User status defines the consumer's relationship with the product and brand. It involves level of loyalty and propensity to become a repeat buyer. Arens et al. (2012) identified six categories of consumers based on user status: sole users, semi-sole users, discount users, aware non-triers, trial/rejecters, and repertoire users.

USER STATUS (BRAND A)

Sole users These individuals purchase only your brand, even when competitors are promoted. They are the most brand loyal and require the least amount of advertising and promotion.

Semi-sole users typically use brand A, but have an alternative selection if it is not available or if the alternative is promoted with a discount.

Discount users are the semi-sole users of competing brand B. They don't buy brand A at full price but perceive it well enough to buy it at a discount.

Aware non-triers are category users but have not bought into brand A's message. Musically speaking, they like the genre but haven't listened to your artist's music and aren't likely to unless they have a big hit.

Trial/rejecters bought brand A's advertising message but did not like the product.

Repertoire users perceive two or more brands to have superior attributes and will buy at full price. These are the primary brand switchers and therefore the primary target for brand advertising.

Neil Howe and William Strauss examined the emergence of the millennial generation, which they define as the generation of young adults, the first of whom have "come of age" around the time of the millennium (2000). This segment includes people born after 1980, according to Pew Research, although Howe and Strauss defined the segment as those born between 1982 and 1995. As of 2019, millennials numbered about 72 million and are the largest generational group in the U.S. population (Fry, 2020). This segment is characterized by a sharp break from Generation X and hold values that are at odds with the baby boomers. Pew Research (Millennials, 2014) described millennials as "relatively unattached to organized politics and religion, linked by social media, burdened by debt, distrustful of people, in no rush to marry and optimistic about the future." Bock (2002) also found them to be optimists and team players who follow the rules and accept authority (2002).

R. Craig Lefebvre, a professor at George Washington University, characterizes the millennial generation as more involved in peer-to-peer communication than previous generations. He states "reliance and trust in nontraditional sources – meaning everyday people, their friends, their networks, the network they've created around them – has a much greater influence on their behaviors than traditional advertising." He goes on to state "the challenge for marketers is how to create peer to group exchanges that feature their brands, products, services and behaviors. The question is no longer 'what motivates someone to change' but rather 'what motivates someone to share something they find intrinsically useful and valuable with their most trusted friends and colleagues?'"

GENERATION-Z

Generation-Z, or Gen-Z, are the post-millennial group born between 1996 and 2010. They are frequently referred to as "digital natives," having grown up with the Internet and cell phones and a greater level of comfort with technology than previous generations. Gen-Z is more entrepreneurial and less likely to identify with a religious institution than their elders. Maturing at a time of rapidly growing minorities, Gen-Z is more accepting of diversity, especially in their musical tastes. Having grown up in time of personal customization, Gen-Z expects their entertainment to be customized for them as well – Spotify rather than radio, Netflix over network television.

TARGET MARKETS

Once market segments have been identified, the next step is choosing the segment or segments that will allow the organization to most effectively achieve its marketing goals – the target markets. Simply stated, a *target market* is a group of persons for whom a firm creates and maintains a product mix that specifically fits the needs and preferences of that group (Dictionary of Marketing Terms, 2008). With the target markets identified, the company begins the task of positioning the product offering, relative to the competition, in the minds of the consumers using the tools available to the market, primarily advertising.

CONSUMER BEHAVIOR AND PURCHASING DECISIONS

According to the American Marketing Association, consumer behavior is defined as "the dynamic interaction of affect and cognition, behavior, and environmental events by which human beings conduct the exchange aspects of their lives." More simply put, *consumer behavior* is the buying habits and patterns of consumers in the acquisition and use of goods and services. Entire courses on consumer behavior exist because it is one of the most interesting areas of marketing. In the following, we will discuss only a few key concepts in consumer behavior as they relate to music purchasing and consumption.

NEEDS AND MOTIVES

When a product is purchased, it is usually to fill some sort of need or desire. There is some discrepancy between the consumer's actual state and their desired state – *I want that!* In response to a need, consumers are motivated to make purchases. **Motives** are internal factors that activate and direct behavior toward some goal. In the case of shopping, the consumer is motivated to satisfy the want or need, as outlined in Chapter 2.

FIGURE 3.6 *Maslow's Hierarchy of Needs*

Understanding motives is a critical step in creating effective marketing programs. Psychologist Abraham Maslow developed a systematic approach of looking at needs and motives. Maslow (1943) states that there is a hierarchy of human needs. More advanced needs are not evident until basic needs are substantially met. Maslow arranges these needs into five categories of physiological needs, safety, love, esteem, and self-actualization. Marketers believe that it is important to understand where in the hierarchy the consumer is before designing an effective marketing program.

Physiological needs are the basic survival needs that include air, food, water, and sleep. Until these needs are met, there is little or no interest in fulfilling higher needs. Survivors in the VALs system fit into this category. Remember, survivors' characteristics include prime concerns about safety and establishing a sense of security. They watch a lot of TV, are more trusting of advertising, are risk averse, and feel powerless. Until these basic physiological needs are substantially met, the needs for love and belongingness are absent. The consumer is in survival mode.

Love, approval, and belongingness are the next level, along with feeling needed and a sense of camaraderie. Products oriented toward social events sometimes appeal to this need, and this includes music, particularly live performances. Esteem needs drive people to seek validation and status. Self-actualization involves seeking knowledge, self-fulfillment, and spiritual attainment.

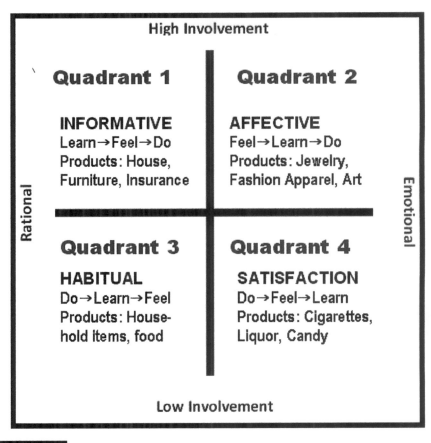

FIGURE 3.7 *The Hierarchy of Effects*

An individual may temporarily sacrifice a lower-level need to achieve a higher-level need, like skipping a meal and donating the money to a worthy cause (like buying tickets for your favorite band's show) or fasting for spiritual reasons, but in the long term, the lower-level needs must be fulfilled to achieve the higher-level needs.

Nobody *needs* music in the Maslovian sense. We can live without it. We might not like it, and certainly life is better with music, but we won't die without it. The challenge to music marketers is to make consumers *feel* like they need music.

CONVERTING BROWSERS TO BUYERS

One of the greatest challenges facing marketers is converting potential customers who are browsing the merchandise into actual purchasers. Retail stores refer to a *conversion rate* – the percentage of shoppers who actually make a purchase. The actual rate varies and depends upon the level of involvement and risk, with small-ticket items having a higher conversion rate than expensive items such as appliances or jewelry. Conversion rates can be improved by reducing the purchase risk through sampling (e.g., listening stations and free downloads; streaming sites also provide this service, and most now offer the option of a permanent download) and liberal return policies, as well as offering a level of service that meets or exceeds expectations. None of this matters, however, if the product is not in stock or can't be found. When shoppers don't buy, it's often because they were unable to find what they wanted or the purchasing process was too difficult. Making the product easy to find is equally important in both the physical and digital world.

Once the shopper has made their choice, then the checkout process must be quick and simple – don't give them a chance to change their mind! People have been known to spend hours shopping for just the right item only to abandon their shopping cart, physical or digital, because the checkout process took too long. This is why many online stores such as Amazon.com offer accounts to customers, with their address and credit card information stored in the system. The consumer can purchase with a single click of a mouse button.

CONCLUSION

This chapter has provided a brief introduction to some of the fundamental theories and concepts in consumer behavior and market segmentation. The music business is just like any other business when it comes to the basics of market segmentation and consumer behavior, but a deep understanding of these basic theories and constructs and being able to apply them to your specific artist or album will help separate you from everyone else in the business. The best music marketers will know their consumers as well as they know their artists.

GLOSSARY

ADI—Area of dominant influence (see DMA).

Affect—A fairly general term for feelings, emotions, or moods.

Behavioral segmentation—Based on actual customer behavior toward products.

Cognition—All the mental activities associated with thinking, knowing, and remembering.

Conation—Represents intention, behavior, and action.

Consumer behavior—The dynamic interaction of affect and cognition, behavior, and environmental events by which human beings conduct the exchange aspects of their lives.

Cookies—A small file sent by a server to a web browser and then sent back by the browser each time it accesses that server. The main purpose of cookies is to identify users and store information about them, including preferences specific to the website.

DMA—Designated marketing area. The geographic area surrounding a city in which the broadcasting stations based in that city account for a greater share of the listening or viewing households than do broadcasting stations based in other nearby cities.

Generation-Z—Or Gen-Z, people born between 1996 and 2010.

Geographic segmentation—Dividing the market into different geographical units such as town, cities, states, regions, and countries. Markets also may be segmented depending upon population density, such as urban, suburban, or rural.

Involvement—The amount of time and effort a consumer invests in the search, evaluation, and decision process of consumer behavior.

Market—A set of all actual and potential buyers of a product. The market includes anyone who has an interest in the product and has the ability and willingness to buy.

Market segmentation—The process of dividing a large market into smaller segments of consumers that are similar in characteristics, behavior, wants, or needs. The resulting segments are homogenous with respect to characteristics that are most vital to the marketing efforts.

Millennials—People born between the years 1982 and 1995. Also called the "Net Generation" because they grew up with the Internet.

Personal demographics—Basic measurable characteristics of individual consumers and groups such as age, gender, ethnic background, education, income, occupation, and marital status.

Psychographic segmentation—Dividing consumers into groups based upon lifestyles, personality, opinions, motives, or interests.

Target market—A set of buyers who share common needs or characteristics that the company decides to serve.

User status—The consumer's relationship with the product and brand. It involves level of loyalty.

Web analytics—The use of data collected from a website and its visitors to assess and improve the effectiveness of the website.

Web data mining—The process of analyzing navigation patterns and other behavior of users on the web in order to improve their navigation and interaction with the website.

BIBLIOGRAPHY

The American Marketing Association. (2004). www.marketingpower.com.

American Marketing Association Dictionary. (2014). www.ama.org/resources/Pages/Dictionary.aspx.

Arbitron Radio Today: How America Listens to Radio. (2010). http://futureofradioonline.com/wp-content/uploads/2010/12/RadioToday_20101.pdf.

Arens, W. F., Weigold, M. F. and Arens, C. (2012). *Contemporary Advertising*, New York: McGraw-Hill Higher Education.

Bloch, P. H. and Richins, M. L. (1983). A Theoretical Model for the Study of Product Importance Perceptions, *Journal of Marketing*, Vol. 47, No. 3 (Summer), pp. 69–81.

Bock, W. (2002). www.mondaymemo.net/010702feature.htm.

Broadcast Yearbook. www.americanradiohistory.com/Archive-BC-YB/2010/D5-2010-BC-YB-7.pdf.

"DMA" *AMA Dictionary*. (2018). www.ama.org/resources/Pages/Dictionary.aspx?dLetter=D.

Evans, J. R. and Berman, B. (1992). *Marketing* (5th edition), New York: Macmillan Publishing.

Finch, J. E. (1996). *The Essentials of Marketing Principles*, Piscataway: Research and Education Association.

Fry, R. (2020). Millennials Overtake Baby Boomers as America's Largest Generation, April 28, 2020. www.pewresearch.org/fact-tank/2020/04/28/millennials-overtake-baby-boomers-as-americas-largest-generation/. Accessed March 26, 2021.

Hall, C. and Taylor, F. (2000). *Marketing in the Music Industry* (4th edition), Boston: Pearson Custom Publishing.

Howe, N. and Strauss, W. (2000). *Millennials Rising: The Next Great Generation*, New York: Vintage Books.

Imber, J. and Toffler, B.-A. (2008). *Dictionary of Marketing Terms*, Hauppauge, NY: Barron's Educational Series, Inc.

Jayson, S. (2009). It's Cooler Than Ever to Be a Tween, but Is Childhood Lost? *USA Today*, February 2, 2009.

Jesdanun, A. (2007). Ad Targeting Grows as Sites Amass Data on Web Surfing Habits, *The Tennessean*, December 1, 2007.

Kotler, P. (1980). *Principles of Marketing*, Englewood Cliffs: Prentice-Hall.

Kotler, P. and Armstrong, G. (2008). *Principles of Marketing* (12th edition), Upper Saddle River, NJ: Pearson Education.

Kotler, P. and Armstrong, G. (2014). *Principles of Marketing* (15th edition), Upper Saddle River, NJ: Pearson Education.

Lefebvre, R. C. (2006). Communication Patterns of the Millennium Generation. http://socialmarketing.blogs.com/r_craiig_lefebvres_social/2006/01/communication_p.html.

Marketing Teacher. (2004). www.marketingteacher.com/Lessons/lesson_positioning.htm.

Marketresearch.com. (2001). *The U.S. Urban Youth Market: Targeting the Trendsetters*.

Maslow, A. (1943). A Theory of Human Motivation, *Psychological Review*, 50, pp. 370–396.

"Millennials in Adulthood: Detached from Institutions, Networked with Friends." Pew Social Trends. Pew Research Center, March 7, 2014. Accessed February 7, 2022. http://www.pewsocialtrends.org/2014/03/07/millenials-in-adulthood/.

Nickels, W. G. and Wood, M. (1997). *Marketing*, New York: Worth Publishers.

Rentfrow, P. J. and Gosling, S. D. (2003). The Do-Re-Mi's of Everyday Life: Examining the Structure and Personality Correlates of Music Preferences, *Journal of Personality and Social Psychology*, 84, pp. 1236–1256.

Resources and Learning. (2021). www.riaa.com/resources-learning/parental-advisory-label/.

Vaughn, R. (1986). How Advertising Works, *Journal of Advertising Research* (February/March). https://en.wikiquote.org/wiki/Henry_Ford www.statista.com/statistics/184685/percentage-of-households-with-computer-in-the-united-states-since-1984/.

Younger, S. (2014). Tweens by the Numbers: A Rundown of Recent Stats. www.chicagonow.com. Accessed June 17, 2014.

Zappa, F. (2012). Frank Zappa at PMRC Senate Hearing on Rock Lyrics. *YouTube*, May 16, 2012, uploaded by Bartek Kaszuba. https://youtu.be/hgAF8Vu8G0w.

CHAPTER 4

Marketing Research

INTRODUCTION

Marketing research is the function that links the consumer, customer, and public to the marketer through information – information used to identify and define marketing opportunities and problems; generate, refine, and evaluate marketing actions; monitor marketing performance; and improve understanding of marketing as a process. "**Marketing research** specifies the information required to address these issues, designs the method for collecting information, manages and implements the data collection process, analyzes the results, and communicates the findings and their implications" (American Marketing Association 2020). Marketing research and **market research** are not the same thing: market research is a subset of marketing research that looks specifically at the size, location, and makeup of a product market.

RESEARCH IN THE MUSIC INDUSTRY

There has been an increased focus on metadata and analytics in recent years. Many of the new hires at record labels have been recent college grads with skills and experience in analyzing social media statistics. The largest labels have also expanded their staff dedicated to consumer and industry research, hiring experienced experts from inside and outside the entertainment industry, creating teams of people dedicated to conducting research. The more sophisticated research departments produce work that can be

DOI: 10.4324/9781003153511-4

used to, for example, persuade major corporations of the benefit of teaming up with a particular artist in pursuit of a common consumer. The labels utilize syndicated data collected and published as reports by companies like MusicWatch, MIDiA, and NPD, but this information is general industry research. Of course, labels collect information on individual artists, but before the artist is signed, that data is often limited to checking with others in the industry, streaming and social media numbers, sales numbers if they are available, attending a few live shows, and then going with their "gut feelings."

Radio stations perform much of the research in the music industry. They purchase syndicated research as well as measuring the appeal of new music through regular online, telephone, or auditorium testing.

TYPES OF RESEARCH

Custom vs. Syndicated: One way to categorize research is by the purpose for which it is collected. **Custom research** is done to answer a specific question, such as "How much are consumers willing to pay for an advertising-free streaming music service?" Since the research is initiated and paid for by a single company or organization, they can customize the research specifically for their needs. Research that is done on an ongoing basis and sold to subscribers in the form of periodic reports is called **syndicated research**. Examples of syndicated research in the entertainment industry include Mediabase (radio listenership), Nielsen (television viewership), Broadcast Data System's (BDS) airplay monitoring, and MRC's retail sales monitoring. Because syndicated research has many users and is often used for longitudinal purposes (comparisons over time), the information is standardized and may be more general. The results are published periodically, ranging from daily to quarterly, as a report; however, subscribers may elect to pay for more detailed data or access to "raw" data from which they can create their own reports. For example, if you want to know what the best-selling song is for this week, you can simply open up *Billboard Magazine*, where the results from the MRC research will be reported in a simple sales chart form, but if you want to know in which areas of the country the song sold best, you would need to subscribe to Music Connect to get the more detailed information.

Purpose of the Research: Research can also be categorized by the kinds of questions the researcher is trying to answer. This is commonly referred to as research design. The three types of research design, exploratory, descriptive, and causal, are discussed in the following.

OVERARCHING RESEARCH ISSUES: VALIDITY AND RELIABILITY

Steps must be taken to ensure that the research results are valid and the process is reliable. There are many levels of validity, the scope of which is beyond the purpose of this text, so for our purposes let us define **validity** as the "truth." That is, we want our measures to reflect true differences among

FIGURE 4.1   *Validity or Reliability?*

the subjects being measured and not some random error or bias. To do this, it is very important to be sure the researcher asks the right questions, starting with the problem definition and ending with the interview, focus group, or survey questions. Subtle differences in the wording of a question can lead to different results: "You don't really like bluegrass music, do you?" will get a different response than the less biased "Do you like bluegrass music?"

Reliability is the researcher's ability to get similar results from repeated applications of the measures or from independent but comparable measures of the same trait or construct. If you measure how tall you are using a tape measure under similar conditions (no shoes, stand up straight each time), you expect the difference in measurements to be attributed to changes in the object being measured – you grew! But if your tape measure was printed on stretchy material, you might get different results each time because you have a bad measuring instrument.

There is a clock on the shelf in my office. It reads 10:35. The battery died years ago, and I just never replaced it. Is the time on clock valid? Reliable? Think about your answer before you continue reading. The clock is very reliable. No matter when I look at it, I always get the same information – "similar results from repeated applications of the measures." 10:35. Unfortunately, it is only valid twice a day! For research results to be useful, they need to be both valid and reliable.

Unlike height, much of what we want to measure in the music business is not physical or real – you can't touch or see it. They are latent or abstract **constructs** such as "engagement," "identification," or "emotional response," (Stewart 2013). These constructs are much harder to accurately measure than streams or spins because they are complex and abstract, making it more difficult to construct valid and reliable measurement instruments.

THE RESEARCH PROCESS

There are six stages to the research process once a problem is recognized, all of which are equally important and demanding: problem definition, determining the research design, choosing and designing the data collection, actual data collection, data analysis, and communicating the results. Although presented as discrete stages, it is important to understand that this is not a linear process. At any stage of the research process, you may have to go back to a previous stage and start over or modify the research due to new information or complications.

Problem definition. The first step is to determine what questions need to be asked. The questions may seem obvious, but the researcher must make sure they are getting to the root of the problem. Let's say that an artist or label manager has come to you and wants to know why the artist's last album didn't sell as well as the one before. The obvious answer is because not as many people bought copies of the album, but the real question is why? The end goal of the problem definition stage of the research process is to determine the objective of the research. What is it that you want to learn by doing the research? The second step in the research process is determining a research design. You could look at the three research designs, exploratory, descriptive, and causal, as a continuum. The less you know about the problem, the greater your need for basic information, the more likely you are to use an exploratory research design. Alternatively, the more you know already, the greater the probability that a descriptive or causal research design will be appropriate.

In some situations, you may not be sure exactly what the problem is or what questions you should be asking. In that case, an **exploratory research** design may be helpful in clarifying and defining the research objectives. Exploratory research may be done to help better understand the situation, screen alternatives, or discover new ideas and results in qualitative data. "The focus of qualitative research is not on numbers but on stories, visual portrayals, meaningful characterizations, interpretations, and other expressive descriptions" (Zikmund and Babin 2007, p. 84).

The primary methods for exploratory research are focus groups and depth interviews, discussed in the following.

Descriptive research is done to describe the existing characteristics of a defined target market or population (Hair, Bush, and Ortinau 2003). The primary tool of descriptive research is the survey because of its flexibility, speed, and costs. As mentioned before, the key to a good survey is asking the right questions. Because the survey may be administered without the researcher present (via email or online, for example) clear, concise questions are essential. Finally, **causal designs** are rigidly specified experiments designed to determine cause and effect and are rarely used in the marketing side of the entertainment business.

Data collection methods. Once you have an understanding of what you want to know (problem definition), you should research what has already been learned on the topic and what information is already available. The answer to your question may already exist thanks to prior research, which will negate the need for further, more difficult, and expensive research. Whenever you use somebody else's research to answer a question, you are doing **secondary research**. Even if the answer isn't already out there, examining existing research will help you avoid repeating the same mistakes and may give you valuable insight into the problem you are investigating. It may also help you determine which research design will give you the best answers to your questions.

If you don't find the answer to your question in the existing research, then you will have to conduct **primary research**. The main methods used to collect data in the music business are focus groups and surveys. A popular choice for exploratory research, the focus group is a face-to-face interview of six to ten people that allows the researcher to delve deeply into participants' responses by asking follow-up questions. Focus groups may appear to be unstructured and free flowing, but the well-planned interview is designed to discover new information while maintaining the flexibility to pursue interesting answers and topics as they arise. Focus groups are good for testing new music and finding out why consumers like or dislike certain songs, but because they involve a small number of people, they are not good for generalizing results or providing statistical data about the market. In other words, you should not make marketing decisions based on a single focus group (although it is often done!). Automakers also use focus groups to get feedback on design changes and new model features, but they conduct dozens of focus groups before drawing any conclusions from them. **Depth interviews** are similar in structure to focus groups but are done on a one-on-one basis.

FIGURE 4.2 *A Focus Group*

Research panels. A research panel is like an ongoing focus group, except in most cases the panel does not meet face to face. Data is collected from the same participants on an ongoing basis, either via mailed surveys or online or perhaps passively by an app on a phone. Panels may be asked to provide more detailed data than can be asked in a one-time survey, or they may be asked to provide data over a long period of time for a longitudinal study. Nielsen Mobile collects data on music and entertainment exposure using a mobile phone app that collects information from the same group of participants on an ongoing basis for as long as they choose to remain a member of the panel. Panel members are recruited to fit the profile desired by the researcher, and if a member of the panel drops out for some reason, they can be quickly and easily replaced.

Survey research may be used to further test information gathered from focus groups or as a stand-alone research tool. Survey research is deceptively difficult. How hard can it be to ask somebody a few questions? The challenge comes in not only asking the *right* questions but asking them the right way to the right people.

It is tempting to take shortcuts and ask questions like "On a seven-point scale, please indicate how much you like the music and lyrics of this song." But how do you answer the question if you love the lyrics but hate the melody? Other problems might not be so obvious. A researcher might focus on the artist's music as the cause of declining sales when the actual cause may come from their appearance or their behavior. This problem could have been solved by good exploratory research and asking the right questions in a focus group. It is also important to make sure you are asking your questions to the right people, the artist's potential audience (target market).

ONLINE SURVEY TOOLS

Several companies offer free survey tools with limited capabilities to entice users to buy the full packages. These free versions may be all you need for a short survey with a small sample size. Some of these tools are linked to or packaged with email services like Mail Chimp or MyEmma to make questionnaire distribution easier. Free survey tools include:

Survey Monkey	www.surveymonkey.com
Survey Gizmo	www.surveygizmo.com
Qualtrics	www.qualtrics.com
Constant Contact	www.constantcontact.com
Instant Survey	www.instantsurvey.com
My Survey Lab	www.mysurveylab.com

Choosing a method. Experiments require that all the variables, the factors which might influence the outcome of the experiment, be controlled. This is a monumental task under the best of conditions. Sometimes variables we didn't even know existed are later discovered to influence the outcomes. Nobody in the music business wants to do experiments with their artists because to control and manipulate the variables (so that you can see the consequences) means some part of the target market isn't going to get the full marketing exposure. If we withhold advertising here (but not everywhere else) or we use in-store promotions in one region but not the others, we may hurt sales in those areas, and nobody wants to do that. It is difficult enough to get record company executives to do any kind of research without the threat of decreased sales. Labels and managers are sometimes willing to do other kinds of research, primarily focus groups and surveys. So, when should each be used?

When focus groups are appropriate. Focus groups are considered exploratory research but can be conducted either before or in conjunction with a survey or any time the company wants to probe an issue more in depth. Focus groups are often conducted in advance of a survey in order to gain a better understanding of the issues or of the target market for the survey, allowing the researcher to better understand the context and vernacular and how survey questions might be worded better. Care must be taken to make sure that the participants in the focus group are representative of the same target market that will participate in the survey. You would not want to conduct a focus group of 8- to 16-year-old girls on the subject of their favorite artists and then give the follow-up survey only to adults 25 years old and older. In short, the focus group should be used to inform the survey research.

Other times a focus group will be conducted without any follow-up survey or after a survey in order to probe deeply into a few research questions or to get consumer feedback to aid in decision making. Management may have been surprised by the results of the survey research and want to find out why consumers answered the way they did. It is unlikely they can go back to the exact same people who responded to the survey, but they will be able to go back to the same target market, the same sampling frame, and draw a representative focus group that should be able to shed light on the answers given in the survey.

Let us caution you once again that a single focus group should not be used to make decisions. The ideal procedure would be to conduct multiple focus groups until no new information was gained, but this isn't the ideal world, and decisions are often made based on the feedback from a single focus group.

FIGURE 4.3 *Data Collection by Survey*

When surveys are appropriate. Surveys are best suited to situations where the research questions can be answered in a straightforward manner, when more information about aggregate consumer groups is needed, and when that information will need to be generalized to a larger population.

Surveys are good for identifying characteristics of target markets, describing consumer purchasing behaviors, and measuring consumer attitudes. Surveys provide a relatively inexpensive, efficient, and accurate means of evaluating information about a market by using a small sample and extrapolating the result to the total population or market.

That said, writing a good survey question can be very difficult. All the ambiguity must be removed because you probably won't get a chance to explain what you are trying to ask. Because of this, entire textbooks have been written just on survey design. Whenever possible, the researcher should use existing scales to measure abstract concepts such as attitude and personality, because these scales will have been validated by previous research. Complete, validated, and reliable scales are published in journals and books like *The Handbook of Marketing Scales*. If a valid and reliable scale is used and an appropriate sampling procedure followed, your survey results should convey, with a high degree of confidence, opinions and characteristics that the entire market or population exhibit.

Simple surveys are conducted on a routine basis in the form of warranty registrations and bounce back or customer response cards. The record labels used to put survey cards inside CD packaging and incentivize responses by entering the respondent into a contest for a t-shirt or something autographed by the artist, but as the industry has shifted towards digital sales, this data collection has, unfortunately, disappeared. Because these cards were small, only a very few questions could be asked, but an online version could go in greater depth. Be careful, though, about overgeneralizing the responses. This is not a random sample, and research indicates that there are real differences between the people who send back the cards and those that don't. Still, the responses are beneficial in gaining some understanding of the market.

Data collection. Hair et al. (2003) defined data as "facts relating to any issues or subject." Data is the answers we get by conducting experiments, focus groups, and surveys. Once the research design (experiment, descriptive, or qualitative) has been determined and the method (focus group or survey) has been chosen, the next step is to figure out the sampling frame and sample size. You would get the best information if you asked everyone in the target population the questions, but a census is not practical or realistic. Instead, we seek to take a representative sample of the population that will allow us to draw accurate conclusions. Keep in mind that we are not talking about the entire population of the United States or the world. We are likely only to be interested in potential buyers of our artists' music or tickets, and that market is a segment of the bigger population (see Chapter 3). So, our sampling frame would be all potential buyers of a music genre or some other target market. From that sampling frame, we would draw a sample, a subset, of people to actually take our survey or be in our focus groups. The exact size of the sample can be calculated depending on the desired precision and confidence, but that is beyond the scope of this text. Suffice it to say that a research company testing songs for a radio station can calculate the sample size needed to give the programmer statistically precise information with great confidence, provided they get the right people to participate in the research, and that is, arguably, more important than the size of the sample. You don't want to ask some 60-somethings their opinion on synthwave – they aren't likely to be the right audience.

Surveys can be administered online if the questions are straightforward and the population you want to sample is small or widespread. Surveys can also be administered in person. If you have the permission of the promoter, surveyors may "intercept" concertgoers entering or leaving the venue. Not exactly a simple random sample, but if you are interested in capturing the opinions of the artist's fans, this is a quick way to reach a lot of them at one time. Surveys may also be included as part of one-on-one, depth interviews. Depth interviews, like focus groups, allow the researcher to dig more deeply into a topic, but they are done with individuals rather than small groups.

Data analysis. Once the data is collected, it must be put into useful form. The first stage of data analysis is to edit and code the data. **Editing** is the process of checking the data forms for errors, omissions, and consistency. The data is then coded. **Coding** refers to the systematic process of interpreting, categorizing, recoding, and transferring the data to the data processing program. Much of the process of editing and coding has been simplified through the use of computers. Online surveys minimize the possibility of coding errors because the human element can be minimized.

When the answers to the questions can be represented by numbers (i.e., for question five, the respondent chose answer two, and for question six, they chose answer seven), powerful statistical programs like SAS and SPSS can be used to look for relationships among the answers or commonalities within the respondents.

Not all data can be reduced to a series of numbers to be put into a computer program for analysis. Focus groups and depth interviews may result in long transcripts that will need to be carefully read and analyzed to identify key results and trends. More and more, even this kind of data can be analyzed using specialized software.

Analysis of the data is not the final step. The results of the analysis must be interpreted and given meaning, and that information must be communicated to the decision makers in the form of **conclusions and recommendations**. This is normally done in a formal **research report**, although in the music business, the report or presentation may be less formal than in other industries. The purpose of the report is to communicate the results, findings, and recommendations to the marketing client. The written report should begin with a one- or two-page executive summary that covers how the research was conducted and the highlights of the findings. The purpose of the executive summary is to give a quick but thorough overview of the research report for a busy executive that does not have time to read the full report. The full report should address each question asked in the research and how the findings are different than expected or otherwise interesting. In the conclusion of the report, the researcher should explain what was found and their interpretation of the meaning of the responses. Finally, recommendations should be made based on the results of the research.

Advantages of Online Surveys One of the advantages of doing surveys on computers is that it reduces error during data collection and coding. There are other advantages and disadvantages to computer-based and online research. Smartsurvey.co.uk gives the following ten advantages to online surveys over paper-and-pencil surveys:

Faster: The time it takes to complete an online survey is one-third of the time it takes to complete a paper-and-pencil survey. Turnaround time is shortened, making the data and analysis timelier.

Cheaper: Online surveys can cut your costs in half. Fewer administrators to train, no printing costs, no travel or telephone costs (the researcher can take a global sample from the comfort of their office), and the responses are collected automatically and immediately accessible.

More Accurate: As stated before, there is less error because "participants enter their responses directly into the system."

Quick to Analyze: Because the responses go directly into a database and less time is needed to fix errors, analysis can begin sooner.

Easy to Use for Participants: Ninety percent of people that have access to the Internet prefer to answer surveys online instead of using the telephone. With an online survey, participants can pick a time that suits them best, and the time needed to complete the survey is much shorter. Questions that are not relevant to a particular participant can be skipped automatically (this assumes, of course, that the respondent has easy access to a computer).

Easy to Use for Researchers: Since the answers are already in the database, the researcher can quickly pull the data into statistical analysis software for more detailed analysis. Charts and graphs can be easily generated for visual presentation.

Easy to Style: If desired, the survey can be branded to match the company doing the research. Graphics and fonts can be easily changed, and audio and video can be easily included in the survey.

More Honest: Market researchers have found that participants in online surveys usually provide longer and more detailed answers. Because participants feel safe in the anonymous environment of the Internet, they are more likely to open up and give a more truthful response.

More Selective: A more thorough screening can be done and only relevant questions need be asked of each respondent thanks to preprogramming of the survey software.

More Flexible: One issue with long surveys is participant fatigue. By the end of the survey, respondents just want to be done and may not think about their responses as much or even just tick off answers just to be done. Online surveys allow the researcher to randomize the order of questions, thus avoiding the possibility that the questions at the end of the survey are never seriously considered (www.smartsurvey.com).

Disadvantages of Online Surveys:

Sample Quality: Have you heard the term "the Digital Divide"? It refers to the difference between those who have access to the Internet and those that do not. This is something that must be considered in any online survey. If your target market is suburban, middle-class, Caucasian teenagers, then you are probably safe doing an online survey, but if your target market is inner-city youth, you may want to reconsider using online surveys because they may not have the same level of access to computers and the Internet.

Clarity and Follow-up: Online surveys don't allow for direct interaction between the researcher and the respondent, so there is no second chance to explain what you are asking or to encourage respondents to stick it out to the end. This makes pretesting the survey for clarity and length even more important.

Perception and Response Rate: Email solicitation of respondents will probably be treated as junk mail and ignored unless there is a third-party endorsement. Getting a known entity, like a well-respected university, a popular blogger, or a celebrity, to endorse the survey will help increase the response rate – the percentage of people invited to participate that actually complete the survey.

Technical Issues: While problems can arise in any type of survey, the more technology involved, the bigger the problem may be. A respondent may lose power or Internet access during the survey and be unable to finish. The researcher will not know this and may think they just quit halfway through unless they are willing to spend the time and money to create a survey that will allow respondents to return to the survey at the same point where they were interrupted. If the proper precautions are not taken, the same person may take the survey repeatedly, thus skewing the responses. Finally, the survey itself must be tested and retested to make sure that every possible combination of responses is glitch free.

www.cvent.com/en/blog/events/advantages-disadvantages-online-surveys

SYNDICATED RESEARCH

The buzzword in research circles and business today is "big data." **Big data** has been defined as "a collection of data from traditional and digital sources inside and outside [the] company that represents a source of ongoing discovery and analysis" (Arthur 2013). Companies like Chartmetric specialize in making the mountains of data, especially Internet data, manageable for the music business industry. They will be discussed in depth in Chapter 12. These companies are collecting detailed information ranging from physical sales to artist likes and mentions on Facebook and Twitter to create charts and for analysis by their subscribers.

MRC Data is arguably the most important company providing research data and information to record labels. They are self-described as "The industry's premier global music measurement platform. Industry leading music measurement and analytics platform that provides the most comprehensive view of streaming, radio airplay and sales data for your artists, albums and songs" (mrcentertainment.com/data). MRC Connect collects weekly sales data from over 14,000 retail, mass-merchant, and nontraditional (online stores, venues, etc.) outlets. Weekly data from sales are compiled by MRC Music Connect and made available every Sunday (albums) and Monday (songs).

Since the introduction of SoundScan and BDS (both now owned by MRC), the use of syndicated research has become a valuable tool for making marketing decisions in the record business. Chapter 12 illustrates how this data can be used as a basis for more in-depth research to detect sales trends and the impact of marketing strategies. Airplay data can be merged with sales data to determine a more precise impact of radio airplay on record sales than was possible before. With the use of MRC Data, primary research such as business reply cards, syndicated research from other sources, and occasional focus groups, is combined for predicting marketplace performance of new releases, tour analyses, and target market definition and to persuade radio stations to increase airplay.

Mediabase is another company that tracks radio airplay (see Chapter 12). A division of iHeartMedia, Mediabase 24/7 monitors and provides research to nearly 1,800 affiliate radio stations in the United States and Canada on a barter subscription basis. The data collected from radio stations is used not only by record labels and radio stations but to compile airplay charts reported in *USA Today*, AllAccess, and Country Aircheck. Mediabase data is also used in countdown shows such as American Top 40 with Ryan Seacrest and CMT Country Countdown USA with Lon Helton.

Nielsen Audio (formerly known as Arbitron) provides information on radio listening audiences, and much of that information is valuable to the record business. Nielsen Audio not only determines how many listeners each radio station has, but they also break down the audience demographically. This information is useful not only to record labels trying to work a new single to the radio audience but to advertisers trying to reach the same target audiences.

TRADE ASSOCIATION RESEARCH

All the major entertainment trade organizations provide research for their members. The Music Business Association (formerly the National Association of Recording Merchandisers or NARM) publishes periodic results of studies they have commissioned. The Music Business Association provides

FIGURE 4.4 *RIAA Logo*

research findings at its annual convention on a variety of current and ever-changing industry topics.

The Recording Industry Association of America (RIAA) provides data on shipments and contracts with an outside research firm such as MusicWatch and MRC to create its annual consumer profile. The International Federation of Phonographic Industries (IFPI) collects data from member countries and publishes an annual report called *The Global Music Report*. The IFPI also releases periodic reports focused on digital music and global piracy. The associations tend to conduct issue-oriented research of benefit to all members of the record industry. Generally, the reports are provided to dues paying members of the trade organizations or sold to anyone willing to pay the substantial fee for them, but like the RIAA, much of the information is available on their website.

CUSTOM RESEARCH FIRMS

Custom research firms collect and provide data and analyses to answer a specific question for a client. These research firms may specialize in a specific area, such as Internet consumers, and several cover the recording industry and technology. These same companies are also contracted by the various industry associations to conduct specialized research that is then made available to association members.

Edison Media Research is a leader in political, radio, and music industry research with clients that include major labels and broadcast groups.

Music Forecasting does custom research projects on artist imaging and positioning.

ComScore offers consulting and research services to clients in the entertainment and technology industries and conducts audience measurements on website usage through its *Media Metrix* division.

CONCLUSION

Quality research is deceptively difficult and requires careful planning. Different research designs are used depending on the research question to be answered and the depth of information desired. Most of the data collected in the music business is for syndicated reports such as Nielsen reports and *Billboard* charts. The Internet has allowed researchers to more easily reach respondents, reducing the costs and time needed to complete the research process.

GLOSSARY

Big data—"A collection of data from traditional and digital sources inside and outside [the] company that represents a source of ongoing discovery and analysis" (Arthur 2013).

Causal design—A research design in which the major emphasis is on determining a cause-and-effect relationship (AMA online dictionary).

Coding—The systematic process of interpreting, categorizing, recoding, and transferring the data to the data processing program.

Construct—A theory or concept representing ideas that cannot be measured directly.

Custom research—Research undertaken to answer a specific question.

Depth interview—A one-on-one, face-to-face interview done out of the researcher's office that allows the research to probe deeper into the respondent's answers by asking follow-up questions.

Descriptive research—Research undertaken to describe the existing characteristics of a defined target market or population (Hair et al. 2003).

Editing—The process of checking the data forms for errors, omissions, and consistency before coding.

Exploratory research—A research design focused on better understanding a situation, screening alternatives, or discovering new ideas, thereby clarifying and defining the future research objectives.

Focus group—A face-to-face, interactive interview of a small group of people that allows the researcher to delve deeply into participants' responses by asking follow-up questions.

Market research—A subset of marketing research that looks specifically at the size, location, and makeup of a product market.

Marketing research—The function that links the consumer, customer, and public to the marketer through information – information used to identify and define marketing opportunities and problems; generate, refine, and evaluate marketing actions; monitor marketing performance; and improve understanding of marketing as a process. Marketing research specifies the information required to address these issues, designs the method for collecting information, manages and implements the data collection process, analyzes the results, and communicates the findings and their implications (American Marketing Association).

Primary research—Information that is collected directly from a source.

Problem definition—Stating what problem is to be solved and what questions need to be asked to obtain the information needed to determine the solution to the problem.

Qualitative research—Research used to gain an understanding of how underlying attitudes, opinions, and motivations impact consumer behaviors.

Reliability—The researcher's ability to get similar results from repeated applications of the measures or from independent but comparable measures of the same trait or construct.

Research design—A plan that guides the collection and analysis of research data.

Research panels—A group of people put together by a researcher from which data is collected on an ongoing basis.

Secondary research—Using existing research data in an attempt to answer your own research question.

Survey research—A research design that collects data using identical questionnaires administered either in person or electronically.

Syndicated research—The information collected on a regular basis that is then sold to interested clients (American Marketing Association Online Dictionary).

Validity—The extent to which differences in results in the measurements reflect true differences among the objects or characteristics being measured rather than constant or random errors.

BIBLIOGRAPHY

American Marketing Association (2020). "Marketing Research." www.ama.org/the-definition-of-marketing-what-is-marketing/. Accessed March 29, 2021.

Arthur, L. (2013). "What is Big Data?" *Forbes.com*, 15 August www.forbes.com/sites/lisaarthur/2013/08/15/what-is-big-data/?sh=35362ccc5c85. Accessed February 7, 2022.

Hair, J., Bush, R. and Ortinau, D. (2003). *Marketing Research Within a Changing Information Environment*, 2nd ed. Smartsurvey.com www.smartsurvey.co.uk/articles/10-advantages-of-online-surveys/#.UgPpeVPpaww. Accessed August 9, 2013.

Stewart, S. M. (2013). "Artist-Fan Engagement Model: Implications for Music Consumption and the Music Industry" (Doctoral dissertation). ProQuest Dissertations & Theses database. (UMI No. 3612137).

Zikmund, W. G. and Babin, B. J. (2007). *Essentials of Marketing Research*, 3rd ed. Mason, OH: South-Western.

CHAPTER 5

Artist Branding

BRANDING BASICS

What Is a Brand?

The American Marketing Association defines a **brand** as a "Name, term, design, symbol, or any other feature that identifies one seller's good or service as distinct from those of other sellers." A strong brand increases a product, company, and artist's marketability and value. Brands like Apple, Amazon, and Google are just a few examples of strong identities whose brand marks alone are valued at hundreds of billions of dollars each. In fact, a brand can be one of the most valuable components on a corporation's balance sheet, even though it is an intangible asset. For example, Flowers Foods reportedly paid bankrupt Hostess Brands $350 million for its bread brands (including Wonderbread) and attributed 55%, or $193 million, of the purchase price to identifiable intangible assets, mainly trademarks.

> **A STRONG BRAND INCREASES AN ARTIST'S VALUE TO FANS, THE MEDIA, AND BRAND PARTNERS.**

Why Does Artist Branding Matter?

A strong brand increases an artist's value to fans, the media, and brand partners alike. By building a strong brand, an artist can develop a deep connection with their fan base that is likely to result in increased streams and merchandise sold. It also enables the artist to leverage their name to generate both increased interest from the media as well as additional sources of revenue – from licensing deals to endorsements to business ventures where the artist owns the company – to help fund an artist's initiatives and to provide additional promotional support for recordings.

History of Branding

Branding is often thought of having originated with cattlemen, who would burn a distinctive symbol into an animal's skin with a hot branding iron as a way of differentiating one farm's cattle from another's. Historically, however, all manner of goods and crafts were imprinted with the marks of the tradespeople who produced or crafted the product.

The modern concept of branding products originated in the 19th century with the advent of packaged goods. Industrialization created the need for manufacturers who were mass-producing goods in factories to associate their trademarks with trust and familiarity for consumers who may have been previously producing or purchasing the same goods locally, so when shipping the items, the factories would brand their logo on the barrels used. A few of the first manufacturers to embrace this "branding" concept were Campbell's Soup, Coca-Cola, and Ivory Soap.

The basic principles of product branding that corporations have been using for years are also used to build powerful brands in the music industry. In fact, branding has become a central focus of marketers across the music landscape. Radio stations, radio personalities, concert venues, musical instrument manufacturers, and artists each have their own unique personality and attributes for which they are or want to be known. Through imagery, words, slogans, logos, and style, these entities can create a powerful brand that consumers recognize and trust.

Artist Branding and Its Benefits

For an artist, branding is about clearly communicating who they are and what they represent to the people they are trying to attract. It is all the associations that come to mind when you see or hear a name, and it is the way that one artist is differentiated from all others in the marketplace. What are the images and words that come to mind when you think of your favorite music artist? What about Lady Gaga? Carrie Underwood? The Rolling Stones? Miley Cyrus? Chris Stapleton? Each of these artists has a strong brand that is vastly different.

When you think about it, each person you encounter is well versed in branding, although they may not realize that is what they are doing when they choose how they will style their hair, what type of clothes they choose to wear, what car they drive, and how they communicate to others. Just like artists, we each have our own brands to manage, which is referred to as **personal branding**.

An individual artist's or group's brand involves how they are perceived by the public, fans, corporate brands and advertising agencies, bloggers, playlist curators, and more. The artist's brand has to do with all aspects of their public image from how the artist's website looks, to the album art, to the style and brand of clothing that is worn. Logos, fonts, the look and feel of marketing materials and press kit, live performances, concert merchandise, hair style, videos, wardrobe, what the artist says and does in public, and the brands with which the artist is affiliated all play a role in the development of the artist's brand.

COALESCING A STRONG ARTIST BRAND

- Artist name
- Genre/type of music
- Artwork such as album art, logos, branding elements, fonts, and merchandise creative
- Physical appearance (clothing preferences, hairstyle)
- Website and social media accounts (colors and other branding elements, tone and voice, images, and videos
- People and brands with whom the artist collaborates)
- Live performance style (energy of the show, lighting, visual elements, special effects)
- Video concepts and aesthetics

Building a strong brand helps an artist develop more meaningful connections with their fan base. Artists who are adept at projecting a consistent and cohesive brand in everything they do are often perceived as being more genuine and trustworthy. Fans value authenticity and often will reciprocate through support of an artist's initiatives, whether that is purchasing tickets to a virtual event, fan-funding of a music video, or an increase in pre-saves of new releases. As noted earlier, artists with strong brands may also be able to leverage their name to generate additional sources of revenue to help supplement low streaming royalties.

The music company also benefits from strong artist brands since it opens the door to branding and positioning opportunities for both the artist and their recorded music outside the traditional music retail space. More exposure in the media and **non-traditional outlets**, usually executed through strategic **partnerships**, creates a broader awareness of the artist which leads to a bigger audience and, in turn, the opportunity for increased streams and

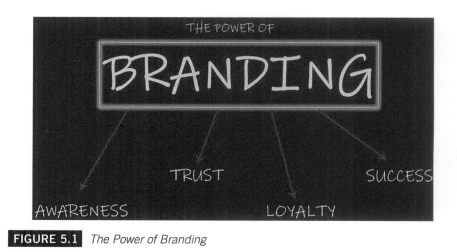

FIGURE 5.1 *The Power of Branding*

overall brand equity for the artist, all resulting in greater profitability for the music company. Thus, our discussion in this chapter will focus not only on the curation and nurturing of an artist's brand but also on being able to leverage the artist's brand to aid in the promotion of recorded music.

SUCCESSFULLY BUILDING A STRONG ARTIST BRAND

Strong artist brands do not happen overnight. It takes years of cultivation on behalf of the artist and the artist's team. To be successful at branding, one must have a firm understanding of how to create the artist's brand, how to promote and build the brand, and how to maintain the brand.

Defining the Artist and Creating a Brand Identity

The first step to developing an artist brand is to truly understand who the artist is, as the branding should reflect their overall persona.

- In what genres does the artist usually play?
- What is their story?
- What makes the artist interesting to fans? What makes them interesting to the media?
- What makes this artist stand out from others in the genre?
- What values do they hold dear? What drives them? Are they focused on family or a free spirit who loves adventure?
- What is their personality and style? Is the artist polished and fashion-forward, or rugged and disaffected?

- Are the artist's live shows full of energy and high production or more laid back and intimate? Does the artist drink alcohol or publicly denounce any kind of vices?

These and other such questions provide insight into the characteristics and personality of the artist. Unique characteristics can often be embraced, as those differentiators help a consumer more easily recognize that brand when compared to others. Defining the artist and being clear about who they are authentically is the first step to building an image and developing a strong brand.

Many artist teams begin this process by identifying approximately ten descriptors that succinctly convey the artist's key attributes. For example, a few descriptors to convey music star and television personality Kelly Clarkson's vibe may be "outspoken, vulnerable, fun, genuine, sociable, humorous, humble, family-oriented, and honest."

Another suggestion to zero in on an artist's core attributes is through the development of an elevator pitch. In other words, how would you describe an artist to a stranger within the time it takes for a brief elevator ride? An elevator pitch is a concise summary of the artist's brand that may be recited at a moment's notice. Elevator pitches vary in length and could be anything from one sentence to 60 seconds in length. It should be memorable and convey the artist's core brand, how the artist is unique, and the value to the artist brings to the table. It should quickly communicate the artist's sound and music by including the genre and a description of the sound using popular artist comparisons or noting artists that were an inspiration.

Consideration must also be given to the artist's logos and creative elements. What type of energy and vibe is projected through the artist's music videos and album art? Is there a consistent theme or brand element that will be carried throughout all projects associated with the artist?

American music artist, YouTuber, and dance star "JoJo" Siwa has a very distinctive brand that separates her from others. Not only does she consistently project a happy, positive, and bubbly attitude, but she reinforces those brand attributes through visual elements such as the bright colors of her clothes, the bow in her hair, and even her famous ponytail. This imagery is carried throughout all facets of the artist's landscape, from social media accounts to video projects to set design at her live shows.

Famous Logo and Brand Element Examples in Music

- The "Metal Lord" font of the Iron Maiden logo gives it a distinct appearance.

- The Grateful Dead's legendary skull and lightning bolt logo was initially created to differentiate the band's road cases on tours but has become one of the most recognizable music logos.
- The tongue and lips logo of the Rolling Stones has become synonymous with the band.
- The popular country duo Brooks & Dunn consistently incorporated a steer head throughout their decades-spanning career.

Creative Elements – Best Practices

- Visual elements should relate to and reinforce the sound of the music.
  - Logos for heavy metal bands often use jagged-edged fonts, while those for a pop band may feature rounded corners and softer imagery.
  - Bold colors such as red and black may be associated with metal and rock, while pink and more subtle hues are often utilized in pop music projects.
- Visual elements should be applied consistently across the artist's landscape from social media to stage signage to merchandise.

Promoting and Building the Brand

Once an artist's brand is clearly identified, the marketing team must then take care to promote and build the brand. This effort involves consistently shaping and promoting all the associations of the brand, both physical and intangible.

Strong Artist Brands Are Authentic and Consistent

Artists must be *clear* and *consistent* about who they are and what they represent. The artist's brand identity should reflect the values, attitude, and personality of the artist in all communication with the public and industry. This step is critically important because of the strong bond an artist has with their fan base. That relationship is built on trust and an understanding of knowing who the artist is. The importance of consistency extends to all creative elements surrounding the artist. Logos, copywriting, and imagery all should remain consistent across all platforms and media. They should be mindful of the tone and voice used in all communications. It should also be consistent with the artist's brand regardless of whether it is the artist himself speaking or a representative. Voice reflects aspects of the artist's brand that are consistent, words and phrases that should never be used, and/or characteristics of how the artist should be conveyed. For example, one artist's voice may be "high energy, upbeat, and positive," while another's is "strong, aggressive, and rebellious." The voice gives the fans and others an idea of who the artist is. Tone, on the other hand, is how the artist's voice is used in various situations. This might be conveyed with the choice of words, punctuation, or even emojis. For instance, adding an angry face emoji to a social media post alters the tone completely from the same message using the laughing emoji.

Strong Artist Brands Are Active and Engaged

Artists should actively communicate who they are. They should engage and interact with their audience on a regular basis and embrace social networks to communicate on a one-to-one basis, particularly while building their fan base.

Labels and artists should actively look for opportunities to showcase the artist's brand, from including the logo and marks across all available media, to walk-on roles in popular television programs, to getting the artist involved in a charitable effort, all of which help build awareness and brand equity.

The best artists know how to meaningfully connect with their fans by sharing a story, a special moment, or just communicating an everyday frustration. These interactions are opportunities to promote the artist's authenticity and create a lasting bond with loyal supporters. The fan base of the band One Direction has often been described by the media and insiders alike as one of the most passionate groups in the music business. The group often sends updates on projects to its social audience, recognizes its strongest supporters publicly, and engages with fans in creative ways like asking fans to submit video of themselves for a chance of being featured in a movie about them.

Maintaining and Protecting the Brand
Strong Brands Monitor What Is Being Said About Them

TweetDeck, Hootsuite, Social Mention, Google Alerts, and Google Trends are just a few of the services and tools that make it easy to monitor social conversations about a brand. Savvy marketers might set up a command post to monitor the artist's name, album names, song titles, related hashtags, and other strategic keywords to engage with consumers who are talking about the artist and to protect the brand by quickly diffusing any negative comments that may damage the brand.

Sinead O'Connor's career was negatively impacted by a Saturday Night Live performance in which she tore up a photo of Pope John Paul II. A video that allegedly featured artist R. Kelly having sex with an underage girl threatened to derail his career with the negative media coverage. Natalie Maines, lead singer of the multi-Grammy winning group formerly known as the Dixie Chicks, announced the group's objection to the Iraq war while performing oversees. This prompted once-loyal country radio stations and fans to have CD-burning parties and resulted in immediate cancellation of U.S. concert tour dates as their conservative country fan base turned against the trio. These are just a few moments that resulted in irreparable damage to the brands of these artists.

While social media has made connecting with fans easier than ever before, today's connected world means artists and the artist's team must be vigilant to ensure that all messaging is "on brand" and that potentially damaging incidents are controlled quickly.

SUCCESSFULLY LEVERAGING THE ARTIST'S BRAND

Partnerships between artists and brands have evolved significantly over the last decade. While fans previously may have referred to a blatant sponsorship as contrived or as an artist "selling out," corporate partnerships are now commonplace and often seen as an authentic extension of the artist that provide much-needed financial and promotional support. Funding from branding deals has often been the factor that enables mid-tier artists to turn a profit for the year vs. just breaking even, and during the COVID-19 crisis, when the live events business came to a screeching halt, brands dedicated to the music space stepped up to keep the music playing – from paying for call-outs or discreet **product placement** during livestreams from artists' living rooms to prominent sponsorship of other virtual events. While deals were much less lucrative than they had been for full tour sponsorships in years prior, the much-needed capital provided by brands kept many artists afloat during a difficult time and no doubt strengthened the relationship between both parties.

And it is not just the superstars who benefit from brand relationships. A strong brand can be leveraged by artists at every level – from local artists who are attempting to grow their audience or secure a partnership with a local retailer to mid-sized acts that are on the cusp of breaking out. Importantly, partnerships between artists and corporate brands can be critical to artists that are not securing prominent exposure on playlists or radio by generating awareness of and interest in the artist and their music.

Examples of Artist Brand Partnerships

Double-platinum-selling artist Kanye West is a rapper, record producer, and businessman. In addition to successfully selling music, the unconventional artist has turned his strong brand into numerous business ventures. *Forbes* has named West as a top-earning celebrity with an estimated $1.3 billion in wealth that is derived from royalties from a partnership with Adidas for his Yeezy branded shoes, a separate Yeezy apparel line, a Yeezy partnership with top clothing retailer Gap, and his Sunday Service events, in addition to music-related royalties.

Platinum-selling artist Jay-Z is consistently on the top of the *Forbes* Top Celebrities list, made possible by the income generated not necessarily by his music sales but from his corporate partnerships and business ventures like Roc Nation, a full-service entertainment and sports management company, a chain of night clubs, and top champagne and cognac brands. The artist also used his strong brand to launch the music streaming service, TIDAL. The hip hop mogul famously utilized his partnership with Samsung for his promotion of *Magna Carta Holy Grail* with the company purchasing

Table 5.1	2020 Celebrity 100 List		
Forbes Celebrity List Rank*	**Artist**	**Net Worth**	**Major Businesses, Partnerships, and Endorsements**
2	Kanye West	$1.3 B	Royalty from Adidas, maker of Yeezy shoes; Yeezy apparel line; Gap/Yeezy partnership; and Sunday Service events
17	Ariana Grande	$72 M	Cloud Eau de Parfum
20	Jonas Brothers	$68.5 M	Villa One, a tequila brand in partnership with spirits giant Stoli and designer John Varvatos
21	The Chainsmokers	$68 M	Largest stakeholders in JaJa Tequila
23	Ed Sheeran	$64 M	Partnership with Heinz ketchup that includes commercial spots and a limited-edition bottle, as well as a branded line with Lowden guitars
25	Taylor Swift	$63.5 M	Keds, Diet Coke, CoverGirl, Capital One, and Apple
28	Post Malone	$60 M	Bud Light Seltzer deal and his own branded rosé, Maison No. 9
37	Sean Combs (formerly Puff Daddy)	$55 M	Partnership with Ciroc vodka, as well as an ownership in DeLeón tequila and Aquahydrate alkaline water
42	Jay-Z	$ 1 B**	Entertainment company Roc Nation, Armand De Brignac champagne, and D'Ussé cognac

*2020 Celebrity 100 List
** Net worth for Jay-Z was updated by Forbes on 2/1/21

1 million copies to give to owners of its smartphones before the album's official release. The artist and album promotion were featured in millions of dollars' worth of advertising across all mediums – from digital to social to radio to television and out-of-home.

Rihanna is another artist who has been very successful in leveraging her brand. Beyond her music-focused endeavors, the superstar has made millions from Fenty Beauty, the makeup brand she co-owns with luxury goods group LVMH.

Artists in every genre are learning to leverage their brand partnerships across multiple **product categories**. Taylor Swift's campaign for her *Red* album was a great illustration of this with the **branded displays** in the front of Walgreen's stores, an album-and-pizza promotion with Papa John's that included album art on box tops as well as a special offer to purchase the CD with every pizza order, her own shoe line for Keds promoted across multiple media outlets, a deluxe exclusive and branded display with Target, a promotion with American Greetings, and a four-sided corrugated cardboard floor display in Wal-Mart's **"Action Alley."** Swift also renewed her presence in fragrance departments with the launch of a new fragrance for Elizabeth Arden to follow the success of Swift's Wonderstruck perfume. The buzz and excitement around the album launch was largely credited to the comprehensive marketing campaign. The exposure value of the partner promotions was worth millions, it helped secure the album's successful launch, and it increased Swift's brand equity. Swift continued to utilize her corporate partners with subsequent releases and was successful in generating enormous buzz with each coordinated effort.

> Product category – a group of similar or related items. Examples of major product categories include apparel, automotive, beauty, banking and finance, entertainment, consumer electronics, mobile and telecom, healthcare and pharma, sporting goods, travel, office supplies, pet supplies, food and beverage, and home improvement.

Kanye West, Jay-Z, Taylor Swift, and many other artists have discovered the power of leveraging their names or brands into successful business ventures where the revenue potential can augment or sometimes dwarf their earnings from recorded music.

The Role and Benefits of Corporate Partnerships in the Music Industry

Music and brand partnerships continue to gain in popularity because of the enormous benefits they offer to each party – the brand, the label, and the artist/management team.

CAMPAIGN SPOTLIGHT: RAM TRUCK BRAND AND FOO FIGHTERS

In 2021, the Ram Truck brand and Foo Fighters partnered to pay homage to everyday heroes with a multimedia campaign, "Spotlight," which included media distributed across television, digital, and Ram Truck brand social media channels. The campaign kicked off just before Mother's Day and coincided with the release of *From Cradle to Stage*, a new unscripted series directed by Foo Fighters' frontman Dave Grohl and inspired by his mother Virginia Hanlon Grohl and her critically acclaimed book. This effort followed the release of *What Drives Us*, a documentary Grohl directed as "a love letter to every musician that has ever jumped in an old van with their friends and left it all behind for the simple reward of playing music."

Olivier Francois, Global Chief Marketing Officer of Stellantis, RAM's parent company, explained that "Foo Fighters' story with the Ram Truck brand started 25 years ago when they piled into a Ram van and headed out on the road in search of a dream. Their ability to pursue those dreams were enabled by the ones who have continued to support the band throughout what has turned out to be a one-of-a-kind journey."

Stretching back to the #RamBandVan initiative launched in 2018 and decades earlier to the 90s and the band's original Ram van featured in *What Drives Us*, the partnership between the Ram Truck brand and Foo Fighters was born out of a shared passion to enable hardworking musicians to get out and tour, knowing the struggles, financial and otherwise, that come with the grassroots touring lifestyle.

Soundtracked by Foo Fighters' "Making A Fire," the campaign launched with an extended 60-second version of "Rock Star" featuring Grohl. As a tribute to the mentors

FIGURE 5.2 *Ram Truck Brand and Foo Fighters*

and leaders who are Ram truck owners, the video showcased everyday people such as parents, teachers, and coaches supporting kids, while Grohl is seen in the spot driving his Ram 1500 and listing all the things it takes to overcome obstacles on the path to rock stardom before stating, "bringing out the very best of them takes the very best of us."

The Ram Truck brand's "Spotlight" campaign was created in collaboration with Austin, Texas-based agency GSD&M with support from G7 Entertainment Marketing, who brokered the Foo Fighters partnership.

FIGURE 5.3 *Foo Fighters Campaign*

(Source: May 6, 2021 Press Release, "Ram Trucks Brand Teams With Foo Fighters to Launch New Advertising Campaign to Shine a Spotlight on Everyday Rock Stars")

FIGURE 5.4 *Dave Grohl and Mom*

From the Brand's Perspective

Corporate brands value partnering with music acts and events to build brand awareness among the artist's fan base, engage with consumers, and build brand loyalty. Corporate brands understand the deep connection artists have with their fans, and they hope that by being involved in the artist's activities and events, they can enhance the brand's image or improve the attitude toward the brand.

In recent years, the increased use of social networks to develop a strong direct relationship with a brand's consumer base has made music partnerships even more attractive. Brands today are embracing **content marketing** strategies at a rapid rate, seeking to build goodwill and drive a deep emotional connection to their customers by creating great content (primarily video) around topics in which their customers are most interested rather than pushing out sales messages. It is a natural choice since music partnerships offer both the connection aspect and video. And brands today see more value in music as a promotional vehicle than ever before. In fact, brands spend billions annually on music sponsorships.

> The energy drink company Red Bull has built a strong brand for more than 20 years with its music initiatives. The company's efforts include a multi-city music festival, a record label that works with up-and-coming artists, and a music publishing arm. For 20 years, skateboard shoe manufacturer Van's has also created a bond with music fans through sponsorship of the Warped Tour.

From the Music Company's Perspective

As the revenue model for music companies has shifted from a focus on physical products to streaming, conglomerates and independents alike have embraced partnering with a corporate brand for a range of reasons – from supplementing the cost of a tour or music video to expanding the promotional opportunities for the music. Consider that marketing budgets for new music largely target advertising and promotional elements such as online advertising and radio promotion because they have been shown to have a strong, proven track record of driving consumer behavior. Ask any music marketer, and they will tell you that they would love to incorporate other mediums and tactics, but their budgets just generally will not allow for it. However, the label may gain access to all those elements through a strategic partnership with a corporate brand if the timeline of the corporate brand aligns with that of the new music being released into the marketplace.

The music company sees value in the significant promotion that corporations can bring to an artist and recorded music. The media and assets provided by the brand partner bring value to the artist to which they might not otherwise have access given the label's limited marketing budget. Many corporate brands spend millions of dollars each year on in-store displays, television ads, radio campaigns, direct mail, outdoor advertising, and digital efforts, dwarfing the amount generally spent annually by the label to promote a recording. By incorporating promotion of the album into the corporate brand's messaging, the corporations offer the opportunity to expose the artist and the music to a wider audience, which translates into greater awareness; increased revenue from ticket sales, merchandise sales, and streams; and the building of the artist's brand.

For example, consider that the cost of a billboard in Times Square is $150,000 to $1.5 million per month; a 30-second spot on a single airing of NBC's "The Voice" costs more than $250,000; and a full-page, four-color advertisement in People magazine has an **open rate** of $420,000, and you can start to see the tremendous value that brands can bring to the partnership table. These high-priced promotional vehicles are more easily attained by an artist with a limited budget when acquired through corporate partnerships.

Labels also look at corporate partnerships favorably because of the potential to generate additional revenue from a sponsorship fee. Notably, most recording contracts now offer the opportunity for the label to share in most of the artist's revenue streams, not just sales and streaming revenue of the recorded product, and that often applies to sponsorship revenue as well.

Although many partnerships do involve a sponsorship fee, some agreements provide only **in-kind support**, where the media and marketing support the corporate brand brings to the table are the only "currency" involved in the deal. These types of partnerships may be particularly important for the "baby" acts who may not yet be selling sold-out shows but can benefit from having some of their costs either covered or offset by a sponsoring brand.

> **In-kind support** – Goods or services are often provided to artists by companies instead of money. Examples include mobile phone services, music equipment, or technological support that companies might trade artists in exchange for brand inclusion on the artist's tour bus or on-stage signage.

From the Artist's and Artist Manager's Perspective

As important as partnerships are for the music companies in promoting the recorded music to the masses, managers and artists are also seeing the value brands provide in generating awareness of the artist, expanding the fan base, and building brand equity. Partnerships can be used to promote the artist in non-traditional product categories and places such as in the cosmetics counter of a department store or on the shelves of a grocery or sporting goods retail store. This not only helps stimulate streams of the artist's music but also provides valuable impressions that help build brand equity. As previously mentioned, partnerships may also provide an alternative source of revenue to help offset low artist royalty payments.

Linking Brands

There are many opportunities to leverage the strength of two or more brands through strategic partnerships. These relationships take different forms, with some of the most popular being product placement, sponsorships, and artist endorsements.

Product placement is a form of advertising that strategically places branded goods or services within a produced work. These efforts most commonly take place within music videos, films, and television programs, with the intention of having the product blend naturally into the scene so the integration is not blatant. In many cases, the integration fee is used to help offset the production costs associated with the video. Prominent examples of product placements in music include Avril Lavigne answering her Sony Xperia phone that was sitting in a glass of water during her video for "Rock and Roll," Britney Spears' integration of a dating site in "Hold it Against Me," EOS Lip Balm in Miley Cyrus's "We Can't Stop," and Lady Gaga's numerous product integrations into her videos.

Sponsorships occur when an organization pays a sum of money toward the cost of an activity or event (such as a music festival, tour, or other event) in return for the right to attach their logo to it. Sponsorships allow companies to connect with consumers by becoming part of an event using its marks being included on stage signage, on bus wraps, and exhibition opportunities at the event. The brand hopes that by being associated with an event that the audience likes and enjoys, the positive feeling about the event may get transferred to the brand. This is referred to as the **association effect**. For example, while Bud Light may sponsor a festival, most fans understand that the support is directly tied to that specific event rather than a statement of support for the product from the artists who perform on the stage.

A celebrity **endorsement** is the act of giving one's public support or approval to someone or something. In essence, the artist is affirming that they use and endorse a product or service exclusively. In this scenario, the artist's name and reputation are more tightly linked to the product. Examples of music endorsements include Ed Sheeran for Lowden Guitars, DJ Khaled for WW (formerly Weight Watchers), and Beyoncé for Pepsi.

Today, "partnership" is often the preferred term in the development of artist and brand arrangements, with the thought being that each group is bringing valuable assets to the agreement, and most contracts involve a mix of cash and other assets.

Many deals developed today enable the brand to become organically involved in the artist's career and initiatives so that there is authenticity in the partnership. Pepsi, a longtime supporter of music initiatives, signed megastar Beyoncé as a "brand ambassador" in a deal that was reportedly worth $50 million. The agreement was heralded as the future of music partnerships because of the star appearing on Pepsi products and commercials; having a guaranteed performance in the Pepsi-sponsored Super Bowl; and being provided a multimillion-dollar fund for Beyoncé to use to generate content from live events, for videos, or humanitarian efforts. The move fosters an even deeper spirit of collaboration between the parties by allowing the artist to drive the decision regarding how the funds will be used, while Pepsi benefits from the brand equity and sales boost generated from the partnership.

FINDING THE RIGHT BRAND PARTNER

Defining the Audience

In addition to needing to know the artist's brand, it is important to understand whom that artist serves. In other words, who is streaming the music? Who is purchasing tickets to concerts or livestreams? Who is engaging with the artist on social media?

Not only is this important information for the artist as they look to build a larger following, but it is critical for the brand. Corporate brands seek music artists who have attributes that align with theirs, but they also want to ensure that the artist's fan base is in alignment with the brand's target audience, which ensures that the partnership is a good **brand fit**; in other words, a collaboration where the audiences of both parties share the same attributes and personality.

Research plays a critical role in this step, and the data may come from numerous sources. One of the easiest ways to gain valuable demographic and geographic information is through the insights available on the artist's social accounts. Spotify, TikTok, Facebook, YouTube, Instagram, and other platforms provide valuable insight that may collectively be used to better understand the characteristics of the audience, particularly demographic and geographic information. Online surveys may also be created and distributed to the artist's email list, mobile list, ticket purchaser database, or social profiles asking lifestyle and behavior questions, and intercept surveys may be conducted as fans enter or exit a concert.

Through this research activity, artists and their teams can find out more about their existing customer base, which in turn helps them better target potential fans. The artist's marketing teams can gain a deeper understanding of what their fans want from them and what characteristics and traits they love about the artist and the artist's music, which will help the artist better meet their needs. With this insight, a **target audience** and **customer personas** can be developed. A customer persona is a profile or description of the ideal customer that includes demographic, geographic, and psychographic characteristics, as well as consumption patterns. Often, multiple personas are created to help visualize multiple segments of the artist's target market. These profiles may be used to identify the best places to place paid advertising promoting new music, as well as to identify which categories and brands they like most.

In addition to primary research on the audience, labels and management groups use secondary tools and services to help them develop the artist's profile and identify corporate brands that make the most sense to target for partnerships. One alternative data collection source is using a third-party provider of consumer behavior. There are companies in the United States that specialize in surveying consumers and selling the results to interested parties such as corporate brands, media outlets, agencies, and other entities. Two of the most widely utilized databases in the advertising and marketing industry are MRI-Simmons and Scarborough. While these two companies provide research across many different sectors, MusicWatch and MIDiA are two additional research providers who specialize in providing insight on the music consumer.

- MRI-Simmons gathers information on consumer behavior through its online National Consumer Study. Data of the study is released quarterly and includes the survey results of approximately 25,000 U.S. adults across more than 8,000 products in approximately 500 categories. Media choices, demographics, lifestyle and attitudes, and a plethora of other information may be cross-tabulated to reveal specific behaviors or characteristics of a target market. MRI-Simmons licenses this database to companies and organizations that can pull data specific to their need. For example, a music company or talent agency looking to secure a brand partnership for an EDM artist might run a report of all adults aged 18–49 who listen to EDM music with the brands and products they purchase, what television shows they watch, and what activities in which they like to participate during their spare time. The reports can be used to identify the most likely prospects for a partnership or to strengthen a partnership proposal. On the flipside, most large advertising agencies subscribe to MRI-Simmons and may use it to identify the best genre with which to align for a tour sponsorship.

- Nielsen Scarborough is another market research service that offers consumer demographics and insight on media consumption, attitudes, and purchase behavior. Nielsen Scarborough captures consumer data on more than 2,000 product categories and brands collected from an annual survey of more than 200,000 adults in the U.S. The main difference between MRI-Simmons and Scarborough is that Scarborough specializes in local market insight. For example, while MRI-Simmons can tell you how many people who go to hockey games stream rock music heavily, Scarborough can tell you how many people who go to Seattle Seahawks games do so. This localized information is particularly helpful to radio and television stations, which often use the data to sell ads to local businesses.
- MusicWatch, Inc. provides custom consumer research for the music and entertainment companies through its proprietary dataset. From studying emerging digital and video trends to fan base profiling, MusicWatch is utilized by labels, trade organizations, and other industry groups to better understand how and where Americans are consuming music.
- MIDiA Research is a research and analysis subscription service that also uses a proprietary database. The company focuses on digital content, including music, online video, and mobile content from which insight is published regularly. They also offer a consulting service for those needing custom research.

The Process

Increasingly, when an artist is signed to a record deal, brand partnerships are a top priority. Early on, the partnerships team at the music company or talent agency will ask the artist a series of questions to gauge his or her affinity for certain products, interests, and values to begin to formulate a list of potential corporate partners for the album launch. The team wants to find a good brand fit for the product because that can help ensure an artist's commitment to the partnership (or buy-in), which is a critical component to its success.

In today's music environment, it is not at all uncommon for partnership pitches to come from either the artist's agency or from an agency that specializes in developing and implementing these types of opportunities, as opposed to the label. The independent agents may work on behalf of a corporate brand that is seeking to secure sampling opportunities by attaching themselves to a nationwide tour or on behalf of the artist who is hoping to find a corporate partner to cover the production cost of a music video.

Often, each party involved will approach a potential partnership by throwing each of their assets on the virtual table and seeing which ones are interesting and valuable to the other party. Any cash payment involved as well as which assets are offered up depend on what the artist is asked to do (private concerts, commercial shoots, meet and greets, etc.). Again, trade-only or in-kind deals are more popular with up-and-coming or mid-tier artists; some sort of cash and trade combination is standard for well-known, mainstream acts.

CREATING AN ARTIST BRAND DECK

An **artist brand deck** (also referred to as an artist presentation, artist brand profile, or brand booklet) is a visual presentation of an artist that is sometimes provided to corporate brands, agencies, and the media to offer some background on the artist's career and personality. The presentation is usually brief (8–10 slides) and includes the following elements:

- Links to the artist's recent and most popular videos.
- Highlights of the artist's career and successes they have obtained – radio, sales, albums released, and so on.
- Social stats and fan reach and engagement.
- Press coverage or other industry quotes that speak to the success or appeal of the artist.
- Upcoming events including appearances in movies or television programs, album releases, collaborations with other artists, tours, and so on. This is important information, as it gives the brand an opportunity to compare their promotional calendar to see at which events they may be able to leverage their involvement.

5 STUDIO ALBUMS
1 GREATEST HITS ALBUM / 1 CHRISTMAS ALBUM
20 MILLION ALBUMS SOLD WORLDWIDE
12 MILLION ALBUMS SOLD IN THE US
10 TOP 10 SINGLES
3 GRAMMY AWARDS
4 AMERICAN MUSIC AWARDS
3 MTV AWARDS
2 ACADEMY OF COUNTRY MUSIC AWARDS
1 COUNTRY MUSIC ASSOCIATION AWARD
12 BILLBOARD AWARDS

FIGURE 5.5   *Kelly Clarkson*

A brand deck may also include product categories or brands that would be a good fit for the artist, as well as research statistics to strengthen and support those assertions. Perhaps it denotes activities in which they like to engage, products they consume regularly, or brands that speak to their personality. It may also include the artist's audience profile.

The information conveyed should help agency and company representatives understand the attitude, style of music, preferred activities, and other characteristics of the artist even if the brand representatives were unfamiliar with the artist previously.

FIGURE 5.6   *Need to Breathe One Sheet*

FIGURE 5.7 *Michigander One Sheet*

A more condensed version of the brand deck is often presented in a **one sheet**. The **one sheet** often serves multiple purposes in that it may be sent to Digital Service Providers (DSPs), radio programmers, or brand agencies. A one sheet includes an artist biography, an embed of the latest music video or other links to music, career highlights, social stats and links, streaming and chart successes, and upcoming tour information.

Cause Marketing With a Brand

Many artists have found success in partnering with a corporate brand to jointly promote a charitable or public relations effort, referred to as **cause marketing**. Corporate brands have money and media assets (magazine ad, television spots, large social networks) that can be used to support an effort about which an artist is passionate. This is often a more comfortable way for artists to create an affiliation with a brand, particularly those that fear their fans will see a corporate endorsement as the artist "selling out." The brand often gets the benefit of creating goodwill among its customers and the artist's fans for its support of a project. It is a win-win-win scenario for the artist, corporate brand, and cause. For example, Big Machine Label Group partnered with General Mills and Feeding America for the Outnumber Hunger campaign, which included special concert events and the label's roster of artists being featured on selected boxes of General Mills cereal.

Artist-Owned Brand Opportunities

Increasingly, rather than an artist renting or licensing their name to an organization through a short-term agreement, the artist sometimes chooses to take more of an ownership role by owning and controlling the brand. While the risk with this approach is greater for the artist, the potential rewards are substantial if the artist has the time and bandwidth to be engaged with a product at a deeper level.

Jimmy Buffett's expansion of the Margaritaville brand is one example of this. Buffett has built an empire from the leveraging of the name, which includes massive merchandising efforts and a national restaurant chain.

Avril Lavigne has spent years building a brand that she was able to leverage into several businesses, including her own line of clothing and a line of fragrances.

Rather than opting for another endorsement deal, multi-platinum artist Kenny Chesney create his own Caribbean premium rum company, Blue Chair Bay, inspired by his music and island lifestyle.

Chesney was able to use his connection with his loyal fan base to promote the brand. In fact, one of the key marketing tactics for the launch of

FIGURE 5.8 *Kenny Chesney's Blue Chair Bay Rum*

(Photo credit: Allison Ann)

Blue Chair Bay was a co-sponsorship of Chesney's No Shoes Nation tour in 2013 where fans could sample the three flavors of rum before the show on the deck of a specially created 20-foot sailboat, reaching an estimated 1.25 million fans in 42 markets nationwide. The rum line was also front and center for the 2015 Big Revival Tour, and in 2021, Chesney continued to integrate his fan base into the promotion of the spirit by asking his supporters to choose the next Blue Chair Bay rum flavor. Today, the spirit continues to be cited as one of the fastest-growing rum brands in the country.

FIGURE 5.9 *SiriusXM NSR Logo*

Chesney also continues to reinforce his brand through No Shoes Radio, an exclusive Sirius XM channel that is curated by Chesney himself. It features his music and exclusive content, and it also highlights his favorite artists across the musical spectrum. As the superstar noted, the channel offered the opportunity to "blur the lines between my records and everything from reggae to rock to bluegrass." In essence, the station and rum line both offer the artist the ability to reinforce the brand he has spent years developing through his tropically inspired life and music.

CONCLUSION

The importance of artist branding and brand partnerships cannot be overstated. Today's environment is challenging, retail space is scarce, and revenue from music consumption is still down from industry highs, resulting in less profit for the label and artists alike.

A strong artist brand:

- Assists an artist in building a passionate following.
- Generates interest among the media, resulting in increased publicity opportunities.

- Allows the artists themselves to capitalize on the brand name they have developed and generate additional revenue.
- Attracts the attention of corporate partners looking to expose their products to a passionate fan base. The increased marketing impressions lead to more brand equity for the artists.

Increased consumer awareness from broader exposure helps build brand equity that can be leveraged in many ways for greater profitability by both the artist and the label.

GLOSSARY

Action Alley—A term used most within the Wal-Mart community of employees and suppliers that refers to the main aisles down the center of the store that separate the departments.

Artist brand deck (a.k.a. artist brand profile, artist deck, or brand booklet)— A visual presentation usually created in software such as PowerPoint or Keynote that is created to provide a snapshot of an artist's background, lifestyle, sales success, career highlights, and brand preferences. This presentation is most often used to educate members of the media, advertising agencies, and brand marketers on the artist's brand.

Brand—A name, term, design, symbol, or any other mark that uniquely identifies a seller.

Branded display—A special section in a retail environment, usually at the end of an aisle or in the middle of a wide aisle that separates departments, that features select products and is branded with that product's logo and photos.

Brand fit—refers to how well two partnership brands align. A good brand fit refers to a partnership where the artist's personality and fan base fit the personality of the corporate brand and vice versa.

Cause marketing—A form of marketing where an artist teams up with a charitable organization to raise awareness or resources for a social issue or problem.

Content marketing—A marketing tactic by which companies create, gather, and distribute valuable and relevant content to its existing and desired target audience. The objective of content marketing is to establish goodwill, which will eventually lead to sales or other customer action.

Customer profile—A "persona" or detailed description of the ideal customer. Whereas a target audience might be 18–21-year-old females who make a salary of less than $25,000 and live in suburban markets, a customer profile might be a 19-year-old female college student who likes Top 40 pop music, shopping at Forever21, and dancing.

Endorsement—The act of an artist giving his or her public support or approval to an individual, brand, or company (e.g. Jessica Simpson and Proactiv).

In-kind support—Goods or services provided to an artist instead of money. Common examples include music equipment, lighting for stages, and travel services.

Non-traditional outlets—Online and offline stores where music is not generally sold, such as grocery stores, sporting goods stores, and clothing retail stores.

Open rate—The highest rate charged by an advertising medium such as a magazine or newspaper for placement in the publication. Contract customers, or those placing multiple ads throughout a specified timeframe, receive quantity discounts, but non-contract advertisers (such as those who only run one ad) are charged the open rate.

Personal branding—The branding and self-positioning of the individual person. A personal brand involves all the attributes and characteristics associated with an artist, including their style, demeanor, and reputation.

Product category—A group of similar, like, or related items; for example, Ford, Toyota and GM all make products that fall under the "automotive" category.

Sponsorships—The act of an organization paying a sum of money toward the cost of an activity or event such as a tour or music festival in exchange for assets surrounding the event like inclusion in advertising, on-site signage like stage banners, and the ability to engage with consumers onsite via booths and product sampling.

Target audience—A group of potential customers usually defined by ranges (age, income, etc.) that is the group of people targeted for a promotional effort.

BIBLIOGRAPHY

Aaker, D., and Erich Joachmisthaler. *Brand Leadership*. New York: The Free Press, 2009. 97–128. Print.

AMA (American Marketing Association). 2021. www.ama.org.

"BGB Archive." *Interbrand*. Web. 24 Mar. 2021. interbrand.com/best-global-brands.

Bieler, Kristin. "High Tide for Blue Chair Bay." *Beverage Media Group Blog*, 1 Apr. 2014. Web. 14 Nov. 2014. www.beveragemedia.com/index.php/2014/04/high-tide-for-blue-chair-bay.

"Big Machine Label Group, General Mills and Feeding America® Partner with Multi-Platinum Artist Thomas Rhett for 2017 Outnumber Hunger Campaign." *Big Machine Label Group*. www.prnewswire.com/news-releases/big-machine-label-group-general-mills-and-feeding-america-partner-with-multi-platinum-artist-thomas-rhett-for-2017-outnumber-hunger-campaign-300413397.html.

Blue Chair Bay. "Blue Chair Bay Rum Launches 25-City Spring into Summer Sampling Tour." 24 Apr. 2014. www.prnewswire.com/news-releases/blue-chair-bay-rum-launches-25-city-spring-into-summer-sampling-tour-256529591.html.

"A Brief Overview of the History of Branding." *AEF Advertising Educational Foundation*. AEF, 2009. Web. 14 Nov. 2014. www.aef.com/index.html.

Coffee, Patrick. "Pepsi and Beyoncé: The New Sponsorship." *PR Newser*, 11 Dec. 2012. www.mediabistro.com/prnewser/pepsi-and-beyonce-the-new-sponsorship-model_b52383.

Dahud, Hisham. "The 3 Cs of Effective Artist Branding." *Hypebot*, 6 Aug. 2012. Web. 14 Aug. 2014. www.hypebot.com/hypebot/2012/08/the-3-cs-of-effective-artist-branding.html.

Diaz, Ann-Christine, and Shareen Pathak. "10 Brands That Make Music Part of Their Marketing DNA." *Advertising Age*, 30 Sept. 2013. adage.com/article/special-report-music-and-marketing/licensing-10-brands-innovating-music/244336.

Dimofte, C. V., and Richard F. Yalch. "The Mere Association Effect and Brand Evaluations." *Journal of Consumer Psychology*, Vol. 21, 2011, pp. 24–37. http://www-rohan.sdsu.edu/~dimofte/jcp11.pdf.

Erdogan, Zafer. "Celebrity Endorsement: A Literature Review." *Journal of Marketing Management*, Vol. 15, No. 4, May 1999, pp. 291–314.

"Forbes World's Highest-Paid Celebrities." *Forbes*. Web. 2020. www.forbes.com/celebrities.

Greenburg, Zach O'Malley. "Rum Diaries: Kenny Chesney Diversifies with Blue Chair Bay." *Forbes*, 1 July 2014. www.forbes.com/sites/zackomalleygreenburg/2013/07/01/rum-diaries-kenny-chesney-diversifies-with-blue-chair-bay.

Hampp, Andrew. "From This Week's Billboard: How Taylor Swift's Red Is Getting a Boost from Branding Mega-Deals." *Billboard*, 22 Oct. 2012. Web. 14 Nov. 2014. www.billboard.com/biz/articles/news/branding/1083319/from-this-weeks-billboard-how-taylor-swifts-red-is-getting-a.

Hayes, Adam. "The High Cost of Times Square Advertising." *Investopedia*, 29 Mar. 2020. www.investopedia.com/articles/investing/022315/high-cost-advertising-times-square.asp#:%7E:text=It%20costs%20between%20%241.1%20and,on%20Time%20Square's%20largest%20billboard.

Hill, Eliot. "Dave Grohl Working on New Documentary, Confirms Foo Fighters Album Is Done." *IHeartRadio*, 12 Feb. 2020, www.iheart.com/content/2020-02-12-dave-grohl-working-on-new-documentary-confirms-foo-fighters-album-is-done.

Houghton, Bruce. "Samsung Buys 1 Million Copies of Jay-Z New Album." *Hypebot*, 17 June 2013. Web. 14 Nov. 2014. www.hypebot.com/hypebot/2013/06/samsung-buys-1-million-copies-of-jay-z-new-album-magna-carta-holy-grail.html.

Huber, Chris. "The History and Meaning of the Grateful Dead Steal Your Face Logo." *Extra Chill*, 29 Nov. 2020. extrachill.com/2020/11/grateful-dead-steal-your-face-logo-meaning.html.

"Kenny Chesney Enlists No Shoes Nation to Choose Next Blue Chair Bay® Rum Flavor: Mocha VS. Mango | Ebmediapr.Com." *EB Media*. Web. 1 Feb. 2021. ebmediapr.com/news/kenny-chesney-enlists-no-shoes-nation-choose-next-blue-chair-bay%C2%AE-rum-flavor-mocha-vs-mango.

Knopper, Steve. "Sponsorships Are Helping Homebound Artists Replace Lost Revenue, but 'Their Leverage Has Diminished.'" *Billboard*, 17 Apr. 2020. www.billboard.com/articles/business/branding/9359953/artist-brand-deals-livestreams-rates-sponsorships-coronavirus.

Lavin, Kevin. "What's in a Name? Just Ask Hostess Brands – The Value of Branding in Bankruptcy." *ABF Journal*, January/February 2014. www.abfjournal.com/articles/whats-in-a-name-just-ask-hostess-brands-the-value-of-branding-in-bankruptcy.

Lukovitz, Karlene. "Subway, Diet Coke Team on Taylor Swift Promo." *Media Post*, 27 Sept. 2014. www.mediapost.com/publications/article/234991/subway-diet-coke-team-on-taylor-swift-promo.html.

Mansfield, Brian. "Taylor Swift's '1989' Sells 1.287 Million in First Week." *USA Today*, 4 Nov. 2014. www.usatoday.com/story/life/music/2014/11/04/taylor-swift-first-week-million-sales/18480613.

McQuicklin, Jeff. "Understanding Image and Branding as a Musical Artist." *Artist Development Blog*, 27 Jan. 2011. Web. 9 Sept. 2014. http://artistdevelopmentblog.com/advice/understanding-image-branding-musical-artist/.

Poggil, Jeanine. "The Most Expensive Shows on TV; Out of 80 Returning Series on Broadcast TV Tracked by Ad Age, 38 Saw the Cost for a 30-Second Commercial Increase." *AdAge*, Vol. 91, No. 18, Nov. 2020, p. 0010.

"MRI Survey of the American Consumer – MRI-Simmons." *MRI*. Web. 2021. www.mrisimmons.com/our-data/national-studies/survey-american-consumer.

"No Shoes Radio | Kenny Chesney." *No Shoes Radio*. Web. 2021. www.kennychesney.com/no-shoes-radio.

PEOPLE.com. "People.Com | Celebrity News, Exclusives, Photos and Videos." *People.Com*. Web. 2021.

Sauer, Abe. "Avril Lavigne May Have Just Pulled off the Greatest (and Worst) Product Placement of All Time." *Brand Channel*, 20 Aug. 2013. www.brandchannel.com/home/post/Avril-Lavigne-Product-Placement-Rock-N-Roll-082013.aspx.

"Scarborough Survey Participant Information." *Nielsen*. Web. 24 Mar. 2021. scarboroughlocal.nielsen.com/scarborough/index.html.

CHAPTER 6

The U.S. Industry Numbers

SALES TRENDS

The last 30 years of music "sales" has witnessed the emergence and disappearance of new formats of music delivery while withstanding some of the most trying economic challenges of our modern era. Between the "Great Recession" of 2007 and the worldwide pandemic brought on by COVID-19 that shut down most international markets in 2020, the music industry has been confronted creatively and financially – and yet has survived. But to fully appreciate its future, let's look at its recent past.

In the early 1990s, the recording industry was in the midst of a *replacement cycle*, where consumers were in the process of replacing their old vinyl and cassette collections with the first digital format – the compact disc (CD). This boom was fueled by discount retailers such as Best Buy and Circuit City aggressively entering the music market, opening up a multitude of retail outlets, and using discounted CD prices as a loss leader to attract customers. These practices, along with the concept of selling "used CDs," threatened the viability of traditional music stores and caused a plateau in the number of units shipped in the mid-1990s.

By 1995, the CD replacement cycle was nearing completion, and the impact of the changing retail landscape was beginning to take its toll on the bottom line. Traditional retail stores were struggling to compete with discount stores, and sales gains in the industry overall were modest (from $12.068 billion in 1994 to $12.320 billion in 1995). Now-defunct music retailer Blockbuster closed hundreds of stores, causing a massive rush of returned product.

DOI: 10.4324/9781003153511-6

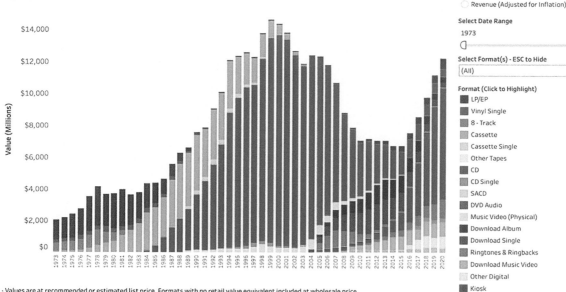

U.S. Recorded Music Revenues by Format

1973 to 2020, Format(s): All
Source: RIAA

Source: RIAA. Permission to cite or copy these statistics is hereby granted, as long as proper attribution is given to the Recording Industry Association of America

- Values are at recommended or estimated list price. Formats with no retail value equivalent included at wholesale price
- SoundExchange Distributions are estimated payments in dollars to performers and copyright holders for digital radio services under statutory licenses
- Paid Subscription includes streaming, tethered, and other paid subscription services not operating under statutory licenses.
- Limited Tier Paid Subscription includes streaming services with interactivity limitations by availability, device restriction, catalog limitations, on demand access, or other factors
- On-Demand Streaming includes ad-supported audio and music video services not operating under statutory licenses
- Other Ad-supported Streaming includes revenues paid directly for statutory services that are not distributed by SoundExchange and not included in other streaming categories
- Kiosk includes Singles and Albums
- Synchronization Royalties include fees and royalties from synchronization of sound recordings with other media
- Ringtones & Ringbacks includes Master Ringtones, Ringbacks, and prior to 2013 Music Videos, Full Length Downloads, and Other Mobile
- Other Tapes includes reel-to-reel and quadraphonic
- Other Digital includes other digital music licensing

FIGURE 6.1 *RIAA Annual Shipment Data*

(Source: RIAA)

The industry experienced a decline in 1997 as "the industry was responding to a smaller but healthier retail base" (RIAA, 1998). Growth returned briefly in 1998, as shipments grew by 5.7%. While shipments of singles dropped, CD units and music video units showed a healthy increase. The RIAA attributed the increase to a steady flow of releases by top artists throughout the year and an increase in the diversity of offerings. The moderate growth continued through 1999, with a 3.2% increase in units, fueled by strong growth in the full-length CD format (12.3% in value), despite a drop-off in music video sales. Credit is given to retailers and suppliers for improved efficiency in inventory management and the emergence of boy bands. As noted by the RIAA, The year 1999 generated the highest total revenues in U.S. history at $14.6 billion.

The new millennium ushered in a period of decline in recorded music sales as the industry struggled against threats on numerous fronts. The market for CD singles plummeted as peer-to-peer file-sharing services were blamed for much of the downturn. Additionally, the year 1999 introduced Napster, and the concept of free downloading rocked the marketplace. Sales began its steep decline, with labels and retailers trying to make up for their losses by increasing the unit cost per transaction, with the height of the practice hitting in 2003 at $15.03 per unit. Adding to the rapid adoption of digital files was the introduction of Apple's iPod in 2001 and its online retail megastore, iTunes, in 2003.

In 2002, the market dropped another 11.2% in units and 8% in value. Every configuration saw a decrease in sales except DVDs. At this point, the RIAA began an aggressive campaign to discourage consumers from using illegal file-sharing services, but other factors also contributed to the downturn. Young consumers were spending their entertainment budget on cell phones, computers, video games, DVD collections, and other forms of entertainment. Consumers began to report (through research studies) that they perceived the cost of CDs to be too high. Upon further analysis, young consumers, who had grown accustomed to cherry-picking songs from albums through P2P services, were opposed to paying retail price for an album just to own the one or two songs they wanted. These consumers were beginning to burn their own personal mix CDs either from their own collection or from P2P services. As a result, they preferred music á-la-carte and were a receptive market for licensed music downloading services. But these services were slow to develop, and the offerings were not sufficient to entice droves of consumers into subscription services.

In 2005, the RIAA added new format categories of mobile, digital subscription, digital music video, and kiosks. "Counting all formats and all distribution channels (retail and special markets distribution), overall unit shipments of physical product decreased by 8.0 percent in 2005" (RIAA press release 2006). Mobile formats (such as ringtones) shipped 170.0 million units, representing $421.6 million in retail value. Full-track downloads were a not significant sector at that time.

In 2007, the world and U.S. market was thrust into what has come to be known as the Great Recession, brought on by several economic factors, including the subprime mortgage crisis, which was attached to a real-estate bubble nationwide. These troubling times magnified household debt, and consumers withheld discretionary income purchases for fear of harder times ahead. For 2008, there was a 14% drop to 428 million albums sold in the United States, while sales of digital downloads increased 27% to just over a billion songs. CD sales were down almost 20% from the previous year (Sisario, 2009).

Music is a luxury good, and during the distressing economic downturn from 2007 to 2012, sales plummeted by 32%. By 2010, the U.S. music industry bottomed out at approximately $7 billion dollars and stayed there for another five years. To think that the industry had been financially cut in half within a ten-year period is staggering, and yet the industry had made steady increases in value since 2015 through innovation and re-invention.

In a relatively short time period, the introduction and adoption of streaming services and their various subscription-based portals means they have basically replaced and "canceled" the CD format. Companies like CD Baby, TuneCore, DistroKid, and Ditto Music help music creators find pathways for consumers to access music through portals such as Spotify, SoundCloud, Deezer, Apple, and Pandora and more. This shift in "accessibility" of both free/ad-supported and subscription-based streaming has massively changed the way the consumers now listen to music. As for physical sales, by 2020, vinyl record sales have outpaced CD sales. But digital streaming accounts for 88% of revenue generated, making streaming the dominant format of choice – full stop.

Sales Trends and Configuration

Since 1991, the most substantial change of the use of SoundScan data (now called Music Connect) in compiling the *Billboard* Charts occurred December 13, 2014, with the modification to the *Billboard* Top 200 to include not only album sales but the additional data of track equivalent album (TEA) and streaming equivalent album (SEA) information. Whereas the RIAA measures shipments for their annual industry numbers, SoundScan measures over-the-counter sales. In an effort to monitor recorded music sales and determine trends and patterns, the industry in general and SoundScan in particular have come up with a way to measure digital album sales and compare them with music sales in previous years. When SoundScan first started tracking digital download sales, the unit of measurement for downloads was the single track, or in cases where the customer purchased the entire album, the unit of measurement was an album. But this did not give an accurate reflection of how music sales volume had changed, because most customers who download buy individual songs instead of complete albums.

In an attempt to more accurately compare previous years with the current sales trend, SoundScan came up with a unit of measurement called *track equivalent albums*, which means that ten track downloads are counted as a single album. This evaluation is based on a financial equivalent, given that a $.99 download × 10 would equal a $9.99 album download or CD. Thus, the total of all the downloaded singles is divided by ten, and the resulting figure is added to album downloads and physical album units to give a project picture of "total" activity reported in *Billboard*'s Top 200 chart.

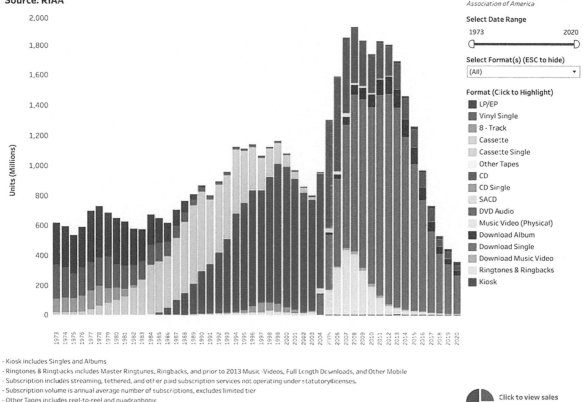

U.S. Recorded Music Sales Volumes by Format
1973 to 2020, Format(s): All
Source: RIAA

Select Date Range
1973   2020

Select Format(s) (ESC to hide)
(All)

Format (Click to Highlight)
- LP/EP
- Vinyl Single
- 8 - Track
- Cassette
- Cassette Single
- Other Tapes
- CD
- CD Single
- SACD
- DVD Audio
- Music Video (Physical)
- Download Album
- Download Single
- Download Music Video
- Ringtones & Ringbacks
- Kiosk

Click to view sales volume for single year

- Kiosk includes Singles and Albums
- Ringtones & Ringbacks includes Master Ringtunes, Ringbacks, and prior to 2013 Music-Videos, Full Length Downloads, and Other Mobile
- Subscription includes streaming, tethered, and other paid subscription services not operating under statutory licenses.
- Subscription volume is annual average number of subscriptions, excludes limited tier
- Other Tapes includes reel-to-reel and quadraphonic
- Total units excludes Paid Subscriptions

FIGURE 6.2 *RIAA Annual Shipment Data*

(Source: RIAA)

The value of consumption does not stop at the download. Streaming songs generate licenses and royalties from the various sites and add revenue to the bottom line of copyright holders. The *streaming equivalent album* was introduced in 2013 to measure streaming consumption. To evaluate streaming equivalents, the general industry standard is that 1,500 streams of any songs from a particular album are counted as a single album. 1,500 streams × the standard royalty generated by this airplay $.005 = $7.50, the wholesale price of an album. All of the major on-demand audio subscription services are considered, including Spotify, SoundCloud, and Tidal, as well as YouTube/VEVO. Combine the streaming activity of all the singles from a particular album and divide by 1,500, and the resulting figure is added to the album downloads and physical album units to give a project picture of "total" activity known as the Top 200.

Since the initial evaluation of streams, further appraisal by Nielsen created a deeper look at streaming equivalents and created a "weighted" measure of each stream based on its source, whether it be a free, ad-supported stream or a premium, subscription stream. Although not revealed in most data, the values look like the following:

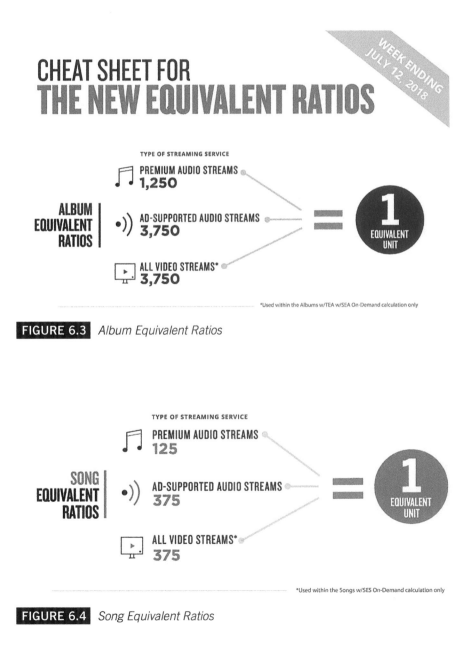

FIGURE 6.3 *Album Equivalent Ratios*

FIGURE 6.4 *Song Equivalent Ratios*

So when looking at data streaming information, it is not clear how many of the streams are ad supported or premium, per se. But looking at raw data, calculating the "total activity" becomes clearer in ranking the top-performing titles for each week.

Here is an example of how the calculations work:

Rank	Artist	Title	TW Total Activity	TW Album Sales	TW Song Sales	TW TEA	TW Audio Streaming Activity	TW Video Streaming Activity	TW TOTAL SEA Audio+ Video	TW Streams Per Album
1	Lil Uzi Vert	Eternal Atake	287,995	8,510	18,641	1,864	375,934,513	24,484 074	277,620	1,442
2	Jhene Aiko	Chilombo	151,972	37,996	19,545	1,954	142,041,951	6,940 709	112,021	1,330
3	Bad Bunny	YHLQMDLG	111,209	7,515	7,534	753	130,923,137	14,739 284	102,941	1,415
4	Lil Baby	My Turn	104,262	2,992	24,340	2,434	127,092,515	18,199 007	98,835	1,470
5	NCT 127	NCT #127: The Final Round	87,207	82,749	3,797	379	5,188,057	919 519	4,077	1,498

Table 6.1 "Total Activity" Chart Calculations

Lil Uzi Vert:
8,510 albums sold
1,864 TEA = (18,641/10 songs)
277,620 SEA = (375,934,513+24,484,074 = 400,418,587/1442 weighted avg streams)
287,995 total activity
(Source: Music Connect Top 200 Chart w/e 3/12/20)

In looking at the data in Table 6.1, the artist NCT 127 *sold* more albums (at 82,749 units) this week than Lil Uzi Vert, but Lil Uzi Vert sold more songs and had near triple the streaming activity, which caused his album project *Eternal Atake* to be Number 1 on the Top 200 for this week. You may notice that Jhene Aiko also sold more albums than Lil Uzi Vert, but streaming and video activity outperformed the entire inventory of releases for the week ending March 12, 2020. Without being able to see the actual detail of what streamed as ad supported vs. premium, it is not possible to derive the SEA exactly. By adding both the audio and video streaming data and dividing by the SEA revealed by Music Connect, an average "streams per album" matrix is calculated. Looking at these five titles, it is understandable where the industry derived the original standard of approximately 1,500 streams equaling an album.

What has been remarkable is the fast pace in which consumers have adopted this technology known as streaming. In less than a decade, the industry has realized almost all its revenue from digital music sources, based on shipment data from the RIAA. The year 2011 became the first to generate more revenue from digital sources than physical sales. And by 2012, digital music revenue exceeded $4 billion for the first time, with $1 billion

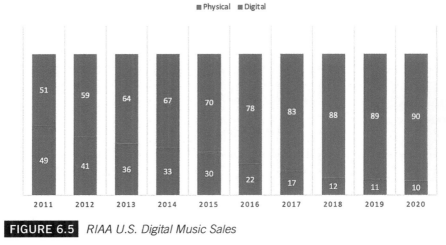

FIGURE 6.5 *RIAA U.S. Digital Music Sales*

(Source: RIAA)

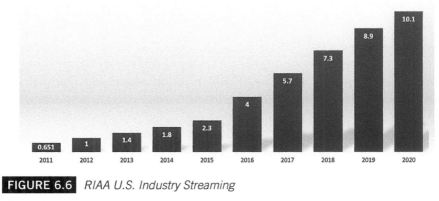

FIGURE 6.6 *RIAA U.S. Industry Streaming*

(Source: RIAA).

of this coming from streaming alone. And within a few short years, 2020 realized shipment revenue of $12.2 billion, nearly its 1999 all-time high of over $14 billion – this in a year of a worldwide pandemic with 90% of the revenue being generated from digital sources, mostly streaming.

Annual Sales Trends

Music sales have historically been seasonal, with the majority of sales in the past occurring during the fourth quarter holiday season. The timing of superstar releases has been reserved for the fourth quarter to take advantage of foot traffic in the retail environment. But thanks to streaming, music consumption practices have dramatically changed, with consumers accessing music more evenly throughout the year. Although seasonality remains, the volatility of these consumption trends have leveled out. Look at 2014: it is easy to spot Valentine's Day at Week 7, which has been a traditional music-giving holiday. Look again at 2014 Thanksgiving and Christmas, with easily recognizable surges in sales. For the year 2019, current streaming practices have radically smoothed the sales model, with a tiny blip for Valentine's at the 7-week mark, a slight bump noticed for the iHeartRadio Music Awards in March, and another rise in week 34 when Taylor Swift released "Lover" in August. The fourth-quarter swell is still noticeable, but is it traditional sales, or do consumers have more time to access holiday favorites and are streaming more? It is both!

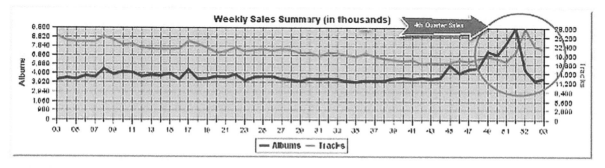

FIGURE 6.7 *SoundScan Weekly Sales – 2014*

(Source: SoundScan)

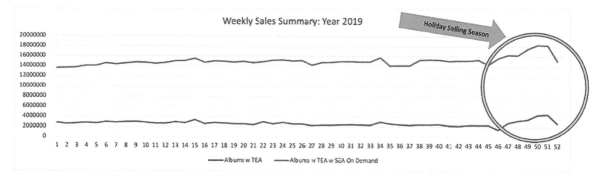

FIGURE 6.8 *Music Connect Weekly Sales – 2019*

(Source: Music Connect)

Genre Trends

2019

Share of Total Volume by Format and Genre
(Selected Top Genres)

Genre	Total Volume*	Physical Album Sales	Digital Album Sales	Digital Song Sales	Total On-Demand Streams	On-Demand Audio Streams	On-Demand Video Streams
R&B/Hip-Hop	27.7%	10.5%	15.9%	20.5%	30.7%	31.2%	29.6%
Rock	19.8%	42.2%	32.4%	22.3%	14.7%	17.1%	10.2%
Pop	14.0%	11.0%	11.0%	18.8%	14.7%	13.7%	16.5%
Country	7.4%	10.4%	8.1%	11.7%	5.9%	7.2%	3.5%
Latin	5.3%	1.0%	1.0%	2.8%	8.2%	4.8%	14.5%
Dance/Electronic	3.6%	1.0%	2.2%	4.0%	3.9%	3.8%	4.1%
Christian/Gospel	2.3%	4.8%	3.2%	3.7%	1.8%	1.9%	1.6%
World Music	1.5%	2.3%	1.7%	1.5%	1.6%	1.3%	2.3%
Holiday/Seasonal	1.4%	3.3%	1.8%	1.0%	1.0%	1.1%	0.8%
Children	1.2%	2.1%	1.9%	0.9%	1.1%	1.0%	1.3%
Jazz	1.1%	2.7%	2.0%	0.9%	0.7%	0.9%	0.3%
Classical	1.0%	2.0%	2.5%	0.7%	0.7%	0.9%	0.5%

Read as: 27.2 of total album sales come from R&B/hip-hop genre *Total Volume = Albums + TEA + On-Demand Audio/Video SEA

FIGURE 6.9 *MRC Year-End Genre Report – 2019*

2020

Share of Total Volume by Format and Genre (Selected top genres)

	Genre	Total Volume*	Total Album Sales	Physical Album Sales	Digital Album Sales	Digital Song Sales	Total On-Demand Streams	On-Demand Audio Streams	On-Demand Video Streams
1	R&B/Hip-Hop	28.2%	13.6%	12.3%	16.2%	20.0%	31.1%	30.7%	33.9%
2	Rock	19.5%	39.5%	44.0%	30.8%	21.4%	15.6%	16.3%	11.4%
3	Pop	12.9%	10.3%	10.0%	10.9%	17.8%	13.1%	13.1%	13.5%
4	Country	7.9%	8.8%	9.3%	7.8%	11.9%	7.1%	7.5%	4.6%
5	Latin	4.7%	0.8%	0.6%	1.3%	2.7%	6.0%	5.3%	10.5%
6	Dance/Electronic	3.2%	1.6%	1.3%	2.1%	3.9%	3.3%	3.4%	2.9%
7	Christian/Gospel	1.9%	2.4%	2.2%	2.9%	3.5%	1.8%	1.7%	1.8%
8	World Music	1.8%	3.5%	4.3%	1.9%	2.1%	1.6%	1.5%	2.5%
9	Children	1.3%	1.6%	1.6%	1.6%	0.8%	1.2%	1.2%	1.3%
10	Jazz	1.1%	2.9%	3.2%	2.3%	1.1%	0.7%	0.8%	0.3%
11	Classical	1.0%	2.1%	1.9%	2.5%	0.7%	0.8%	0.8%	0.4%

*Total volume = Albums + TEA + on-demand audio/video SEA

FIGURE 6.10 *MRC Year-End Genre Report – 2020*

In the past, the category of rock music dominated the marketplace for many years, but in 2017 the R&B/hip-hop category took over the top of the genre charts and has remained there ever since. For 2020, artists such as Drake, Lil Uzi Vert, Cardi B, and The Weeknd continue to rule the Top 200 as well as the R&B and hip-hop charts. Rock and pop continue to lose in market share, but only slightly. Rock continues to wax nostalgic with perennials such as Elton John, Journey, and the Eagles as continual featured artists. Country gained a bit of market share with artists such as Luke Combs, Marren Morris, Gabbey Barret, and Morgan Wallen.

The take-away for each genre should be how each category is being accessed. To look at the market share of each music type, check out the combination of formats and how each creates their overall "footprint" of the genre. Do these consumers stream video more than audio? How active are these genre consumers in purchasing albums? When targeting to R&B/hip-hop consumer, creating and utilizing a music video is an essential marketing tool based on the format information as to how R&B/hip-hop patrons access this music, with 33.9% viewing videos within this genre – a significant trend not to be ignored.

DEMOGRAPHIC TRENDS

The RIAA measures age and gender for their consumer profile studies. Current data is available from 2019 Music Watch Report.

Age

Table 6.2 looks at the Internet population that is represented by people over the age of 13. When identifying strong-performing groups, look to those numbers that outperform the average Internet population percentage. As you see, active buyers look to be between the ages of 18 and 44, with CD buyers aging older, which shouldn't be a surprise since this is part of their culture and habits. Interestingly, free streaming skews older too, but it could be accounted for by one portal – Pandora. Pandora was the first streaming service that was free and allowed listeners to "design" their own station with artist and genre preferences. And many aging users have never left. Based on this data, P2P file sharing seems to have lost favor among the young, with messaging about this "illegal" practice finally sticking, along with access to free listener services that allow for "dial up" selection of songs and artists.

Gender

It's difficult to correlate access of music with the popularity of certain genres, the adoption of music consumption technology, and gender and age.

Table 6.2 MusicWatchInc Report 2019

MUSIC CONSUMER PROFILE – 2019

	Total Internet Population 13+	Music Buyers	Cd Buyers	Digital Buyers	Vinyl (New)	Music Streamers	Paid Subscribers	Free Streamers	P2p Downloaders	Streamrippers
GENDER										
Male	49%	53%	50%	45%	52%	49%	56%	44%	67%	62%
Female	51%	47%	50%	55%	48%	51%	44%	56%	33%	38%
AGE										
13–17	9%	6%	4%	8%	7%	10%	7%	8%	1%	10%
18–24	13%	15%	8%	12%	14%	14%	18%	5%	11%	14%
25–34	18%	23%	13%	22%	21%	20%	28%	11%	28%	24%
35–44	17%	22%	13%	19%	19%	18%	26%	10%	35%	28%
45–54	17%	17%	25%	21%	19%	16%	13%	22%	15%	14%
55+	27%	17%	36%	18%	21%	22%	9%	43%	9%	9%
RACE & ETHNICITY										
Black/African American	13%	13%	6%	8%	6%	14%	16%	10%	16%	22%
Not Black/African American	87%	87%	94%	92%	94%	86%	84%	90%	84%	78%
Hispanic, Latino or Spanish origin	17%	17%	10%	12%	19%	18%	20%	10%	14%	16%
Not Hispanic, Latino or Spanish origin	83%	83%	90%	88%	81%	82%	80%	90%	86%	84%
FAVORITE GENRES (RANK)*										
#1	80s-'90s Hits	80s-'90s Hits	Classic Rock ('60s-'80s)	80s-'90s Hits	Classic Rock ('60s-'80s)	80s-'90s Hits	Rap/Hip-Hop	Classic Rock ('60s-'80s)	80s-'90s Hits	Rap/Hip-Hop
#2	Classic Rock ('60s-'80s)	Classic Rock ('60s-'80s)	80s-'90s Hits	Classic Rock ('60s-'80s)	80s-'90s Hits	Classic Rock ('60s-'80s)	80s-'90s Hits	80s-'90s Hits	Rap/Hip-Hop	80s-'90s Hits
#3	Country	Rap/Hip-Hop	Country	Pop/Top40/Current Hits	Alternative/Modern Rock	Rap/Hip-Hop	Pop/Top 40/Current Hits	Country	Rap/Hip-Hop	Classic Rock ('60s-'80s)

Source: MusicWatch/2019 Annual Music Study

DEFINITIONS:
Music Buyer: Purchased at least one CD, digital track/album, or vinyl record or paid to listen to online radio or on-demand music services in the pst year
CD Buyer: Purchased at least one full/single CD in the past year
Digital Buyer: Purchased at least one digital track/album in the past year
Vinyl Buyer: Purchased at least one new vinyl album in the past year
Music Streamer: Listened to music via free/paid online radio or on-demand services in the past year (i.e., Pandora, Spotify, YouTube)

For more information contact, MusicWatch, Inc. www.musicwatchinc.com [Insert 15031-4988-UnFigure-001 Here]

Paid Subscriber: Personally paid for an on-demand music subscription service (not including Amazon Prime subscriptions).
Free Streamer: Stream music but did not use a paid subscription service
P2P Downloader: Downloaded at least one track for free from a file-sharing service in the past year
Streamrippers: Streamripped at least one song in the past year

***QUESTION:**
Of the types of music listed below, which are the three that you are most interested in? That is, you not only listen to them occasionally but you actually buy the music, pay to go to concerts, etc. (Select up to 3)

The overall population of potential music buyers is split nearly equally, with women edging out men, 51% to 49%. Historically, as the technological gateway to music has advanced, men out-engaged women, specifically in peer-to-peer exchanges, paid subscriptions, and streamripping such as Mediafire, and the percentages of usage verify this practice, with men using paid subscriptions 56%, P2P downloading 67%, and streamripping 62% – outpacing their female counterparts.

MARKET SHARE OF THE MAJORS

Label market share can be measured in a myriad of ways. The term "overall" business includes all format sales, but to see how well a music company is doing with new releases, one can isolate catalog sales from currents (current releases) to reveal the "state" of the competition's business. (See more data in the following.) Universal has dominated the market since acquiring Poly-Gram Records in 1998. It is impressive that Universal has managed to grow additional market share since this acquisition, and they have continued

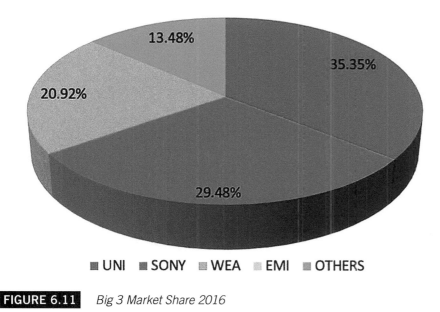

FIGURE 6.11   *Big 3 Market Share 2016*

(Source: Music Connect)

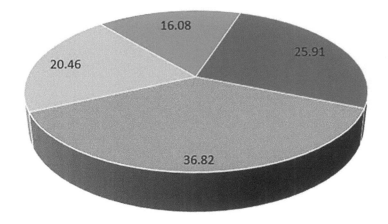

WEA   Sony   UNIV   Other

FIGURE 6.12   *Big 3 Market Share 2017*

(Source: Music Connect)

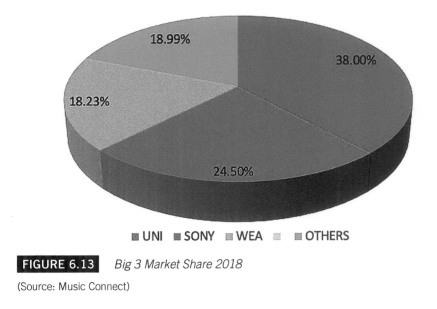

UNI   SONY   WEA   OTHERS

FIGURE 6.13   *Big 3 Market Share 2018*

(Source: Music Connect)

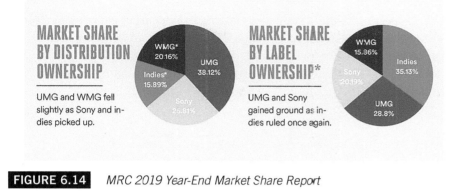

MARKET SHARE BY DISTRIBUTION OWNERSHIP

UMG 38.12%
WMG* 20.16%
Indies* 15.89%
Sony 25.81%

UMG and WMG fell slightly as Sony and indies picked up.

MARKET SHARE BY LABEL OWNERSHIP*

WMG 15.36%
Indies 35.13%
UMG 28.8%
Sony 20.19%

UMG and Sony gained ground as indies ruled once again.

FIGURE 6.14 *MRC 2019 Year-End Market Share Report*

on that trajectory since the turn of the century. Imagine the powerhouse labels of Universal's Interscope, Def Jam, and Island Records merging with EMI imprint Capitol Records, among many more. No wonder the company enjoys near 40% market share.

At one time, there were six major music conglomerates that controlled the U.S. music distribution landscape. The 2012 EMI sale was preceded by the 2004 merger of Sony and BMG, which created the second-largest music company in the industry, behind Universal. Before the merger, BMG held a strong market position in the late 90s and early 2000s due in part to their partnership with Jive Records who, at that time, was responsible for the teen hit sensations 'N Sync, Britney Spears, and Backstreet Boys. Sony enjoyed success in the late 90s with Celine Dion, Mariah Carey, and other pop artists but saw their market share slip when the pop movement subsided. Sony swallowed BMG creating Sony Music Entertainment.

Importantly, as market shares vary from hit to hit and year to year, remember the importance of the independent label and their relationship with the music group. Each of these music companies has "independent distributors" as part of its business model. Sony Distribution distributes records for truly independent labels through their Sony Red and The Orchard distribution arms. Universal's indie distributor is Caroline Distribution, and Warner's is ADA. If the labels that were distributed through these major distribution arms were re-categorized with their percentage of business moved to the "Independent" category, the market shares would look very different. As generated by *Billboard* in 2019, the value of the independents was measured by doing just this: look at the graphic. Independents generate over 35% market share if isolated by themselves. This revenue is "recognized" by the major distributor, but the independent label is generating the money.

Comparison of All and Current Albums

In the following charts, a comparison is made for market share of all albums and market share of new releases for each label. An increase or decrease in current album market share is one indicator of the company's health, but a comparison of all albums vs. current and catalog albums indicates how much a label relies on catalog sales compared to sales from its new releases.

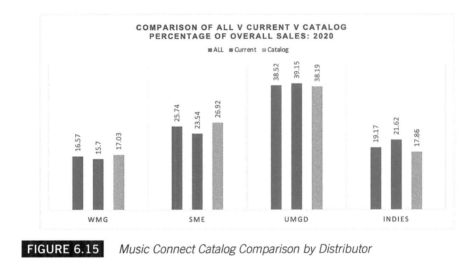

FIGURE 6.15   *Music Connect Catalog Comparison by Distributor*

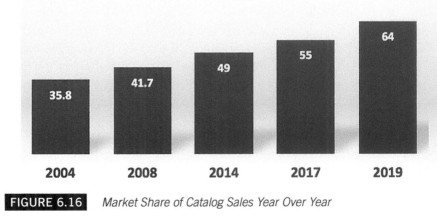

FIGURE 6.16   *Market Share of Catalog Sales Year Over Year*

(Source: 2019 Nielsen Music U.S. Report)

Of the major labels, only UMG has a significantly larger market share for current releases than for all album releases, although new independent releases have also fared well during 2020. This larger piece of the pie indicates an entity's power to release and sell new hit records in the current climate. When one slice grows bigger, another slice must shrink. In 2020 Warner's current releases suffered, but their healthy catalog helped to maintain a 16.57% market share. When reading this chart, the "current" and "catalog" numbers combine to create the "all" share.

CATALOG SALES

Catalog sales are defined as sales of records that have been in the marketplace for over 18 months. Current catalog titles are those over 18 months old but less than 36 months old. Deep catalog albums are those over 36 months since the release date.

When the compact disc was first introduced in the early 1980s, it fueled the sales of catalog albums as consumers replaced their old vinyl and cassette collections with CDs. This windfall allowed labels to enjoy huge profits and led to the industry expansion of the 1980s and early 90s. However, catalog sales started to diminish in the mid-1990s as consumers finished replacing their collections. The closure of traditional retail stores also contributed to the decline in catalog sales, with customers having fewer opportunities to be exposed to the older titles.

Music companies have always looked to maximize their catalog's potential by utilizing various pricing strategies. Throughout the years, labels sought new ways to promote catalog sales through reissues, compilations, and looking at new formats. During the mid 2000s, catalog sales declined overall, but the introduction of digital downloads saw a spurt in catalog sales as consumers sought to fill in their collections with catalog albums and singles that had not been available for some time.

And now with the endless availability of whatever artist and title from whatever era as accessible through the various streaming portals and curated playlists, catalog releases are the largest source revenue for the industry. Current releases may dominate the sales trends, but catalog sales in both physical and digital formats have seen steady growth over the past 15 years.

CONCLUSION

This chapter presented tools and measures used to evaluate the health of the labels and the industry. We used these assessments to illuminate some trends and shifts within the U.S. recording industry. By looking at trends in music genres, the demographics of buyers, configurations, label market share, and the proportion of catalog to current, one can make inferences about the current state of the industry. Armed with an understanding of these tools and measurements, the reader will be able to draw their own conclusions in the future as new data becomes available.

GLOSSARY

BMG—Bertelsmann Music Group. Used to be one of the U.S. major labels; merged with Sony and now has publishing entities in the United States and global presence worldwide.

Catalog—Older album releases that still have some sales potential. Recent catalog titles are those released for over 18 months but less than 36 months. Deep catalog: those titles over 36 months.

Dollar value—The monetary worth of a stated quantity of shipped product multiplied by the manufacturers suggested retail price of a single unit. The value of shipments is given in U.S. dollars.

EMI—Electric and Musical Industries. One of the major music conglomerates. Also known as EMI Music Group.

Market share—A brand's share of the total sales of all products within the product category in which the brand competes. Market share is determined by dividing a brand's sales volume by the total category sales volume.

P2P (peer-to-peer)—Electronic file swapping systems that allow users to share files, computing capabilities, networks, bandwidth, and storage.

Product configuration—Any variety of "delivery system" on which prerecorded music is stored. Various music storage/delivery mediums include the full-length CD album, CD single, cassette album or single, vinyl album or single, DVD audio, DVD, mp3, or streaming audio and video.

Replacement cycle—Consumers replacing obsolete collections of vinyl records and cassettes with a newer compact disc format.

Streaming equivalent album (SEA)—An industry standard used to derive an album equivalent. 1,500 streams equals one album count.

SMG—Sony Music Group. One of the major music conglomerates.

Track equivalent album (TEA)—Ten track downloads are counted as a single album. All the downloaded singles are divided by ten, and the resulting figure is added to album downloads and physical album units to give a total picture of "album" sales.

UMG—Universal Music Group. One of the major music conglomerates.

Units shipped—The quantity of product delivered by a recording manufacturer to retailers, record clubs, and direct and special markets, minus any returns for credit on unsold product.

WEA—The distribution arm of WMG. (Stands for Warner, Elektra, Atlantic.)

WMG—Warner Music Group. One of the major music conglomerates.

BIBLIOGRAPHY

2014 Nielsen Music U.S. Report.

Billboard, Staff. "Billboard 200 Makeover: Album Chart to Incorporate Streams & Track Sales." *Billboard*, 19 Nov. 2014. Web. 26 Jan. 2015. www.hollywoodreporter.com/thr/article_display.jsp?vnu_content_id=2059949.

MRC Year-End Report 2019.

MRC Year-End Report 2020.

MusicWatchInc.com, 2019.

Nielsen & Billboard's 2018 U.S. Music Report.

The NPD Group/2013 Annual Music Study.

RIAA, "1997 Yearend Marketing Report on US Recording Shipments," *RIAA Press release*, 1998.

RIAA, (2020). U.S. Sales Database, Shipment Report.

RIAA, (2020). U.S. Sales Database, Revenues by Format Report.

Sisario, Ben. "Music Sales Fell in 2008, but Climbed on the Web." *New York Times*, January 1, 2009.

SoundScan, (2005). aud.soundscan.com.

CHAPTER 7

Label Operations

LABEL OPERATIONS

In an era when every level of the food chain within the entertainment industry is being scrutinized as to its value, record labels are being squeezed from both sides of the equation. Record sales rely on streaming and artists are expecting more value from their relationship with their labels – more so than ever before. As technology advances, artists have seen the opportunity to completely circumvent the "label deal" and market directly to their fans. The challenge for labels is to be a significant contributor for both sides of the equation: create various "products" that draw consumers to the artist via streaming and radio on a consistent basis *and* create a loyal fan base to help market the artist through various outlets such as ticket sales and cross promotional partnerships – which is not as easy as throwing a site onto the world wide web and expecting magic to happen! It takes the creativity and business acumen of a team that is savvy about today's social media and analytics but understands longstanding business tactics that can endure strong competition and challenging economic situations to win in today's music business.

Every record label is uniquely structured to perform at its best. Oftentimes, the genre of music along with the "talent pool" of actual label personnel dictate the organization and inner workings of the company. As talent is signed to a label, the "artist" will come in contact with nearly every department in the process of creating a music product for the marketplace.

A typical record label has many departments with very specific duties. Depending on the size of the label, some of these departments may be

DOI: 10.4324/9781003153511-7

combined, or even outsourced, meaning that the task that the department fulfills is hired out to someone not on staff of the label. But the end result is to be the same – create a viable music product for the marketplace. In the structure as follows, there usually is a general manager/sr. VP of marketing who coordinates the all the marketing efforts.

GETTING STARTED AS AN ARTIST

As talent is being "found" or developed, the first contact with a record label is usually the A&R department. The artists & repertoire department is always on the hunt for new talent as well as songs for developing and existing artists to record. But before the formal A&R process occurs, the talent has to be signed to the label.

BUSINESS AFFAIRS

The business affairs department is where the lawyers of a label reside. Record company lawyers are to negotiate in the label's best interest. Most often, new talent will have a manager and lawyer working on their behalf, with the contract in the middle. Clearly, the label wants to protect itself and hopes to reduce risk by maximizing the contract.

Besides being the point person for artist contracts, label lawyers negotiate and execute many other types of agreements, including:

- License of recordings and samples to third parties
- Negotiate for the right to use specific album art
- Point person when an artist asks for an accounting or audit of royalties
- Renegotiates artist contracts
- Contractual disputes such as delivery issues
- Conflicts with contract such as guest on another
- Vendor contracts and relations

Oftentimes, the accounting department falls within business affairs, since the two are related regarding contractual agreements and financial obligations. The accounting department is the economic force driving all the activity within the record company. It takes money to make money, and the accounting department is responsible for the budgets for each department as it aligns with the forecast of releases. Although the function and need of the accounting department is still needed, many of the tasks that this department has performed is now sourced to the parent company and/or the distribution company as "back office" support. Sophisticated forecasting models are generated at the parent-company level and assist with making signing and release decisions that impact the profitability of the company. Value of product is discussed in the distribution chapter.

On a local level, the accounting department acts as an accounts payable/receivable clearinghouse, managing the day-to-day business of the company such as HR/payroll, leasing of the building, and keeping the lights on.

Artists & Repertoire: How Labels Pick and Develop Artists

Being the A&R person is the job most people want at a record label. They see the glamour and the glory of being the person who discovered the next big act that changes the music scene, hanging out in the studio or backstage with big-name acts, and getting their picture in *Billboard* and *Rolling Stone*. What they do not see is the countless hours and late nights listening to bands that are not that good and never get signed, the sore backs and headaches from endless hours screening videos, or the high turnover rate because you are only as good as the last act you signed. Or the repertoire part of the job – helping the artists find just the right song for their next album and, in some cases, having a hot debate with the label president over which song should be the first single. As the title indicates, the position has two different tasks: signing new talent and helping the label's artists select the best songs to advance their careers.

Discovering Artists – The A in A&R

The single most important talent of a good A&R person is to identify and sign successful artists. Their bosses are likely to overlook a project going over budget if it is a big hit, while bringing the recording in on budget will not advance your career if it the record does not sell. A good A&R person is such a valuable asset that they are often considered in the valuation of a record label in a merger or acquisition. While other departments in the combined companies will probably be downsized after a merger, successful A&R persons will be retained from both labels (Krasilovsky and Shemel 2007). As long as they continue to be successful.

Different labels put more or less emphasis on different resources for discovering potential new artists. Historically, artist were discovered in bars or clubs or brought to the label by a publishing company, but now social media, from TikTok to YouTube, is more likely to be the source of discovery than a bar or club.

More and more, the decision of who gets signed is driven by the data. In recent years, many artists have been discovered on the Internet and gone on to great success, and labels are putting greater emphasis on data analytics,

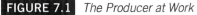 *The Producer at Work*

social media likes and followers, and numbers of both audio and video streams. It's been documented where label presidents have set aside entire days each week just to look at videos of artists recommended by his staff – and this was before TikTok. But this is only some of the many sources used by A&R to find new talent. During the COVID-19 pandemic, quarantined A&R persons rediscovered a "secret weapon" in their own homes – their children. While scouring the internet for the next big act, they often found that their children had already discovered those same acts on social media.

A&R is shaped by both the music and technology. For pop and hip-hop music, discovering new talent may be as simple as an internet search. Other genres, like rock or Americana, may still require some hanging out in bars or utilizing your industry contacts to find the best acts to sign. The importance of social media analytics is further discussed in Chapter 14.

Producers as A&R Scouts

Independent and staff producers are another way that artists land recording contracts. Because producers often hire unsigned talent as musicians or background singers for other artists' projects, they are in a unique position to discover and promote these artists to A&R at the record labels they are working for. An early example of this is John Hammond who, as a talent scout and producer for Columbia Records, discovered and produced Bob Dylan, Aretha Franklin, Pete Seeger, and many others (John Hammond Biography n.d.). Current examples of this kind of connectivity would be Mike Serletic, who signed what would become Matchbox Twenty to a production deal, and Serletic then secured their record deal with Atlantic post-recording sessions, after the record was "in the can." Joey Moi of Big Loud Records worked with Florida Georgia Line pre-record deal, knowing that their songwriting publishing rights were secured with Big Loud Shirt publishing company. The FGL record deal has been a (very successful) co-venture with Big Machine Records. (www.allmusic.com)

Attorneys and A&R

An established entertainment attorney with connections in the industry may be willing to take on an unsigned artist in a "shopping deal." For a fee, ranging from an hourly charge plus expenses to a percentage of the artist's advance when a deal is signed, the attorney will pitch the artist to their connections in the industry. The artist is paying the attorney for her industry contacts. Before signing such a deal, the artist needs to do their research and determine the attorney's stature in the industry. Have they successfully made such deals in the past? Do they represent a number of artists, and are they well respected in the industry? Do they really have the contacts they claim, and can they do the job you are hiring them to do?

The attorney has to believe the artist has potential because their reputation is at stake, and that is no small risk. Like any person doing a sales job (like a song plugger pitching songs to an artist or producer), their credibility is their calling card, so if the attorney is not passionate about an artist and the artist's music, keep looking. Another major deal point is the duration of the shopping arrangement: assuming the artist has entered into an exclusive deal with the attorney, negotiating an end date so that if the attorney is not successful, both can move on.

Publishers and A&R

The A&R department's strongest allies are the publishers who sign and promote their singer-songwriters to the labels in hopes that a record deal will result. Publishing companies want their writers to have recording deals – it is a guaranteed outlet for getting their catalog recorded. So publishers have their own A&R people and may sign artist-writers that they believe have the potential to be recording artists. Examples of successful artists that were developed through publishing companies include Florida Georgia Line (Big Loud Shirt) and Taylor Swift (Sony/ATV), who was signed as a songwriter at age 14.

The publisher will finance the recording of demos and pitch them to record labels. These development deals usually give the publisher 6 to 18 months to secure a recording deal for the artist-writer (Brabec and Brabec 2018). The artist and their attorney should negotiate minimum standards for the deal so they are not forced into a less than desirable deal with a record label. Even if the artist doesn't get a record deal, they may get their songs recorded by other artists, which will generate some income while they continue to search for a record label.

Reality TV

American Idol, launched in 2002, has been an absolute television phenomenon, placing no fewer than 64 finalists on the *Billboard* Charts 345 times in the first ten years (Bronson 2012). Kelly Clarkson, Carrie Underwood, and Daughtry, along with Jordin Sparks, Scotty McCreery, and others have added to those numbers since (Wikipedia 2021). Although they are often criticized for their iron-clad, all-encompassing long-term contracts (Kelly Clarkson's was for seven albums, and it took her 13 years to complete [Newman 2015]), would-be contestants wait in line for hours for an opportunity to audition. By the time a contestant reaches the finalist stage, they have hours of television exposure, thousands of fans, and probably a single playing on the radio. It is an A&R department's dream. Winners receive a recording contract, and other finalists may be picked up, at the label's option, for a development deal or a full contract. Other successful alums of American Idol include Kimberly Locke, Adam Lambert, and Clay Aiken, among many others. Even some non-finalists, like William Hung, have managed post-show success (Bronson 2012).

The Voice's most successful alum is Cassadee Pope, who had a #1 record on the country charts, was nominated for a Grammy, and won a CMT award for her video for "Wasting All These Tears," which first aired in April of 2011. Part of American Idol's success is probably because it was the first show of this type (at least in this century), but more likely it is the fact that it provides a more complete development package for its contestants, with competing contestants having singles sold on iTunes and played on the radio. After the season is over, finalists will go on tour, performing both as a group and solo on as many as 40 dates during the summer and mostly in smaller markets, where fewer shows may be available.

How Artists Get on A&R's Radar

Prior to the internet, there was only one way to get discovered – get out and perform! Here are some of the common ways new talent is discovered in the music business today.

FIGURE 7.2 *Social Media Apps*

Build a Social Media Presence

Be K-Pop sensation BTS, ex-One Direction member Louis Tomlinson, or perennial Taylor Swift! All have been on the *Billboard* Social 50 chart for over 150 weeks or more. These artists are now megastars, but building a social media presence at any level is an essential part of making yourself attractive to a label or publisher, too. How many Facebook friends you have; how many active followers you have on TikTok – how you interact with your fans on social media and whether you are adding fans at an increasing rate is part of the measurement of getting signed.

As for TikTok, 2020 has proven this social media platform can convert songs into viral hits and those songs' performers into prominent major label artists. According to TikTok, over 70 artists have broken on their platform and signed major deals, like 24KGoldn, whose single *Valentino* became a viral hit not long after signing a record deal with Columbia Records, and Fousheé, whose vocals were first heard on the platform on the track *Deep End* and who has recently signed with RCA. Other artists who have jumped the divide from TikTok to label deal include Claire Rosinkranz, Dixie D'Amelio, Powfu, Priscilla Block, and Tai Verdes. A&R personnel look at other streaming numbers on platforms such as Spotify and Instagram as well. These portals tend to validate the TikTok phenomena, for viewers will jump from

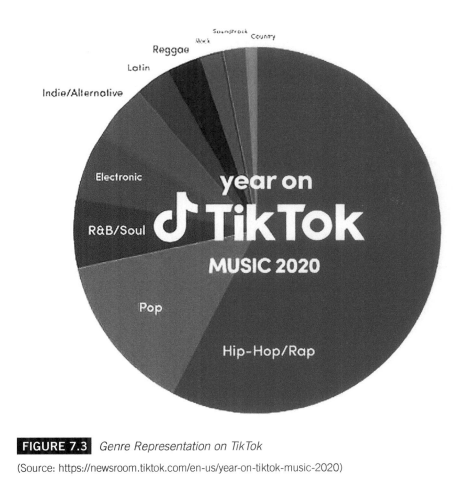

FIGURE 7.3 *Genre Representation on TikTok*

(Source: https://newsroom.tiktok.com/en-us/year-on-tiktok-music-2020)

the short "taste" of TikTok to hear the entire version of the artist's song and continue their search for information on new music and the artists making it (Stassen 2020).

Perform, Perform, Perform. Then Perform Some More

What happens after you get signed? More than one social media sensation has been signed to a million-dollar deal before the label realized they had no depth. No stage presence, no writing talent, no future! One thing is guaranteed in the music business: if you don't put yourself out there, if you never perform in public (live or recorded), you will never get discovered. Performing your own song in a theater or arena full of living, breathing fans is a very different experience than doing a cover song in

FIGURE 7.4 *Live Performance*

front of your iPhone camera. So, play your songs, play somebody else's songs, perform the national anthem at the local high school game, audition for American Idol, and post it all on your social media. Play for free in restaurants and small bars until you don't have to. Give away your music and play *live* at every opportunity. Just get yourself out there and create a buzz about how good you are. When your social media numbers are through the roof and the bars and clubs start filling up with fans, the industry will notice.

Give Away Free Music

Giving music away free gives the artist "social currency" and may attract more fans as friends of friends like their Facebook page or follow them on social media. It can also build a fan base that rewards the artist's generosity by buying concert tickets or merchandise. Sure, some people will take advantage of the situation and never give the artist a dime. Sometimes it is because they are using the free download as a trial and they are never converted to a fan. In other cases, they are just freeloaders.

A legendary example is Radiohead's *In Rainbows*.

The band gave away the entire album in a "name your own price" deal and fans, after a minimal service fee, paid whatever they wanted. According to Comscore.com, 62% of downloaders paid nothing beyond the access fee, but the average payment was $2.26, and the promotion grossed $1.36 million in three weeks ("For Radiohead Fans"). Piracy of the album

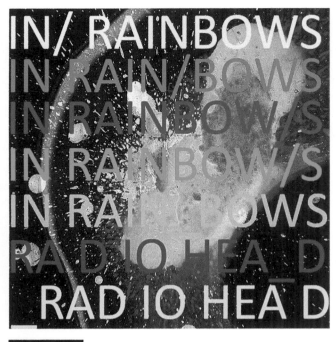

FIGURE 7.5   *Radiohead's* In Rainbows

was rampant, and the number of units actually sold is disputed (Music Connect lists the ATD number at 1,032,063 as of March 26, 2021), but one thing is clear: the publicity from the promotion was followed by a world tour of mostly sold-out shows.

In 2016, Chance the Rapper gave away his *Coloring Book* album, saying that he "never wanted to sell his music." He felt like it was "inhibiting him from making a connection (with his fans)." Chance makes money exclusively from going on tours and selling his merchandise. He feels that music is for the people; therefore, music should be free. *Coloring Book* made it to number one on the *Billboard* Charts, was the first-ever album to do so from streaming alone, and as a bonus, won three Grammys – a stunning feat (Scott-Healey 2019).

Make Professional Connections

Most publishers and record labels do not take unsolicited submissions because of copyright infringement lawsuits, so it helps to get to know somebody on the inside or somebody who knows somebody on the inside. Playing a lot of shows increases the chances of catching the attention of the A&R department, but hiring a lawyer, publisher, or manager may be the ticket in. As noted, an established entertainment attorney with connections in the industry or a publisher with whom you have already developed a relationship may be willing to take on an unsigned artist in a "shopping deal." Join organizations of other writers and artists and build a network of connections.

Fitting In

It is important that every artist have their own unique persona, their own brand, but if a band and its music are cutting edge, then finding a major record label deal may be long in coming. Depending on the music, niche music can find homes on not major labels but independent imprints that cater to a specific listener. Chances are better with a smaller, more forward-thinking labels. Consider this fact: no major label has ever created a new genre. Independent labels like Sun, Chess, and Atlantic brought us the first R&B and rock 'n' roll music; rap was first recorded on labels like Polydor, Sugar Hill Records, Profile Records, Ruthless, and Roc-A-Fella. Regardless of where they are on the scale of musical tastes, artists must capture the attention of A&R through their music, personality, and work ethic. Labels have limited budgets, and they will invest in the artists that they feel are most likely to lead to a commercially successful recording.

Labels also must consider how each additional artist signed impacts the existing roster of artists. If the label already has several solo female acts in the early stages of their career, they are not likely to sign another. The label can only support a limited number of artists before they begin to cannibalize their own sales. There have been cases where an artist had to wait years before a label decided it was the "right time" to sign them (Catino). Once signed, most contracts do not guarantee the label will ever actually release an album. Sadly, some artists have been signed and kept on label rosters for years, then cut without ever releasing an album.

Repertoire – A&R After the Signing

As alluded to previously, these days, many artists come to the label not as a result of the label's own scouting and research but because somebody else discovered and signed them first. A&R is not a traditional marketing function, but, depending on the level of involvement and input after the signing, A&R has a lot of impact on the marketer's job. A&R's primary role after the artist is signed is to find the best possible songs for the artist – to develop their repertoire – but because artists arrive at the label with wide differences in training and experience, they may have to help develop the artist into a marketable product (as discussed in the chapter on branding), perhaps arranging for the artist to get training on how to perform in front of a live audience and how to give interviews.

The A&R person that brought the act to the label will heavily influence song choices and branding. Songs, like artists, may be discovered anywhere. The most common sources are the artists themselves or a publisher. Songs have also been discovered from other artists (covers), producers, and unsigned writers. Sometimes that means that the artist-writer, the one with the record contract, must be professional enough to accept that somebody else's song may be the better choice for their album than their own song.

Conclusion

One of the pluses of the A&R job is that every day is different. The primary tasks of A&R are to discover and sign new artists to the label and help all the artists find good songs. Different labels put more or less emphasis on the Internet as a tool for discovering potential new artists. Although some artists that have been discovered on the Internet have gone on to great success, this is only one of the many sources used by A&R to find new talent. Perhaps their strongest allies in the job are the publishers who sign and promote their singer-songwriters to the labels in hopes that a record deal will result. Other sources include open mic nights, college and industry showcases, booking agents, managers, and producers. Once identified, most labels use Internet analytics and research to confirm that the artist's career is on an upward trajectory and that they have a good relationship with their existing fans.

Timing is also an issue. No matter how talented an artist may be, they may just not be a good fit for the label at that time. The artist must persevere and make himself or herself as visible and attractive as possible until the time is right.

PRODUCT MANAGER/ARTIST DEVELOPMENT

Sometimes called the product development department, this department manages the artist through the maze of the record company and its needs from the artist. Madelyn Scarpulla describes them as, "the product manager in effect is your manager within the label" (Scarpulla). Product managers/artist development specialists hold the hand of the artist, helping them clarify their niche within the company. Artist development usually develops strong relationships with the artist and artist manager, with other departments in the company looking to artist development as a clearing house, helping to prioritize individual department needs with the artist.

Artist development not only manages the artist through the process such as the delivery of the recording, photo and video shoots, and promotional activities but looks for additional marketing opportunities that maximize the unique attributes of the act. And in some labels, the AD can assist in the actual grooming of the talent – taking the act from lounge singer to star quality. But the role consists of being a "hub of the wheel" just as the rubber meets the road and the artist gains traction inside the record label company. The AD can act as scheduler of the artist and head of logistics for all the players and keeping the calendar for the act while carving out their place within the company.

CREATIVE SERVICES

Depending on the company, the creative services department can wear many hats. Artist imaging begins with creative services assisting in the development of style and how that style is projected into the marketplace. Special care is taken to help the artist physically reflect their artistry. Image consultants are often hired to assist in the process. "Glam Teams" are employed to polish the artist, especially for high-profile events such as photo and video shoots, as well as personal appearances.

The creative services department often manages photo and video shoots, setting the arrangements and collaborating on design ideas and concepts with the artist. Once complete, images are selected to be the visual theme of the record, and the design process begins. In some cases, creative services contains a full design team that is "in house" and a part of label personnel. Such "in house" teams can ensure quality and consistency in imaging of the artist – that the album cover art is the image used on promotional flyers, sales book copy, and advertisements. When there isn't an in-house design crew, design of album cover art and support materials is farmed out to outside designers. Interestingly, the use of subcontractors can enhance unique design qualities beyond the scope of in-house designers, but there can be a lack of cohesive marketing tools if not managed properly.

PUBLICITY

The priority of the publicity department is to secure media exposure for the artists that it represents. The publicity department is set into motion once an artist is signed. The biography of the artist via an interview is created. Other tools such as photos from the current photo shoot, articles and reviews, discography, awards, and other credits are collected into one digital "folder," creating a press kit for each artist. These press kits are tools that are used by the publicity department to aid in securing exposure of their artists and are often sent to both trade and consumer outlets.

Pitching an artist to different media outlets can be a challenge. As an artist tours, the ideal scenario would be that the local paper would review the album and promote the show. Additional activities would be to obtain interviews with the artist in magazines and newspapers that can also be used as incremental content for websites by these same entities. Booking television shows and other media outlets falls to the publicity department as well.

As artists become more established, many acts hire their own independent publicist to enhance their profile. This additional media punch is usually coordinated between label and indie publicist so that redundancy is avoided as well as the being an efficiency of best-efforts among staff and hired guns.

RADIO PROMOTION

In most record companies, the number one agenda for the promotion department is to secure radio airplay. Although streaming has become a huge way for consumers to find new music, surveys continue to show that a significant number of listeners still learn about new music and artists via traditional radio. Typically, radio promotion staffs divide up the country into regions, and each promotion representative is responsible for calling on specified radio stations in that region, based on format. Influencing the music director and radio programmer is key in securing a slot on the rotation list of songs played. These communications often take place on the phone, but routinely, radio promotion staffs visit stations, sometimes with the artist in tow, to help introduce new music and secure airplay.

With the consolidation of radio stations continuing within large communication conglomerates, developing an influential relationship with individual stations is getting harder and harder. But there still exists a level of autonomy within each station to create its own playlist as it reflects their listenership. To strengthen the probability of radio airplay, promotion staffs conceive and execute radio-specific marketing activities such as contests, on-air interviews with artists, listener appreciation shows, and much more.

The sales and marketing department have had to radically refocus their energies as traditional retail outlets have all but disappeared. Besides the independent music store outlets, music stores and dedicated music departments have been eliminated since consumers no longer want to purchase – but stream. Although selling product is still a part of the job of the label, the

FIGURE 7.6 *Promoting Music Through Radio Sales and Marketing*

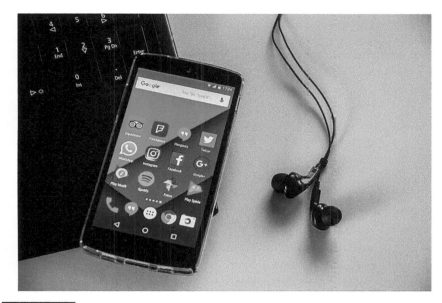

FIGURE 7.7   *Mobile Streaming and Social Media Apps*

"accounts" have drastically changed and labels are pitching music not only to outlets like Apple iTunes, Amazon, and Target but now also streaming services like Spotify, Pandora, Sirius XM, and YouTube.

In the past, sales departments looked to their distributor to be extensions of the label brand and assist regarding the sell-in of new releases. But with the advancement of streaming, the function of distribution companies have also drastically changed, acting instead as an information portal, a "back-office" support team for the labels needing accounting services, and representation of all the labels in coordinating large marketing efforts across label brands. Much more information is discussed in the distribution and sales chapter.

DIGITAL MEDIA/MARKETING

Digital media/digital marketing has become many things for record labels. For many labels, this department handles the social portals for the imprint while pulling the analytics and streaming insights to share with the company, providing valuable information to personnel, including the radio promotion and sales/marketing teams. The digital marketing group should be the megaphone for the label, promoting their artist and label activities through the various social media outlets as an extension of publicity efforts and branding for the imprint. These efforts can include events with artist fans, sponsoring various shows or events by creating online giveaways or other awareness activities. Not to be confusing, song placement on digital portals such as a Spotify playlist is usually reserved for sales, where ongoing marketing efforts through digital media and the collection of strategic data falls under the digital group. Remember that most artists manage their own social media outlets, either themselves or through their management teams. Much more is discussed in the digital marketing chapter.

INDEPENDENT LABELS

In an era when album sales continue to decline, the Internet has fragmented the market, and major record labels have tried to maintain their prominence as "creators of superstars," consumers have been looking elsewhere to satiate their burgeoning musical tastes – and they're eating from the independent labels' table. Historically, independent labels have been relatively small in sales stature and were genre or region specific, tending to react quickly to marketplace trends. Some independent labels strike deals of distribution with the major conglomerates, whereas others find their way to consumers through independent distributors – but all are looking to create access to their music, either literally or virtually to an ever-fractionalizing marketplace. And this new age of consumership has brought with it an entirely new dimension for "indie" labels – one that can include the surprise of a new big artist with big revenue along with even bigger opportunities.

To look at their organizational chart, **Big Machine Records** in Nashville, TN, is structured similarly to that of any major record label, including their size. But what makes them unique is that they are truly independent, meaning that they are not owned by a major conglomerate but act in their own interest and are in charge of their own destiny – and they seem to know where they're going. They broke one of the largest-selling artist of all times, Taylor Swift, who continues to dominated the *Billboard* Top 200 charts even though she is no longer with the label but is now on Republic Records, an imprint of Universal, who distributes Big Machine. The label negotiated "free agent" status and can sell directly to consumers and was also the first label to bargain with the DSPs exchange rates that were more favorable to the label. Big Machine administers several imprints of varying genres: Big Machine Records, the Valory Music Company, BMLG Records, and John Varvatos Records – all with their own personality and focus. But Big Machine does not stop there and successfully runs a full-service music publishing company, along with Big Machine Radio, which streams exclusively Big Machine music. Big Machine has worked on branding itself so that consumers will believe in the imprint when the label introduces new artists into the marketplace (www.bigmachinelabelgroup.com).

Glassnote Records, located in New York City, has garnered many Grammy awards by its eclectic roster of artists, including Childish Gambino, Mumford & Sons, and Phoenix. Its extroverted artistic roll of performers range from indie rock to alternative hip-hop and have been successfully distributed worldwide through AWAL, who was a leader in indie distribution but was purchased by Sony to become another leading arm focusing on the independent supply chain through a major conglomerate. Like other successful indie labels, Glassnote operates Insieme Music Publishing as well as Present Artist Management, supplying services for their artists, allowing the artists to focus on their areas of expertise while letting Glassnote get the business done on their behalf (https://glassnotemusic.com/about/).

So, what is considered "success" in the music business today? Some indies have hit it big and compete with majors head on. But there are many smaller independent labels that are making a go at not just surviving but creating a new model of business where incremental sales on a smaller level are enough to not only keep the lights on but serve as a creative outlet for many more artists who would otherwise not have a voice to sing. If selling 50,000 records at $10 a pop makes a half a million dollars and keeps an artist on the road and his small indie label in business for another year – is this success? To the consumer looking for the indie-club songster whose record may never be heard on mainstream radio – you bet.

REFERENCES

Amanda Palmer quote from www.nme.com/blogs/nme-blogs/did-radioheads-in-rainbows-honesty-box-actually-damage-the-music-industry#MkgIOiqOxXZIMCl1.99.

Brabec, Jeffrey, and Todd Brabec. *Music, Money and Success*. 8th ed. New York: Schirmer Trade Books, 2018. Print.

Bronson, Fred. "Ten Years of 'American Idol' Dominance: Clarkson, Underwood, Daughtry, Fantasia, More." *Billboard Biz*, 11 June 2012. Web. 2 Nov. 2014. <www.billboard.com/biz/articles/news/1093819/ten-years-of-american-idol-chart-dominance-clarkson-underwood-daughtry>.

Catino, Jim. Personal interview, 15 Oct. 2014.

John Hammond Biography. "Rock and Roll Hall of Fame." n.d Web. 25 Oct. 2014. <https://rockhall.com/inductees/john-hammond/bio/>.

Krasilovsky, M. William, and Sydney Shemel. *This Business of Music*. 10th ed. New York: Billboard Books, 2007. Print.

List of American Idol alumni single sales in the United States, Wikipedia, 1 Mar. 2021. <https://en.wikipedia.org/wiki/List_of_American_Idol_alumni_single_sales_in_the_United_States>.

Newman, Melinda. "Now Free From Her 'Idol' Contract, What's Kelly Clarkson Worth?" *Billboard.com*, 3 Apr. 2015. Web. 4 Nov. 2014. <NoiseTrade.com/info/questions>.

"For Radiohead Fans, Does 'Free' + 'Download' = 'Freeload'?" *Comscore Insights*. Ed. Andrew Lipsman. Comscore, 5 Nov. 2007. Web. 26 Oct. 2014. <www.comscore.com/Insights/Press-Releases/2007/11/Radiohead-Downloads>.

Scott-Healey, Rebekah|. "Should Artists Give Away Music for Free?: Artists That Gave Away Music." *Open Mic UK*, 9 Aug. 2019. <www.openmicuk.co.uk/advice/artists-give-away-music-for-free/>.

Stassen, Murray. "TikTok Says Over 70 Artists That Broke on the Platform This Year Have Signed Major Label Deals." *Music Business Worldwide*, 20 Dec. 2020. <www.musicbusinessworldwide.com/tiktok-says-over-70-artists-that-broke-on-the-platform-this-year-have-signed-major-label-deals/>.

TikTok. "Year on TikTok: Music 2020." *Newsroom*, TikTok, 16 Aug. 2019. <newsroom.tiktok.com/en-us/year-on-tiktok-music-2020>.

CHAPTER 8

The Marketing Plan

A **marketing plan** is a single- or album-specific plan that describes activities selected to achieve specific marketing objectives for that release within a set period of time.

Record labels create an initial marketing plan like you might see at any other company. The document allows the label to develop a timeline and coordinate the plan with the artist, their management, and other external partners. For a record label, the marketing plan is the blueprint for each release and is a living, breathing document that is to be adjusted in response to the market. Every record released has its own unique marketing plan based on expectations of sales performance and tailored to the target market.

There are two basic target markets for the marketing plan: the consumer and the trade. The trade consists of those people within the industry for whom business-to-business marketing is done. This includes DSP playlisters, radio programmers, social media and traditional journalists, editors, distributors, retail buyers, and others involved in the **push strategy**. End consumers are the other target, where a theme or message regarding the artist and release is being projected. The hope is that these targeted consumers are motivated to seek out and access the new music by streaming, downloading, purchasing, and supporting the artist through other means such as concert ticket purchases and buying of merchandise. This is known as a **pull strategy** through the various distribution channels.

CONTENTS

DOI: 10.4324/9781003153511-8

WHO GETS THE PLAN?

The plan is distributed to everyone involved in marketing the record, including people both inside and outside the label. The label will provides a version of the marketing plan to trade partners to indicate the seriousness of their effort to create access this particular musical release.

Internally, it is important that everyone at the record label understand what is being done in other departments and how synergy is created when all of the elements come together. The plan acts as a contract between departments – a commitment to deliver specific elements to ensure success of the release. The chart in Table 8.1 gives some examples of who may be involved in the execution of the plan.

There are other departments that benefit from and are included in the plan, including grassroots marketing (street teams), merchandising, and tour support. Coordination is necessary for synergy to occur. The publicity department may develop materials necessary to send to retail accounts and radio (such as press kits). Advertising needs to be coordinated with retail

Table 8.1	Departments Involved in the Marketing Plan		
Function	**Internal**	**External**	**Goals**
Publicity	Label publicist	PR firms, press and media outlets	Getting reviews, features, interviews, photos, and appearances
Radio	Promotion department	Indie promoters, radio programmers	Work with radio to get airplay
Sales	Sales staff	Distribution, DSP, and traditional account buyers	Works with distributors and DSP/retailers on exclusives and promotions
Digital media	Digital media and internet specialists, street team coordinators	Label social media portals, DSPs, and cross-promotional opportunities	Online promotion and access points
Video	Creative, promotions, coordinated by creative department focusing on music and extra content creation	Various outlet uses such streaming services, websites, promotional uses via social media	Production is external to the label, while promotions may be either
Touring	Publicity, retail, radio	Production, distribution, and promotion	Label to ensure there is press coverage and potential cross-promotional partners at events

accounts for sales in the marketplace and with the promotions department for trade advertising aimed at radio. Advertising and publicity go hand in hand, sometimes combined in a "media plan," because in many instances, the same media outlets are targeted for both. Outside of the company, summaries of marketing plans may go to DSPs and retailers, program directors in radio, and the artist's booking agency, and they are always presented formally to the artist and their manager. Often, the manager will contribute to the plan if corporate sponsors are involved or if there are special events that could be mined for extra marketing impact.

The Classic Business School Marketing Plan

To review a classic business school marketing plan concept, companies need to evaluate their "product" as it compares to the competitive landscape and overall marketplace. Evaluators would include:

- Where are we now?
- Where are we going?
- How are we going to get there?

There are eight steps to developing a company's marketing plan.

1. Internal Marketing Audit – what are our internal strengths and weaknesses? Product, distribution, and so on.

2. Environmental Scan – what are the external opportunities and threats?

Table 8.2 SWOT Analysis	
Strengths	**Weaknesses**
• Strong roster	• Underperforming artists on roster
• Strong catalog	• Operating expenses are too high
• Skilled label personnel	• Unhappy artists leaving label (or suing)
• Effectiveness of distribution	• Artists legal problems
• Financial status	• Loss of label personnel
• Reputation of label and/or artists	
Opportunities	**Threats**
• New technologies	• Piracy and P2P file sharing
• New markets (international or domestic)	• Censorship
• Piracy controls	• Competition for entertainment dollars
• Trends swing your direction	• Economic upswings or downturns
• International trade agreements	• Changes in retail and music access
• New ways to monetize music consumption (streaming)	• New technologies

For the record label and its artists, this might be an academic exercise, but all companies, including music business companies, must be honest in their assessment of its current state, its artists' strengths and weaknesses, and how the label will approach the marketplace competitively – not from just a "gut" position.

3. Many labels, both in a formal and informal setting, will "do" a SWOT analysis for their company and artist, evaluating what makes each act uniquely qualified to release music in the current climate. Things to consider are:

4. Marketing objectives – where we can and want to go.

- Improve sales of current products – market penetration (reach potential buyers), develop new markets, increase consumption among users.
- New product development.

5. Marketing strategy – (involves a specified target market).

Specific products are developed to reach specific target markets.

The way in which the marketing mix is used helps to satisfy the needs of the target market and achieve organizational goals.

For the record label, these items would include the sales forecast, the target market by demographics, strategies to increase consumption (streaming), and possible new product development that might include brand partnerships such as artist involvement with gaming or synching placement of songs with artists involved in movie production.

6. The marketing program: the day-to-day operational (tactical) decisions – a label's "marketing plan."

For the record label, the release plan for an artist might include a calendar that highlights the various activities by department as each single is being released and as these releases build towards the project street date. These items might include:

- Promotion – consumer and trade
- Sales – distribution, pricing
- Publicity – where, how to communicate
- Advertising – where, how to motivate

7. Sales forecast and marketing budget.

- How much product do we expect to sell?
- How much should we spend on marketing, and how should we spend it?

For the record label, this information is usually reserved for record label personnel only and is not discussed outside of the company. It is important information in that each department needs to operate within budgets to be sure that the release will be profitable.

8. Control and evaluation measures (the feedback and re-adjustment).

> Are we meeting our sales goals? Are there any unforeseen opportunities or threats?

Records can take a long time to fully develop and realize their potential. Looking backward to evaluate "best practices" can be a great tool in realizing how to replicate success or avoid mistakes in the future.

What's in the Record Company Plan?

The plan provides basic information about the release and a description of what is occurring in each department or area. A timetable is often included to ensure that timing and coordination are optimal. Financial specifics are often eliminated from the copies circulated at the marketing meetings but are an integral part of the plan. Financial sales goals are set but are not always included in the external plan. Table 8.3 shows some basic information included in a marketing plan.

Sections of the Plan

The information section of the plan contains specific technical data about the release, including the configurations available, pricing information, initial shipment numbers, **SKU** numbers, list of tracks, production information, release date, contact information, and a description of the project and goals. Plans usually contain a biography and/or an album highlights section updating the reader about the new music. Elements such as digital EPK, images, and music access are also featured in the information section.

The plan usually moves into singles promotion and contains information on the singles that have been selected to release. Since radio and streaming promotion of the first single(s) usually precedes the release by at least 8–12 weeks prior to project street date, this information needs to be presented early in the plan. The goal is to get substantial radio and streaming airplay for each single released, creating awareness for the album project. The radio and streaming promotion section will also feature any special campaigns geared toward radio, such as promotional touring, promotional contests, or showcases. Specific listening markets may be targeted in this section, although national efforts are also created but concentrated on specific formats. Subsequent marketing efforts such as advertising and publicity

Table 8.3 Record Company Marketing Plan Elements

- Project title and artist name
- Street date
- Price, dating and discounts
- Number of songs and selections
- Overview or mini-bio
- Format configurations and initial financial goals/streaming goals/shipments
- Selection number and bar codes
- Marketing goals and objectives
- Target market
- Promotion (DSPs/radio)
    o Names and release dates of singles
    o Promotion schedule
    o Promotional activities
- Publicity
    o Materials: Electronic Press Kit (EPK)
    o Target media outlets
    o Specific goals (interviews, appearances, reviews, etc.)
- Advertising
    o Materials
    o Target media outlets and coordination with retail accounts
    o Ad schedule
- Sales
    o Sales forecast
    o Specific promotions by outlet
    o Marketing materials: digital/physical
- Video
    o Production information
    o Distribution and promotion
- Social media/Internet promotions
- Street teams/grassroots marketing
- Lifestyle/promotional tie-in with other companies/products
- Tour support
- Tour dates – help with spread of product at retail and tie-ins with radio
- International marketing
- Artist contact information (manager, publicist, and so on)
- Comprehensive timetable
- Budget (not included in copies that are distributed to employees or contract agents)

(sometimes grouped together in a media plan) and distribution promotions follow.

The publicity section will have goals, materials being created, and targeted media outlets, complete with a timeline. Remember that some publications, particularly print magazines, have a long lead time and must be worked several months in advance of the placement you are seeking. Publicity materials, including photos and a bio, are usually created well in advance of the release date. A press release is sent out closer to the release date to generate additional media interest.

The distribution and sales section of the plan includes targeted DSP programs as well as physical retail outlets for album releases and may include particular information about costs for these programs, as well as specific features such as exclusive releases or content, endcap placement or other positioning, and newspaper and coop advertising.

An advertising section may have information on specific media being targeted for advertising and information on advertising materials being generated (such as TV commercials, radio spots, etc.). Different labels have different philosophies on advertising. Some focus almost exclusively on advertising to consumers, while others budget for a healthy amount of trade advertising. Sometimes, radio and TV ads will be developed so that they can be produced on short notice to give the record a boost in the marketplace or capitalize on some unforeseen opportunity.

The section on video may include production information and promotional plans, including placement on national, regional, and local television shows and any digital marketing campaigns.

A tour section will include a list of tour dates and venues and any local media and retail partners that will be included in the plans. A lifestyle section contains information about sponsorships and co-branding with other products. A timeline of all events and deadlines can sometimes be found at the end of the marketing plan, allowing for a quick glance to reassure that all events are coordinated and taking place in a timely fashion. This timeline may not be shared with external partners. A detailed marketing plan for a major artist is included in the appendix of this chapter.

Digital media encompasses all areas of marketing. Digital marketing may include information on specific online promotions through streaming radio, DSPs, and targeted web partners (such as video channels, online media publications, and online retailers). Online consumer promotions, contests, email campaigns, and online street teams are a part of this section.

Timing

Timing is crucial to the success of an album, including the time of year the album should be released. The album release date is called the **street date**. Albums are strategically released on Fridays to align with worldwide street dates and keep piracy from bleeding across international lines. Release dates are subject to change for many reasons, not the least of which is what other artist might release a new album on that particular Friday. Historically, there has been a glut of releases at the end of the year when retail stores were already full of shoppers and the lack of releases in other times of the year when a strong release my lure shoppers into stores. New artists were advised to release during those times of the year when competition is less, but then so was the number of shoppers. New data reveals that seasonality of access to music and releases of new music has "smoothed" – meaning the listeners have evened out their consumption and continually access music, no matter the season. An album may be strategically timed to coincide with a concert tour or a major media event for the artist, with the holiday season still seeing a surge in streaming sales, but revenue generated by streaming has remained relatively steady throughout the calendar year.

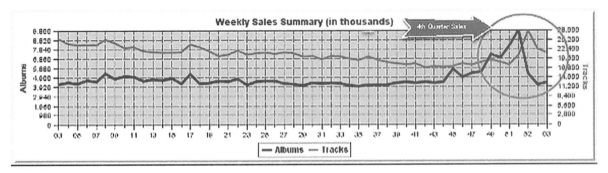

 **FIGURE 8.1** *Weekly Sales for 2014*

(Source: SoundScan)

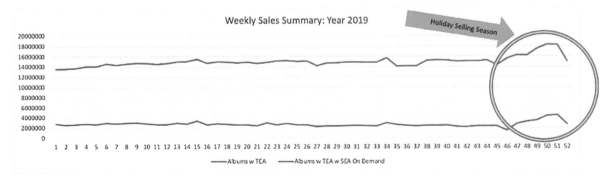

 **FIGURE 8.2** *Weekly Sales for 2019*

(Source: Music Connect)

The Importance of Street Date

The value of the street date has changed in recent times. For many labels, "streeting" an album is not as important as launching a series of singles. For other imprints, the artist's project may release several singles prior to the traditional album street date and generate oodles of streaming activity that translates to revenue. Many consumers are very interested in the body of work the artist will deliver and has in process and will stream the entirety of the project, increasing the stream count for the project just that much more – even though an official single has not been designated. From an administrative standpoint, the official "street date" does generate internal deadlines the label needs to hit: when master recordings need to be complete; when artwork needs to be designed and label copy submitted; and setting of the calendar for which to launch tours, set single releases, and start conversations regarding publicity and social media campaigns to discuss the release of new music. Setting the "street date" has value.

CONCLUSION

Lewis Carroll said, "If you don't know where you are going, any road will get you there." If you want to get to the top of the music business, it is best to have some goals and a plan to get there. In this chapter, we have reviewed the basics of marketing planning used by record labels to promote artists and their recordings. Every label will bring its unique philosophies, strengths, and weaknesses to the process, but they will all have the same goal – to have a hit record. The plan gives the marketing team a starting point and an adaptable framework to achieve that goal. In the chapters that follow, we will discuss the specifics of each of the major marketing tools in greater detail.

GLOSSARY

Marketing plan—A single or album specific plan that describes activities selected to achieve specific marketing objectives for that product within a set period of time.

Push strategy—Pushing the product through the distribution channel to its final destination through incentives aimed at retail and distribution.

SKU: stock-keeping unit—An inventory ID that represents one or more items sold as a single unit. A vinyl and a CD version of an album would each be a separate SKU and each have their own barcode. If the CD and vinyl disk were offered together as part of a box set, that would be a third SKU.

The trade—Those people within the industry (middlemen) for whom business-to-business marketing is done.

Trade advertising—Paid promotions targeted at industry middlemen and designed to stimulate purchasing and, in turn, promotion to consumers.

APPENDIX: MENDING BROKEN HEARTS – INDIE ARTIST MARKETING PLAN

The following is a detailed marketing plan for an indie artist. The names and dates have been changed, but everything else is real.

<u>Dax Carter – **_Mending Broken Hearts_**</u>
Digital Territories: Worldwide
Physical Territories: U.S.
Street Date: April 9, 2021
Label: Lonely Highway Records

Image by Benjamin Balazs from Pixabay

LISTEN

https://open.spotify.com/artist/6vCtweS8UWAXTyau2j0rDT
https://soundcloud.com/lonelyhighwayrecords/dax-carter

All Retailers' Link: https://orch.com/mendingbrokenhearts

Digital UPC: 123456789232
CD UPC: 876345923452
LP UPC: 1234567899325

TRACK LISTING

1. Mending Broken Hearts

2. Roots of the Family

3. Get on the Bus

4. Ride from Chicago

5. Tempting Fate

6. Face of Challenge

7. My Soul

8. Climbing the Hill

9. Riding Alone

10. Broken Heart

FIGURE 8.3 *Artist Image for Marketing Plan*

11. Grand Canyon

12. Travellin' Home

LP ONLY

Only the Winners (Acoustic)
Night to Run (Acoustic)
Wake up Tomorrow (Acoustic)
Climbing the Hill (Acoustic version)

One Sheet Due: 2/12
CD Due in Warehouse: 3/10

Hometown: Austin, TX

Top Consumption Markets (Music Connect): Chicago, Philadelphia, Nashville, Austin, Raleigh, NC, Seattle, Oklahoma City.

Manager: Jane Poulte – jane@poultemanagement.com
Project Manager: Amanda Heartnell – amanda@lonelyhighwayrecords.com

HIGHLIGHTS

- This is Dax Carter's debut album.
- Dax was runner-up on the 2019 season of American Idol.
- Dax is an accomplished songwriter who has penned hits for artists in many genres, including Dawes, Kelsea Ballerini, Pink, and John Legend.
- Mending Broken Hearts features Dax's childhood friend, Jason Isbell, on two of the tracks.

ABOUT THE ALBUM

Having spent a lifetime of love and loss, Dax Carter's *Mending Broken Hearts* tells his personal story from his early childhood into his adult life. He sings the depths of the Grand Canyon on this self-titled track and yet shares the highs of every emotion throughout this album, with the main message being one of hope for the future and a vision to see the positive in the roughest patches of life.

TOUR: Amy Sutherland @ I Love Touring – amysutherlandlovestouring.com.

Full listing found by clicking here.

- Will be in New York in February and LA in March
- Listening party in Austin (early April)
- Performance with Jason Isbell will be livestreamed on Twitch on April 30

Assets: (click hypertext to access)

- Initial press release

- Bio
- Approved publicity photos
- Cover art
- Behind-the-scenes video (ten segments showing the making of the album)
- Q&A with Mumford & Sons (about collaborating on the single "Yes, I Am")
- "Yes, I Am" has been placed in a new television series, which will begin airing on ABC in May

PUBLICITY: Cassidy Potts, Fired Up Media. Cassidy@firedupmedia.com

Key Dates
- **2/15** – Sent press release for first single
- **Week of 4/10** – Album premiere

PRESS SO FAR

Spotify

- Playlist: added to Indie Folk Top Tracks playlist 4/13 (42.425 followers)
- The Sounds of Spotify: added to The Sound of New Americana playlist on 4/11 (1,893 followers)
- Particle Detector: added to The Edge of Roots Rock playlist 4/8 (257 followers)
- Jason Isbell: added to Walk on Songs playlist 4/6 (1,753 followers)
- Particle Detector: added to *The Edge of Indie Folk* playlist 3/31 (2,020 followers)

Others being pitched currently:

The Sound of Indie Folk
Indie Folk and Friends
Acoustic Love
Peaceful Songs
No Country for New Nashville
New Folk Friday

APPLE MUSIC

- Added to Southern Craft playlist 4/12

AMAZON MUSIC

- Included on <u>Fresh Folk and Americana</u> playlist (4/12)
- Included on <u>New Americana</u> playlist (4/5) Anticipated add to <u>Brushland</u> playlist

PUBLICITY HITS

<u>American Songwriter</u> – Facebook livestream aired on album release day
<u>Pitchfork</u>
<u>The Boot</u>
<u>Paste</u>
<u>Americana Music Association</u> (featured in 3/31 newsletter)
<u>Billboard</u> (included in New Releases calendar)
<u>Americana Highways</u> – Facebook livestream aired on release day
<u>Folk Connection</u>
<u>Consequence of Sound</u>
<u>Glide</u>
<u>Take Effect Reviews</u> – album review

CONFIRMED PRESS

Rolling Stone Country – 6/12 issue

No Depression – confirmed inclusion in special Collaborations issue (alongside Jason Isbell) – 6/2–6/7)

OTHER TARGETS PITCHED

Stereogum
Garden & Gun Magazine
Acoustic Guitar Magazine
Premier Guitar

PROMOTIONS

- Instagram filter launched 4/1
- Music.com promotion ran 4/1–4/4

- Partnership with brand partner Yeti featured Dax in their ad campaign running across social networks 4/1–4/27. Dax also performed the single during a livestream on Twitch hosted by the brand 4/10.

RADIO (Art Silas @ Lonely Highway Records | art@lonelyhighwayrecords.com)

Formats: AAA/Non-Comm/Americana
Focus track: "Bending Broken Hearts"
Impact Date: 3/15
Album Servicing: early April

ADVERTISING/MARKETING/SOCIAL MEDIA (tammy@lonelyhighway-records.com)

Social Channels (each is hyperlinked)

Facebook: 193k
Twitter: 32.5k
Instagram: 52.4k
YouTube: 4.2k
Website:

- **Website will promote the new album beginning 3/23**
- **Add tracking pixel**
- Website will highlight the new album on the homepage leading up to the release (along with a countdown clock). A pivot smartURL link will be utilized so visitors may listen to the track on the platform of their choice. The website will also feature a sign up for email.

ADVERTISING

Social Media:

- Facebook, Instagram, YouTube
- 4/8–4/15
- Targeting lookalike audiences on each platform

Digital:

- Google Search, Google Display ads

Publications:

- Stereogum, No Depression, NPR

TARGETS

Demographics/Insights (FB)

- 10% 18–24
- 11% 25–34
- 18% 35–44
- 25% 45–54
- 31% 55–64
- 5% 65+

Artist's fans also love:

- Jason Isbell, Dave Barnes, Drew Holcombe and the Neighbors, John Prine, Shovels and Rope, Mumford & Sons

Sales Goals:
- Focus on building a larger presence with each streaming service and asking fans to engage (saves on Spotify and getting fans to add to their playlists, in particular)
- Amazon will be a key partner to support this release

DIGITAL RETAIL (Sal@LonelyHighwayRecords)

Genre: Americana/Folk
Territories: Worldwide
Pricing: $9.99

LINK: https://orch.com/mendingbrokenhearts

ROLLOUT

- **2/20 –** Pre-order at $6.99 With 2 IGs
- **4/9 –** STREET DATE
- **5/4 –** Revert to $9.99

Apple Music

- Apple for Artists (now integrated with Shazam)
- Apple Music highly values social support driving traffic to their service
  - Correct tagging for socials: @AppleMusic
- We will provide social graphics/assets from Apple
  - Playlisting assets as we find support
  - Instagram story assets (if needed)

Spotify:

- We need to be pitching all focus tracks at least 1 week out from release via the Pitch Tool to take advantage of Release Radar and editorial consideration.
- We should create a discography list to help drive consumption and build profile
- We can utilize "Action Page" link to increase followers

Pandora

- Liners to record (generic/tour)
- Utilize Artist Analytics/AMP Cast:
  - Amp.pandora.com
  - Featured track
  - New release call outs
  - Pin singles to artist page
- Submit for Artists Playlists based on affinity artists

YouTube

- Update banners + artist/album information
- If submitted via The Orchard, we can pitch for channel cards and slates
- Get vertical video call-outs for Amazon, Apple, Spotify

Any assets (video content) we can share ahead of release will be useful

PHYSICAL RETAIL

Pricing:

CD – $13.98
LP – $29.98

Amazon

AMS ads

Will pitch for:

– TIM liner for Alexa
– Originals
– Prime inclusion on street week

In addition to pitching for place listing and program placement, we will pitch for as much digital and physical placement including:

PRE ORDER SUPPORT

– Pre-order Digital Music Newsletter
– Digital homepage pre-order album slider

STREET WEEK SUPPORT

– Amazon Music Twitter Post (1.5m followers)
– Digital Music Newsletter: New Releases
– New release album slider

**Make sure to drive folks to Amazon pre-order*

AEC – 10% discount on Initial Buy
Baker & Taylor/Ingram – Spotlight magazine feature
Barnes & Noble – P&P program: April New Release End Cap
Indie Coalitions
CIMS/AIMS/DORS – P&P and Listening Stations for 60 Days

150 key stores

ADDITIONAL MARKETING

Waterloo Records – Austin TX (hometown) End Cap/LS Featured Artist with signed poster
Grimey's – pre-order with discount to attend Basement East Show
Newbury Comics – signed booklets for online pre-orders

Targeted Accounts:

Electric Fetus, Rough Trade, Music Millennium, Amoeba Music, Silver Platters, Zia Records, Bull Moose, Rasputin Music, Exclusive Company, Everyday Music, Down in the Valley

CHAPTER 9

Music Distribution and Music Retailing

INTRODUCTION

By 2020, over 90% of recorded music revenue has been generated from streaming outlets. Is streaming now retail? So is streaming the "radio" of our times? How does the term "distribution" apply in the music business today? Although the lines are blurry, let's attempt to clarify these issues as they pertain to the business since **access** to music drives revenue (RIAA 2020 Report).

Traditionally, music distributors have been a necessary conduit in getting music product from record labels' creative hands to consumers through brick-and-mortar stores. As the marketplace has shifted to more digital consumption, distribution companies have re-evaluated their value in the food chain and developed sales models by adding services that reflect both direct sales opportunities to music consumers as well as models that supply to various digital portals.

Yes – by 2020, 90% of recorded music revenue has been generated from streaming outlets. This includes services such as Spotify, Apple Music, and Amazon, as well as revenues distributed by SoundExchange, who collects from music portals such as Pandora, SiriusXM, and ad-supported on-demand video services such as YouTube, Vevo, and ad-supported Spotify (RIAA 2020 Report). What a massive shift from traditional music distribution services of just ten years earlier, where each major music conglomerate manned enormous distribution companies with warehouses, distribution centers, and large sales forces that marketed to a vast array of

DOI: 10.4324/9781003153511-9

brick-and-mortar stores – all of which have either vaporized or are reduced to shadows of themselves today.

But to best visualize the current state of "distribution," what we are talking about is the process to facilitate access to music (streaming services) and/or purchase of music (retailers) by consumers via distribution channels. These distributors, both major conglomerate owned and independents, focus their attention on two main outlets that service end consumers: digital and physical. For the purpose of this chapter, let's narrow the focus of our conversation to traditional retail outlets, which still exist, and aggregators that service digital service providers, which generate streaming revenue, known as DSPs.

Traditional Distribution

Traditional physical retail is now a very small portion of the revenue generated in the music industry and yet is still a viable touch point for independent artists, especially those for those on tour. Figure 9.1 reflects a traditional distribution function and shows the points of contact between labels and outlets. In today's business model, record label sales executives communicate with distribution as a conduit to the marketplace, but labels also have ongoing relationships with the various points of consumer contact – whether that be a retailer or a digital service provider. Depending on the importance of the outlet, the label rep will often communicate with the outlet significant releases and marketing plans directly. And as deals are struck, both marketing plans including exclusive releases, playlist placement, and other activities can then be implemented by the label with the help of the distributor.

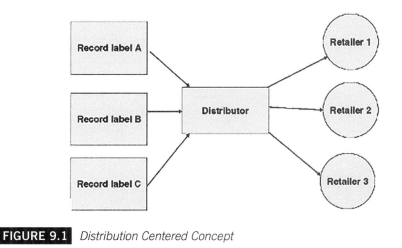

FIGURE 9.1 *Distribution Centered Concept*

THE BIG 3 CONSOLIDATION AND COMPETITION

Within the last 20 years, the Big 6 conglomerates have consolidated into the Big 3, reducing their number of employees to reflect the ever-decreasing size of the music sales pie. But the mergers began prior to the explosion of file sharing with the combination of Universal with Polygram in 1999, the same year as the emergence of Napster. This merger created Universal Music & Video Distribution (UMVD), now Universal Music Group Distribution (UMGD), who has maintained their market share position at #1 since inception. In 2013, most of EMI's stable of labels and artist were added to the Universal fold, increasing the strength of this company's musical power in the marketplace.

Another combination of conglomerates includes the blending of Sony and BMG in 2004, who consolidated their various profit centers as well as their distribution workforces but maintained much of the integrity of their imprints. They, too, gained market share, garnering the #2 position simply by merging. In 2008, Sony purchased the remaining 50% stake held by BMG in the Sony/BMG merger. The merged companies, now wholly owned by Sony, were renamed Sony Music Entertainment Inc. (SME) and include all the labels that were Sony, as well as BMG.

WEA maintains a separate distribution function as part of Warner Music Group (WMG). This company is positioned as #3 of the major music conglomerates.

The Big 3 own their catalog, meaning that the deals that they have struck with the artists signify that the sound recordings are the label's intellectual property. This ownership allows the Big 3 to be able to negotiate directly with digital service providers and their editorial staffs, bargaining for playlist positioning along with exclusive deals and representation on their platforms globally. As of the end of 2020, the Big 3 market share including their independent label representation was 80%. To help with some of the outsized advantage that the majors have grossed, Spotify has created a unified tool for playlist submission to aid independent labels in helping create a more balanced approach to hearing new music on their platform.

As noted in the market share data, independent labels continue to be a force within the music business, and by the end of 2020, independents held an approximately 19.2% market share. With burgeoning digital storefronts, any label can now have an instant "sales" point by which to connect with customers directly. With the creation of "independent" arms within their distribution division, all three majors have taken advantage of this independent surge. By contracting this function of distribution to independent labels, these "independent distributors" can assist the independent label in placing their music in the mainstream marketplace. But there are many truly independent distributors that are *not* tied to the Big 3 that function similarly.

Universal Music Group Distribution

Caroline Distribution represents many independent labels.
INGrooves Distribution represents many independent labels.
Sample labels that it distributes:
Interscope
Geffen
Island/Def Jam
Universal
UMG Nashville
Hollywood
Disney/Buena Vista
Capitol Records
Big Machine
CURB

Sony Music Distribution

Sony's Red and The Orchard Distribution represent many independent labels.
Sample labels that Sony Music distributes:
Columbia
Epic
Arista/Arista Nashville
J Records
RCA
Jive
LaFace
Razor & Tie
WindUp
RCA Label Group Nashville

WEA

WEA's ADA Distribution represents many independent labels.
Sample labels that WEA distributes:
Warner Brothers
Atlantic Records
Bad Boy
Roadrunner
WSM/Rhino
V2 Records
WEA/Fueled By Ramen

VERTICAL INTEGRATION

The Big 3 entertainment conglomerates share profit centers that are vertically integrated, creating efficiencies in producing product for the marketplace. To take full advantage of being vertically integrated, labels looking for songs would tap their "owned" publishing company. (Each of the major entertainment conglomerates has a publishing company. If they only recorded songs that were published by their sister company, more of the

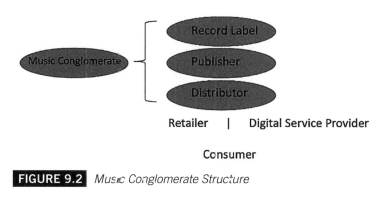

FIGURE 9.2 *Music Conglomerate Structure*

money would stay "in the family.") Once recorded, the music is represented and distributed into retail and the DSPs, using the conduit of the distribution function located within the family (Figure 9.2).

BIG 3 MARKET SHARE

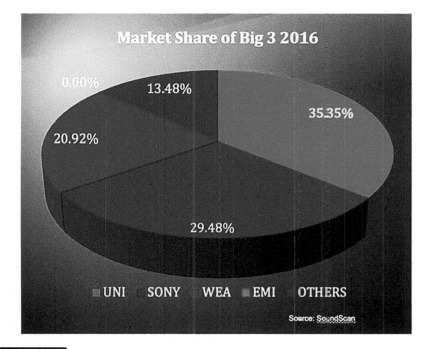

FIGURE 9.3 *Big 3 Market Share 2016*

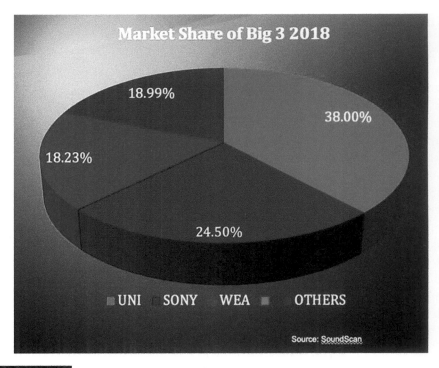

FIGURE 9.4 *Big 3 Market Share 2018*

FIGURE 9.5 *Big 3 Market Share 2019*

MAJOR DISTRIBUTION ORGANIZATIONS

Major distributors such as UMGD, Sony, and WEA have pared back their sales and marketing teams as a result of digital sales dominating the bottom line. There was a time where regional offices were located nationwide, acting

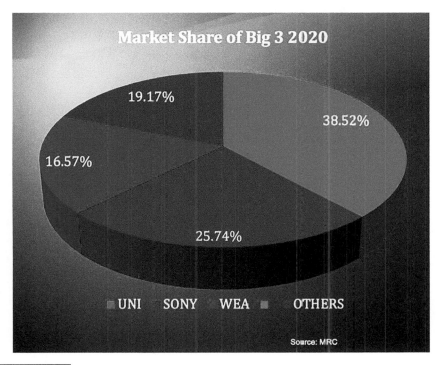

FIGURE 9.6 *Big 3 Market Share 2020*

as sales and marketing conduits to major retail outlets such as Target or Musicland and servicing various retail storefronts with marketing tools such as posters and flats for in-store presentation. In tandem, labels had sales teams that would interface with these distribution teams to transmit sales priorities to the field, but again – sales have been replaced with access, and retail marketing has been traded for social media efforts.

Now, marketing and sales agendas from labels are communicated to the national distribution office. The sales forecasts are generated from the national office, who are the "back office" staff servicing all the labels in the conglomerate family. In many cases, the major labels no longer are stand-alone entities but utilize the talents of the distribution company as an accounting office extension.

The Big 3 have representation in creative hubs and large commerce areas to ease transactions with both labels and marketers. New York, Los Angeles, and Nashville are major creative centers that foster artistic production. As an example, Sony Distribution has representatives in Minneapolis where both Target and Best Buy's headquarters reside. Miami also retains Sony representatives, with this area being rich in musical output. Seattle is both a

FIGURE 9.7 *Sample Major Distribution Office Locations*

hub to several outstanding labels, including Barsuk and Sub Pop while also serving online retail giant Amazon.

INDEPENDENT DISTRIBUTORS – THE AGGREGATORS

The bigger news is the emergence of the independent label in partnership with the independent distributor. The market share of this segment of business is growing as demand from consumers continues to increase. In addition to the Big 3 companies' independent enterprises such as Universal's ADA Distribution or Sony's The Orchard, there are many independent music distributors that are contracted by independent labels to do the same job. Many of these independent distributors, known as aggregators, have developed a niche in marketing unique and diverse products. Ideally, the distribution function is not only to place music into retail – physical and digital – but also to assist in the sell-through of the product throughout its lifecycle.

As mentioned, all of the major conglomerates have their own independent-focused distribution arm within their companies. These divisions supply various functions to each label, as constructed with the contracts, and can assist with just distribution, which can be both physical and digital or can be robust marketing plan development and implementation, including radio and streaming strategies as well as social media campaigns, tour support, and more.

Major Distribution-Owned Companies

Alternative Distribution Alliance (ADA) – owned by WMG
Caroline International – owned by UMGD
INgrooves – owned by UMGD
The Orchard – owned by Sony
AWAL – owned by Sony

Independent Distribution Partners

Believe Digital
Idol
Redeye Worldwide
Ditto Plus
Stem
Symphonic

Most independent deals include payments based on recording royalties that can be as high as 50%. Artists can receive an advance that is recoupable. But know that a dedicated marketing and promotion team will activate strategies focused on the artist/label release (Ucaya "SoundCharts").

White Label Distribution

Consolidated Independent
Sonusuite
FUGA

This type of distributor is for the label/artist who has a full marketing and promotion staff but needs the technical help of getting the record into the consumer pipeline such as supplying audio and metadata to DSPs and collecting and distributing royalties back to copyright holders – which is a huge job.

Open Distribution Platforms

CD Baby
Distrokid
Tunecore
Ditto

These distribution platforms appeal to the masses with simple upload of a single and/or album option, and music is "distributed" to hundreds of DSPs. The first level of service is nominal considering the range of exposure: a flat per-song/album fee, an annual recurring subscription, a percentage based commission, or a combination of the three.

The "premium artist" level could include playlist pitching, publishing administration, physical distribution, and other artist services.

Semi-Label Distribution Services

Amuse

This model is new to the market, taking an independent artist without a label, distributing the music to the DSPs; collecting the artist's consumption data; and, if successful, signing the artist to a real record deal based on the analytics.

(SoundCharts by Dmitry Pastukhov, https://soundcharts.com/blog/music-distribution#5-types-of-music-distribution-companies.)

A League of Its Own

BandCamp

Known to be the "gap filler" between the better-known aggregators and the independent arms of the majors, BandCamp has been known to break artists by being a highly efficient one-size-fits-all online store for independent artists needing a digital streaming service, downloader, merchandise fulfillment, and physical distribution.

Music Distribution Flow Chart

The following chart shows the flow of music distribution with **samples** of companies that represent the category.

Music Supply to Retailers

Once in the distributor's hands, music is then made available to consumers. Varying access portals and retailers acquire their music from different sources. Most mass merchants are often serviced by outside vendors who maintain the store's music department, including inventory management, as well as marketing of music to consumers. Retail chain stores are usually their own buying entities, with company-managed purchasing offices and distribution centers (DCs). Many independent music stores are not large enough to open an account directly with the many distributors but instead work from a *one-stop's* inventory as if it were their own. (One-stops are

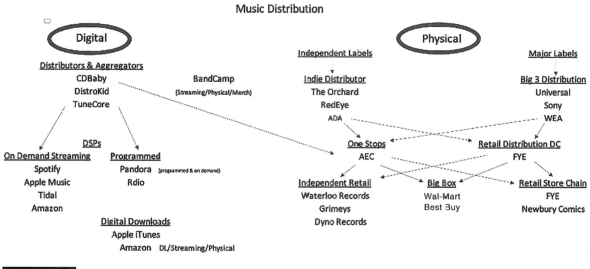

FIGURE 9.8 *Distribution Flow Chart*

wholesalers who carry releases by a variety of labels for smaller retailers who, for one reason or another, do not deal directly with the major distributors.) Retail chain stores and mass merchants will, on occasion, use one-stops to do "fill-in" business, which is when a store runs out of a specific title and the one-stop supplies that inventory.

Retail Store Profiles

A music store's target market or consumer generally dictates what kind of retailer it will be. To attract consumers interested in independent music or to attract folks who are always looking for a bargain determines the parameters in which a store operates. Music retailers have traditionally been segmented into the following profiles:

Independent music retailers cater to a consumer looking for a specific genre or lifestyle of music. Generally, these types of stores get their music from one-stops. Independent stores are locally or regionally owned and operated, with one or just a few stores under one ownership.

> The **Mom & Pop** retailer is usually a one-store operation that is owned and operated by the same person. This owner is involved with every aspect of running the business and tends to be very passionate about the particular style of music that the store sells. This passion can be interpreted as being an expert in the knowledge of the genre and can be a unique resource for the consumer looking for the obscure release.

Mom & pop storeowners tend to have a personal relationship with their customer base, knowing musical preferences and keeping the customers informed about upcoming releases and events.

Alternative music stores profile very similarly to mom & pop stores but with the exception that they tend to be lifestyle-oriented. An electronic dance music (EDM) retailer may have many hard-to-find releases along with hardware offerings such as turntables and mixing boards.

Chain stores tend to attract music purchasers who are looking for deep selection of releases along with assistance from employees who have strong product knowledge. These stores can be found in malls and cater to a broad spectrum of purchasers. These stores have been studied and replicated so that entering any store with the same name in any location feels very similar. Often they have the new major releases upfront with many related items for sale, such as blank media and entertainment magazines. Chain stores traditionally buy their music inventory directly from music distributors, with warehousing and price stickering occurring in a central location.

Electronic superstores do not make the bulk of their profits from the sale of music but rather use music as an attraction to bring consumers into their store environments. Music has historically created traffic to the store in order to make money from the sale of all the other items offered such as electronics, computers, televisions, and so forth. An example is Best Buy, which sources its music from Alliance Entertainment.

Mass merchants use the sale of music as event marketing for their stores. Each week, a new release brings customers back to their aisles with the notion that they will purchase something else while there. Music has become less of an attraction in recent years since demand is being met through streaming, yet an example such as Wal-Mart is using Alliance's direct-to-consumer supply portal to meet the needs of consumers online – all within the Wal-Mart online site.

Role of Physical Distribution

Although a very small part of a music distribution today, physical distribution companies have three primary roles: the sale of the music, the physical distribution of the music, and the marketing of the music, although these roles have been dramatically modified as the shift from physical to digital has advanced. The national staff of distribution companies now focus as brand extensions of the individual labels with an executive staff that acts as a wheelhouse that manages the functions of product information dissemination such as street date and sales information, manufacturing and product management, and implementation of national and regional marketing efforts to be executed at various levels. Sales information consists of setting sales goals, determining and setting deal information such as discount off of wholesale and dating of product, and soliciting and taking orders of the physical product from retail. Additionally, the sales administration department should provide and analyze sales data and trends and readily share this information with the labels that they distribute.

The marketing division assists labels in the implementation of artists' marketing plans, along with adding synergistic components that will enhance sales. For instance, the marketing plan for a holiday release may include a contest at the store level. Distribution marketing personnel would be charged with implementing this sort of activity. But the distribution company may be selling holiday releases from other labels that they represent. The distribution company may create a holiday product display that would feature all the records that fit the theme, adding to the exposure of the individual title.

The physical warehousing of a music product is a big job. The conglomerates as well as independents have consolidated the warehousing of music with positioning of their distribution centers centrally in the United States. For instance, leading one-stop Alliance Entertainment's central warehouse is located near Louisville, Kentucky, which is just next door to the UPS hub. Incorporating sophisticated inventory management systems where music and its related products are stored, retailers can place an order, and it is the distributor's job to pick, pack, and ship this product to its designated location. These sophisticated systems are automated so that manual picking of product is reduced and that accuracy of the order placed is enhanced. Shipping is usually managed through third-party transportation companies.

Because physical music sales have been reduced and many stores have either closed or minimized store floor space dedicated to music sales, music companies and their distributors have outsourced the inventory management, fulfillment, and rack-jobbing to a known entity within the industry. Alliance has stepped up its role as "sales rep" for both Wal-Mart and Best Buy, supervising vendor-managed inventory as well as taking care of the in-store floor space in place of the distributors who would normally assist these accounts with sales and marketing efforts.

As retailers manage their inventories, they can return music product for a credit. This process is tedious, not only making sure that the retailer receives accurate credit for product returned, but the music itself has to be retrofitted by removing stickers and price tags of the retailer, re-shrink wrapping, and then returning into inventory.

How the Money Flows

Interactive Streaming Services – Subscription-based examples: Spotify, Tidal, Apple Music

Knowing that over 85% of revenue is generated from digital sources, the negotiation between record label and distributor with the DSP is essential, be it a major or independent. For interactive streaming services that require a subscription to access, each of these negotiations results in various royalty rate amounts, some more beneficial to the label and distributor than others. Although the value varies, a single standard stream can generate anywhere between $0.0035–$0.0049. On the surface, this rate may seem low, but when multiplied by millions of streams, a hit song can generate tens of thousands of dollars in streaming revenue, which, in this business model, replaces traditional record sales.

Non-Interactive Streaming Services – free streaming examples: Pandora, Rdio

To complete the streaming royalty conversation, in non-interactive streaming that does not require a subscription but makes revenue from advertising to free listeners, royalties are collected by an entity known as SoundExchange. This royalty value is standardized at $0.005 per stream. Each royalty is then split between artist and label, with a small "administration fee" withheld, so that artist and label each receive $0.0023 per stream.

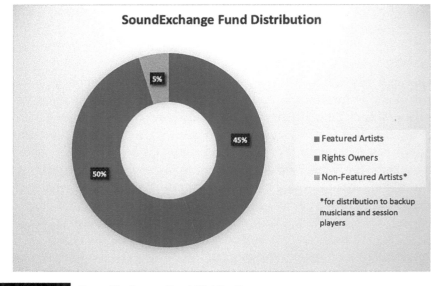

FIGURE 9.9 *SoundExchange Fund Distribution*

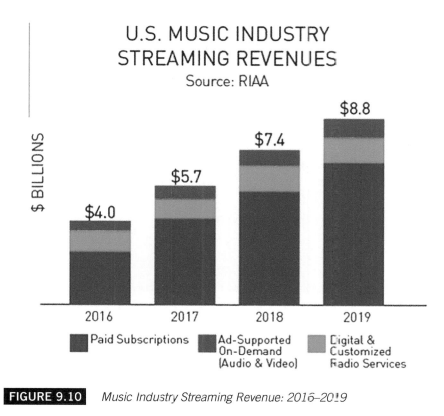

U.S. MUSIC INDUSTRY
STREAMING REVENUES
Source: RIAA

FIGURE 9.10   *Music Industry Streaming Revenue: 2016–2019*

Again, this rate looks small, but added with the other streaming royalty, can accumulate to a large sum of revenue (soundexchange.com RIAA.com)

INDEPENDENT NEGOTIATION REPRESENTATION

Independents, both labels and distributors, can need assistance in finding their voice when negotiating with large DSPs. Using strength in numbers, coalitions have been created to assist with this mediation process. As an example, Merlin is a 20,000-strong member-owned coalition of independent labels and distributors who negotiate and collect royalties from digital service providers worldwide. Its members come from 63 countries and represent 12% of the global digital music market. Members include the Beggars Group, Dim Mak, Domino, Eleven Seven Music Group, Entertainment One, Epitaph Records,!K7, Kobalt Label Services/AWAL, Mad Decent, mtheory, Mom + Pop, Ninjo Tune, [PIAS], Secretly Group, Symphonic Distribution, Sub Pop, and Warp Records.

FIGURE 9.11   *Merlin Represents Independents*

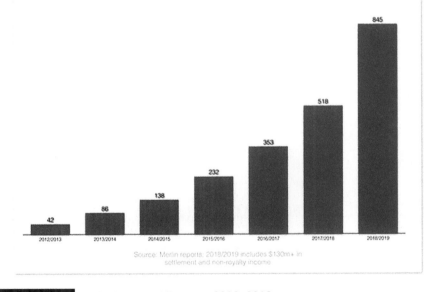

FIGURE 9.12   *Merlin Annual Payouts: 2013–2019*

Merlin doesn't distribute revenues from Apple Music or Amazon Music. However, it *does* license Facebook, YouTube Music, Spotify, Deezer, Pandora, Alibaba, NetEase, and Tencent Music Entertainment – to emphasize worldwide reach (www.musicbusinessworldwide.com/indie-label-annual-revenues-hit-845m-via-merlin-showing-strong-growth/).

FORECASTING

Record labels plan for their financial future by forecasting their revenue generation based on releases that are scheduled throughout a fiscal year. Most labels do this planning a year in advance and, in many cases, do not have under contract the very artists that will be releasing records in the year that they are forecasting. The forecasting process is very difficult, and it takes a skilled sales and financial team to understand the dynamics of the business to fully realize the "bottom line" of this task. Considerations include artist activities, genre timing, other artist releases on the label, the competitive marketplace, musical trends, and other various internal and external issues.

To forecast a release, a label needs to determine if the artist is releasing songs only, an EP (3–5 song mini release), or a full album with 10 songs on it. The process can be looked at several ways, but since so much of revenue is being generated via streaming, the appraisal the label needs to look at is based on money generated. How the money is generated can be produced through various means with different evaluations that include physical releases as well as streaming. Streaming valuation is complicated by the fact that its royalty varies by DSP as well as interactive vs. non-interactive type:

1,250 Premium Audio Streams	= 1 Album Unit
3,750 Ad-Supported Audio Streams	= 1 Album Unit
3,750 All Video Streams	= 1 Album Unit (www.billboard.com)

360 Deals/Multiple Rights Deals

Another source of income for labels is negotiated through the contract with the artist in the 360 Deal or Multiple Rights Deal. Revenue generated from other sources by the artist is shared with the label through monies made via live ticket sales, merchandise, movie and/or book endorsements, product alignments, and sometimes music publishing. The label makes a percentage of these revenues after reaching a certain performance plateau with the artist's music in the marketplace.

Subscription vs. Ad-Supported Streaming: A Weighted Average

To find a weighted average between subscription service and ad-supported streaming, industry research has revealed that approximately 1,500 streams equals 1 album unit. This equation is where the value of $0.005 was derived by SoundExchange to collect as its streaming royalty. 1,500 streams × $0.005 = $7.50, which is the standard wholesale value of a record album. Interactive streaming valuation can vary from $0.0035–$0.0049, depending

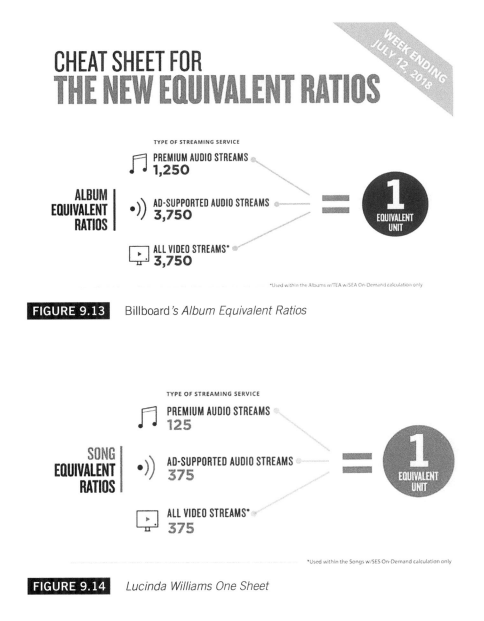

FIGURE 9.13   Billboard's *Album Equivalent Ratios*

FIGURE 9.14   *Lucinda Williams One Sheet*

on the arrangement. The amount of money generated can be significantly different, based on the negotiated contract.

In a competitive landscape, industry norms require that releases generate a certain amount of revenue to cover the expenses of making and marketing the record – known as breaking even. Depending on the genre, these expenses vary. A singer-songwriter may not require as much studio time as that of a full-production pop album by a platinum performing artist. But doing the math using a predictor equation utilizing a profit and loss pro forma will help the label "predict" if the market will stream the music of an upcoming release enough to generate monies to cover expenses – and then go on to make money for the company.

To generate approximately $250,000 in streaming revenue for the label, a release would need to stream:

70,000,000 Non-Interactive Streams × $0.0023	= $161,000
30,000,000 Interactive Streams × $0.0035	= $105,000
100,000,000 Streams	$266,000

Where the Money Goes

Based on a wholesale album unit, a way to see where the money "goes" is to look at percentages based on contracts. Currently, the average wholesale value of an album is between $7.00 and $8.00. Most contracts that labels construct with artists are based on album units of consumption and whole-sale value of the album unit.

Levels of percentages:

- In major record deals, distribution takes 12%–14% of wholesale. Independent distributors take between 10% and 15%.
- Design and manufacturing is based on how many physical units are made, but basic cover art is always required.
- Artist royalties is contractual pivot point. This example reflects an 18% all-in royalty of wholesale.
- Mechanical is based on a ten-song album using the statutory rate of $0.091 per song × 10 songs = $0.91 per unit album.
- Recording costs can be recouped once the album is successful. But early in the calculations, the label will "pay themselves back" until this cost is reimbursed.
- Marketing and promo costs are not recoupable and can be more or less, depending on the artist and genre.

So if a retailer, be it online or physical store, sells a ten-track album for $9.99, and they purchased it for $7.50, the retailer pockets $2.49.

Top 10 Revenue Generators

Every label has its own top accounts that stream and/or sell the greatest amount of their music which, in turn, generate the most revenue for them.

Table 9.1 Money Allocation of a $9.99 Music Sale		
Where the Money Goes **$7.50 Wholesale**		
Distribution	$0.90	12%
Design/Manufacturing	$0.75	10%
Artist Royalty	$1.35	18%
Mechanical Royalty	$0.91	.091 × 10 Tracks
Recording Costs	$1.00	13%
Marketing/Promotion	$0.90	12%
Label	$1.69	22.50%
Retailer	$2.49	33.20%
	$9.99	

This money is based on genre and consumer preference and can vary within the distribution family. The top ten accounts usually include:

Apple iTunes
Spotify
Amazon
Pandora
SiriusXM
AEC – Alliance Entertainment Corporation
YouTube/VEVO
Google
Target

Metadata

Metadata is often referenced when discussing publishing and copyright ownership of songs. The word metadata means "data about other data." Music metadata is the details about the song file about the artist, songwriter, producer, song title, release date, track duration, and more. Initially, the value of this data told who owned the copyright, which included both the musical composition (the publisher) and the sound recording (the label), how much of the copyright was owned by whom, and how to "split" the royalty due to each owner.

But the value of the metadata makes digital music "searchable," meaning if the lyrical content, title, beats per minute (BPM), and/or artist have relevance to a music supervisor looking for a particular song for a movie or commercial, the metadata is the element that will allow this track to be discovered quickly. Metadata is a management tool – like search engine optimization (SEO) for songs, allowing the unique qualities of a track to be discovered through computer searches. A sample form is included here, but other fields that can be inserted in metadata include mood, instrumentation, musicians, and cover versions of existing songs. Accuracy is critical in the uploading of metadata so that search engines can accurately find the information (www.musicgateway.com/blog/how-to/music-metadata-for-sync-the-music-gateway-bible).

Additionally, having this information complete is essential to fully compensate all copyright owners in the current international business climate. Metadata is embedded into the musical file itself during the mixing and mastering process and aids with royalty distribution. Although perceived as a musical production element, one can fully appreciate the value of this data as it pertains to marketing opportunities and corporate alignments (https://artists.spotify.com/blog/metadata-what-it-is-and-why-it-matters).

Table 9.2 Metadata Distribution Form – A Sample Form

Metadata Distribution Form

Artist	
Album	
Format	
Label	
Cat No	
UPC/Bar Code	
Physical Release	
Digital Release	
Artist Website	

Track No	Time	Track Name	ISRC	Primary Genre	Secondary Genre	©Year	©Holder	Composer(s)
1								
2								
3								
4								
5								
6								
7								
8								
9								
10								
11								

Timeline

The communication regarding a new release begins months prior to the street date. Although there are varying deadlines within each distribution company, the ideal timeline is pivotal for the actual street date of a specific release. For most releases, street dates are on Friday worldwide. This is to combat piracy. Working backwards in time, to have product on the shelves by a specific day, physical product has to ship to retailers' distribution centers approximately one week prior to street date. And digital service providers need audio files and metadata information to be uploaded to portals that are to push new releases into the digital market space. Physical distributors need the orders one week prior to shipping. The marketing process of specific titles occurs during a period called *solicitation*. All titles streeting on a particular date are placed in a solicitation book, where details of the release are described. The solicitation page, also known as *sales book copy and/or one sheet*, usually includes the following information:

- Artist/title
- Street date
- Genre category
- Information/history regarding the artist and release
- Marketing elements:
  - Social media
  - Publicity activities
  - Tour and promotional dates
  - Bar code

The one sheet typically includes the album logo and artwork, a description of the market, street date, contact info, track listings, accomplishments, and marketing points. The one sheet is designed to pitch to streaming playlisters, retail buyers, and other marketing outlets. The product bar code is also included to assist in plugging the actual release into the inventory management system.

Retail Considerations

How does a retailer learn about new releases? Distribution companies are basically extensions of the labels that they represent. To sell music well, a distribution company needs to be armed with key selling points. This critical information is usually outlined in the marketing plan that is created at the record label level. Record labels spend much time "educating" their distributor partners and retailers about their new releases.

FIGURE 9.15 Billboard's *Song Equivalent Ratios*

Distribution companies set up meetings with their *accounts*, meaning the retailers. At the retailer's office, the distribution company shares with the *buyer* – that is, the person in charge of purchasing product for the retail company – the new releases for a specific release date, as well as the marketing strategies and events that will enhance consumer awareness and create sales.

Purchasing Music for Physical Stores

When making a purchase of product, the buyer will take into consideration several key marketing elements: airplay, both streaming and terrestrial; media exposure; touring; cross-merchandising events; and, most critically, previous sales history of an artist within the retailers' environment or, if a new artist, current trends within the genre and/or other similar artists. The ultimate goal is to increase the purchasing decision while creating marketing events inside the retailer's environment. Most record labels, along with their distributors, have agreed on a forecast for a specific release. This forecast, or number of records predicted to sell, is based on similar components that retail buyers consider when purchasing product. Many labels use the following benchmarks when determining forecast:

> *Initial orders or IOs*: This number is the initial shipment of music that will be on retailers' shelves or in their inventory at release date.
> *90-day forecast*: Most releases sell the majority of their records within the first 90 days.
> *Lifetime*: Depending on the release, some companies look to this number as when the fiscal year of the release ends, and the release will then roll over into a catalog title. But on occasion, a hit release will predicate that forecasting for that title continues, since sales are still brisk.

INVENTORY MANAGEMENT

Larger music retailers have very sophisticated purchasing programs that profile their stores' sales strengths. Using the forecast, as well as an overall percentage of business specific to the label or genre, the retailer will determine how many units it believes it can sell. This decision is based on historical data of the artist and/or trends of the genre along with the other marketing components.

Keeping track of each release, along with the other products being sold within a store, is called *inventory management*. Using point-of-sale (POS) data, the store's computer notes when a unit is sold using the **UPC** bar code. This min/max inventory management system profiles physical store inventory by how many units by titles should be on hand. Orders and re-orders are generated through an electronic data interchange (EDI) to its supplier, and the product is then shipped to the retailer within a few days. To avoid waiting for the product to arrive, a retailer may opt to drop-ship product directly from the distribution company or a partnering one-stop supplier, avoiding the delay of processing at either the headquarters' distribution center.

TURNS AND RETURNS

Know that a store's success is based on the number of units it sells within a fiscal year. Clearly, the size of the store dictates how much product or inventory it can hold. To manage the real estate within the store, the best-selling product should receive the best space. To keep a store performing well, the inventory management system should notice when certain titles are not selling. Music retailers have an advantage over traditional retailers in that if a product is not selling, they can send it back for a refund, called a *return*. The refund is usually in the form of a credit, and the amount credited is based on when the product is returned, along with other considerations such as if a discount had been received. It has long been an industry standard that the return average is noted at 20%. To put this statistic another way, for every 100 records in the marketplace, nearly 20 units are returned to the distributor.

Retail Promotion

Featured titles within many retail environments are often dictated from the central buying office of the retailer. As mentioned earlier, labels want and often do create marketing events that feature a specific title. This is coordinated via the retailer through an advertising vehicle called *cooperative advertising. Co-op advertising*, as it is known, is usually the exchange of money from the label to the retailer so that a particular release will be featured.

Virtual end caps and digital dashboards drive online buyers to specific releases once "inside" a store's online presence. This real estate is paid for by the label, highlights new releases, and focuses attention on music that the label wants the consumer to notice. Just like online, promotional efforts in-store help to highlight different releases that should aid consumers in purchasing decisions. These marketing devices often set the tone and culture of the store's environment.

Pricing and positioning (P&P) – P&P is when a title is sale-priced and placed in a prominent area within the store.

End caps – Usually themed, this area is designated at the end of a row and features titles of a similar genre or idea.

Point-of-purchase (POP) materials – Although many stores will say that they can use POP, including posters, flats, stand-ups, and so on, some retailers have advertising programs where labels can be guaranteed the use of such materials for a specific release.

Print advertising – A primary advertising vehicle, a label can secure a "mini" spot in a retailer's ad (a small picture of the CD cover art), which usually comes with sale pricing and featured positioning (P&P) in-store.

> **In-store event** – Event marketing is a powerful tool in selling records. Creating an event where a hot artist is in-store and signing autographs of his or her newest release guarantees sales while nurturing a strong relationship with the retailer.

Trade Associations

Music Business Association

Founded in 1958, the National Association of Recording Merchandisers (NARM) was conceived as a central communicator of core business issues for the music retailing industry. This trade organization has had to evolve to "nurture a global music community by connecting, empowering and

FIGURE 9.16 *Logo of Music Business Association*

educating the industry's many players." Rebranded as the Music Business Association, this organization looks to be a central resource for all things music commerce via the various delivery models: digital, physical, mobile, streaming, and more. For more information regarding the Music Business Association and its activities, visit www.musicbiz.org.

GLOSSARY

Big 3—These are the three music conglomerates that maintain a collective 80%–85% market share of record sales: they are Universal, Sony, and Warner.

Box lot—Purchases made in increments of what comes in full, sealed boxes receive a lower price. (For CDs with normal packaging, usually 30.)

Brick and mortar—The description given to physical store locations when compared to online shopping.

Buyers—Agents of retail chains who decide what products to purchase from the suppliers.

Chain stores—A group of retail stores under one ownership and selling the same lines of merchandise. Because they purchase product in large quantities from centralized distribution centers, they can command big discounts from record manufacturers (compared to indie stores).

Computerized ordering process—An inventory management system that tracks the sale of product and automatically reorders when inventories fall below a preset level. Reordering is done through an electronic data interchange (EDI) connected to the supplier.

Co-op advertising—A co-operative advertising effort by two or more companies sharing in the costs and responsibilities. A common example is where a record label and a record retailer work together to run ads in local newspapers touting the availability of new releases at the retailer's locations.

Discount and dating—The manufacturer offers a discount on orders and allows for delayed payment. It is used as an incentive to increase orders.

Distribution—A company that distributes products to retailers. This can be an independent distributor handling products for indie labels or a major record company that distributes its own products and that of others through its branch system.

Drop ship—Shipping product quickly and directly to a retail store without going through the normal distribution system.

Economies of scale—Producing in large volume often generates economies of scale – the per-unit cost of something goes down with volume. Fixed costs are spread over more units, lowering the average cost per unit and offering a competitive price and margin advantage.

Electronic data interchange (EDI)—The inter-firm computer-to-computer transfer of information, as between or among retailers, wholesalers, and manufacturers. Used for automated reordering.

End cap—In retail merchandising, a display rack or shelf at the end of a store aisle, a prime store location for stocking product.

Fill-in—One-stop music distributors supply product to mass merchants and retailers who have run out of a specific title by "filling in" the hole of inventory for that release.

Free goods—Saleable goods offered to retailers at no cost as an incentive to purchase additional products.

Indie stores—Business entities of a single proprietorship or partnership servicing a smaller music consumer base of usually one or two stores (sometimes known as *mom & pop* stores).

Inventory management—The process of acquiring and maintaining a proper assortment of merchandise while keeping ordering, shipping, handling, and other related costs in check.

Loss leader pricing—The featuring of items priced below cost or at relatively low prices to attract customers to the seller's place of business.

Mass merchants—Large discount chain stores that sell a variety of products in all categories, for example, Wal-Mart and Target.

Min/max systems—A store may have ten units on hand, which is considered the ideal *maximum* number the store should carry. The ideal *minimum* number may be four units. If the store sells seven units and drops below the ideal inventory number of four, as set in the computer, the store's inventory management system will automatically generate a reorder for that title, up to the maximum number.

One-Stop—A record wholesaler that stocks product from many different labels and distributors for resale to retailers, rack jobbers, and juke box operators. The prime source of product for small mom & pop retailers.

Playlist—Streaming portals create curated song lists that are "recommended" new music for their listeners and subscribers to hear

Point-of-purchase (POP)—A marketing technique used to stimulate impulse sales in the store. POP materials are visually positioned to attract customer attention and may include displays, posters, bin cards, banners, window displays, and so forth.

Point-of-sale (POS)—Where the sale is entered into registers. Origination of information for tracking sales, and so on.

Price and positioning (P&P)—When a title is sale priced and placed in a prominent area within the store.

Pricing strategies—A key element in marketing, whereby the price of a product is set to generate the most sales at optimum profits.

Returns—Products that do not sell within a reasonable amount of time and are returned to the manufacturer for a refund or credit.

Sales forecast—An estimate of the dollar or unit sales for a specified future period under a proposed marketing plan or program.

Sell-through—Once a title has been released, labels and distributors want to minimize returns and "sell through" as much inventory as possible.

Shrinkage—The loss of inventory through shoplifting and employee theft.

Solicitation period—The sales process of specific titles occurs during a period called *solicitation*. All titles streeting on a particular date are placed in a solicitation book, where details of the release are described.

Vertical integration—The expansion of a business by acquiring or developing businesses engaged in earlier or later stages of marketing a product.

Universal Product Code (UPC)—The bar codes that are used in inventory management and are scanned when product is sold.

REFERENCES

"Metadata: What It Is and Why It Matters – Spotify for Artists." – *Spotify for Artists*, artists.spotify.com/blog/metadata-what-it-is-and-why-it-matters. Thirtytigers.com

Ingham, Tim. "Merlin Paid Indie Labels and Distributors $845m in the Past Year, Showing Strong Growth." *Music Business Worldwide*, 9 Jan. 2021, www.musicbusinessworldwide.com/indie-label-annual-revenues-hit-845m-via-merlinshowing-strong-growth/.

Jimmy Wheeler – Story House Collective/General Manager: Personal Interview.

RIAA.com "YEAR-END 2020 RIAA REVENUE STATISTICS" https://www.riaa.com/wp-content/uploads/2021/02/2020-Year-End-Music-Industry-Revenue-Report.pdf

SoundExchange.com

Music Gateway. "Music Metadata How to: What Is Metadata?" *Music Gateway*, 22 Aug. 2019, www.musicgateway.com/blog/how-to/music-metadata-for-sync-the-music-gateway-bible.

Ucaya. "Market Intelligence for the Music Industry." *Soundcharts*, soundcharts.com/blog/music-distribution#5-types-of-music-distribution-companies.www.musicconnect.comIngham.

CHAPTER 10

Streaming

In the previous chapter, we discussed the quick rise of streaming and how it is currently dominating the music landscape. This chapter dives into the topic in more detail with a look at the business of streaming, top streaming services, playlisting, and the role streaming services are playing in serving as the leading provider of not only music but also of all information about the artist from discographies to tour itineraries.

THE BUSINESS OF STREAMING

In this digitally focused world in which we live, consumers are looking for ease of access to a long list of products and services, from ordering take-out from finding the title and artist of a song through an app like Shazam to ordering any product they wish to buy through hands-free speakers like Amazon's Alexa.

Music services use technology to digitally deliver songs to consumers, whether that be in a downloadable format (iTunes and Amazon Music) or via streaming (Spotify, Apple Music, YouTube Music, Amazon Music, Deezer, Soundcloud, and Tidal). These streaming services are referred to as **Digital Service Providers**. DSPs provide a valuable service to consumers in that they help provide the delivery mechanism for millions of tracks of music to get to consumers. Regardless of the consumer's need, the right music is available at their fingertips. Also, the algorithms used by the services facilitate music discovery in an exciting way that is even more swift and effective than radio.

DOI: 10.4324/9781003153511-10

Streaming services provide an open door for an artist to find new audiences who may not have ever known about them otherwise, which helps them grow their social media audiences, increase tickets sales during their tours, and generate more revenue through merch sales. Streaming has also offered some new artists the chance to get the attention of a record label or even an established artist looking for an opening act. Martin Johnson of the Night Game put a ballad, "The Outfield," on Spotify in 2017. Just weeks after the release on Spotify, the song caught the attention of Grammy winner John Mayer through exposure on Mayer's customized Discover Weekly playlist, and he subsequently invited Johnson to open for him on his "The Search for Everything" tour. Music supervisors are also actively searching for just the right music to fit the movie or television series they are working on, and that type of exposure may open many doors for artists in addition to driving stream counts.

Market Share

Although there are many streaming services globally, Spotify has quickly emerged as the undisputed leader, amassing 32% of the market share as of 2021. Other leaders include Apple Music at 16%, Amazon at 13% and China-based Tencent at 13% (MIDiA). It is easy to see, then, why artists and their teams focus their promotional efforts primarily on Spotify, Apple Music, and Amazon, as collectively they represent 61% of the global market.

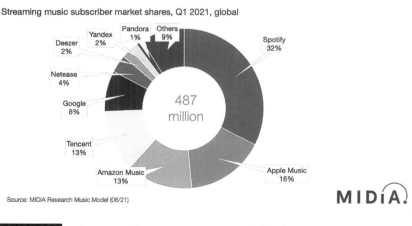

FIGURE 10.1 *Streaming Services Market June 2021 Market Share*

As reported by MIDiA, segmentation in streaming is now starting to take hold. Although each service offers the same catalogue, each service seems to be finding strength within certain demographics:

- YouTube Music appeals to Gen Z and Millennials.
- Amazon Music is converting older audiences to streaming as they move from physical to digital consumption.
- Spotify and Apple Music are top mainstream choices.
- Deezer is seeing growth in emerging markets such as Brazil and with pre-pay mobile bundles.

The Business Model

The streaming services each have different revenue models but are generally either subscription based or supported by advertising. Spotify, YouTube, and iHeart, for example, offer free accounts that provide basic services, with Spotify offering a basic free service as well as premium subscription options. To understand how this **freemium model** works, consider that Spotify has 345 million total monthly active users in approximately 170 markets, of which 155 million are premium subscribers (*The Verge*, 2021). Subscribers to the free version of Spotify music listen to ads every 15 minutes to unlock additional airplay time, while the premium subscribers who pay $9.99 per month for a single user have access to higher-quality music as well as unlimited listening without ads. Premium users are also able to download songs to any device with their app.

Even with continued growth in subscriptions, Spotify has yet to become profitable financially. Although overall Spotify revenue was reported to have declined 8% in 2020 due largely to discounted plans to pull in new subscribers, more than 155 million paid for some level of access. Even though revenue growth will continue to see pressure in future years, it seems advertising with the platform is flourishing as more brands seek to engage with a music-focused audience. In fact, ad revenue was up in 2020, accounting for 13% of total revenue compared to 9%–10% in the few years prior. For its part, Spotify continues to invest in music but has also expanded into the book and podcast sectors.

How Playlists Work

While each platform has playlists that move the needle in terms of increasing awareness and generating music streams, Spotify has become the major portal on which music marketers are focused. Within Spotify, there are several major types of playlists:

1. Algorithmic Playlists

These playlists are customized for the user and automatically generated by Spotify's algorithm. Spotify uses a listener's music listening habits to identify the tracks it feels would be most appealing for that user. Two of the most important playlists that are algorithmic are "Discover Weekly" and "Release Radar."

"Discover Weekly" is updated every Monday. The playlist features music the algorithm thinks the user might enjoy based on artists the user likes, shares, and saves to their playlist. It also considers songs the user skips to identify tracks that should not be included.

"Release Radar" is updated each Friday and features recently released music by artists Spotify thinks the fan would enjoy based on their activity on the platform.

While music cannot be pitched for inclusion on algorithmic playlists, marketing professionals attempt to increase engagement by encouraging fans to pre-save the release and add the songs to their playlists. This increased activity is favored by the algorithm and may increase the chance of the track being featured.

2. Editorial Playlists

Editorial playlists are those curated by a team of music specialists who are hired by Spotify. These lists often have millions of followers and, as a result, can really move the needle for artists who earn a spot on them. For this reason, placement on editorial playlists is coveted by labels and artists alike.

Generally, these playlists either focus on the top music across all genres, are genre specific, or focus on moods and activities that might be appealing to the user at any moment in time such as those featuring dance hits, sad songs, or tracks ideal for weddings.

Importantly, Spotify often tests new tracks on those with smaller followings to determine how they perform prior to placing them on their top lists. For example, tracks that gain traction on a playlist like "Most Necessary" (2.5 million likes) might then be moved to the larger playlist "Rap Caviar," with an audience of more than 13.5 million. Likewise, a country track may be featured on "Indigo" with 320k likes before being considered for Hot Country (6.3 million likes).

3. Other

This category includes branded playlists that feature those created by corporate brands and industry leaders such as trade organizations, publications, blogs, and record labels. Examples include Sony Music's "Filtr" or "Pitchfork's Best New Music." Even Kentucky Fried Chicken and Victoria's Secret are jumping on the bandwagon with their branded playlists.

Rank	Playlist	Followers
1	Today's Top Hits	26,875,333
2	Global Top 50	15,662,686
3	RapCaviar	13,417,265
4	Viva Latino	10,689,669
5	Baila Reggaeton	10,126,752
6	Songs to Sing in the Car	9,726,522
7	All Out 00s	8,731,912
8	Rock Classics	8,457,824
9	All Out 80s	7,758,552
10	Beast Mode	7,498,596
11	All Out 90s	6,138,130
12	Chill Hits	6,038,700
13	Peaceful Piano	6,013,510
14	Hot Country	5,942,899
15	Get Turnt	5,927,111

FIGURE 10.2   *Top Streaming Playlists*

(Source: Hypebot via Chartmetric)

Artists also will develop their own personal playlists that feature their favorite songs – from those they release to songs from their musical influences and more. Another strategy is for artists to build a following for a playlist of other music they curate, then include one of their own songs in the mix as well.

Listener playlists are often referred to as user-generated playlists. These lists are developed and maintained by Spotify users who might be people who work independently in the music industry or just fans who build a playlist as a form of personal expression. These listener playlists are often the most receptive to pitches for new music.

To earn a spot on one of the third-party playlists, marketers would want to pitch the potential placement to the playlist's curator. That task is often more difficult than it might seem, however, as many curators keep their contact information hidden. Some individuals provide a button with the profile to submit music, while others have a link to their Spotify profile that includes social media links (such as an Instagram or Twitter account). If a curator uses their full name in the playlist, those names may be searched on Google or Bing or searched on the social media sites.

There are also submission platforms that submit the music for the artist camp in exchange for compensation. Submit Hub and Groover are top choices for this service, although there are many others. Also, many artists find contact information simply by asking within their community of industry contacts.

Even though it is against Spotify's policy for curators to accept payment for placement, some playlist operators actively engage in the practice of adding or bumping up a song on their playlist in exchange for compensation of some sort. The term "**playola**" was originally used to describe this type of activity, which was a derivative of the radio term "**payola**," the practice of paying on-air personalities and radio station executives in cash or goods in exchange for adds to the station's playlist or increased spins (more on this subject is covered in the radio chapter). Although Spotify prohibits playola by its users, the company interestingly announced in 2020 a new program that would let artists opt in to a lower royalty rate in exchange for increased visibility on some of the algorithmically generated playlists. The move was immediate criticized by industry leaders as a form of payola, and it is unclear where and how this issue might be resolved.

Although playlists are enormously helpful to some artists in amplifying their profile and finding new fans, one must understand that gaining placement on major playlists should not be relied upon for marketing new music.

Charles Alexander, CEO of Outside the Box Music, suggests that artists must embrace a holistic approach to streaming. The streaming specialist notes that playlists may be utilized as a first step to visibility, but it is incumbent upon the artist team to build upon that recognition. If a song gets added to a popular playlist, it should be supported with publicity and through social messaging to help keep up the momentum.

Alexander also notes that streaming is only "one leg of an artist's stool" and that livestreaming, in-person concerts, social media strategy, and radio, as well as television and film placement all are part of raising an artist's visibility and building a fan community. The following is more on streaming strategy from Alexander.

CONTRIBUTED BY CHARLES ALEXANDER

Historically, major labels were responsible for marketing and building audiences for artists. An artist's live performances and tours then capitalized on those efforts and on exposure from radio promotion. Radio has historically been the biggest driver for music consumption.

In the present era, labels provide venture capital and marketing services to launch artist careers. As such, the pyramid has been inverted. It is now incumbent upon artists and artist teams to achieve some level of minimum traction to be considered for label deals. Often these minimum benchmarks are arbitrary and vary depending on the label or related partner such as booking agents and managers, but in most cases, these metrics and key performance indicators (KPIs) revolve around streaming and live performance ticket sales. Some of those KPIs may revolve around the following:

HYPE/BUZZ/PR

Streaming Music Consumption

- Pitching in Platform on Spotify
- Direct conversations with DSPs
- Playlisting

Existing Music Consumption Terrestrial Radio

Social media engagement (especially on TikTok, Instagram, and YouTube)

- Content marketing
- Paid ads on social media – Facebook/Instagram; Google Ads – search and YouTube; TikTok
- Social content marketing tools such as:
  - Ads Manager (for Facebook and Instagram)
  - ToneDen
  - found.ee
- Live performances (COVID-19 exceptions notwithstanding)
- Livestreaming audiences
  - Facebook
  - Instagram
  - TikTok
  - Twitch
  - Reddit/RPAN
- Sentiment analysis
- Analytics tools
- Spotify for Artists
- Chartmetric
- SpotOnTrack
- Pandora AMP
- Dashboard for Pandora
- YouTube Analytics

Spotify is the most dominant platform for audio streaming consumption. The only exception to that fact is perhaps YouTube, possibly the largest video consumption platform on the planet. But most importantly, YouTube is the world's primary music search engine and provides democratic access across the globe.

So even though YouTube serves up video content, one could argue that the primary reason users end up on the platform is because they want to "hear" the music or the artist on a low-barrier platform. Spotify is only available in 93 countries and most recently in India, Russia, and South Korea. It is still not officially available in China and many other economies.

That being said, traction and consumption on Spotify are coveted by most artists and labels, whether they be on major imprints or operate as independents. Even then, it is difficult to draw clear lines of demarcation for what constitutes a major label vs. what is an independent. Many independents now have budgets that rival those of major labels. Some former independents such as AWAL are now part of larger major-label ventures such as Sony. In either case, a large part of the marketing efforts of these entities revolves around getting music consumed on streaming platforms in general, with Spotify and Apple Music being of particular importance.

Spotify

How It Works

At its core, Spotify is a recommendation engine built on a concept called "collaborative filtering." Collaborative filtering collates and aggregates the tastes, opinions, decisions, and preferences of different groups of people. The more a specific song or artist is preferred or consumed by a group or cohort of music fans, the more that piece of music or artist is recommended to a larger group of music fans that the algorithm has predicted will have a predisposed affinity for said music. The inverse is also true. The more a piece of music is skipped or rejected altogether, the smaller the circle of fans that would have the possibility of even hearing the music. The music then proceeds to "fall off" the algorithmic cliff.

In the first scenario, once the consumption reaches a critical point, the song or the artist is considered to have then over-indexed. The song will then be added to an official Spotify **editorial, personalized or algorithmic playlist**. If the song was already added to a playlist upon release due to editorial or algorithmic considerations, the song is then "promoted" to larger and more influential playlists.

It is important to note that collaborative filtering is not the sole criterion in determining if a song has traction. Other factors, such as natural language processing, as well as the music's inherent qualities, such as danceability, energy, key, tempo, and so on, also play a part in how these songs are evaluated.

In the case of most marketing efforts, the idea is to get ahead of the curve in terms of preparing for a release and pitch a minimum of four weeks prior to the release date. The music is pitched in platform on **Spotify for Artists**, a Spotify artist platform site, that enables the pitching of the music to Spotify's editorial staff. The music must be uploaded and ready for release via an artist's distribution partners. The first step, though, is to get the artist's account verified.

It is not unusual for a label marketing person, artist manager, or stakeholder to concurrently have direct conversations with artist and label marketing representatives at Spotify to communicate release priorities. During that process, the conversations center around marketing drivers such as the following:

- Existing fan base
- Social media metrics
- Industry buzz
- Advertising spend on the Spotify Ad Studio platform or off the Spotify platform with social marketing ads in Instagram and Facebook or Google Ads
- Video assets related to the release
- Other efforts by brand partners and sponsors, such as any inclusion in broader advertising campaigns
- Or other forms of cultural relevance

The most influential off-platform consumption driver is TikTok.

However, currently the single most influential platform in terms of fan building and driving streaming consumption off platform is TikTok, the short-form video content app. Streaming consumption conversion from TikTok to Spotify currently ranges from 10% on the low end to 60% or higher on the top end. This is because TikTok takes a very egalitarian approach in terms of visibility on the platform. If a large number of people like the content, the more people see it. Fans' access to the content is not artificially gated (at least for now).

TIKTOK BEST PRACTICES

1. General rule of thumb: Consistency wins. Keep in mind the 3Cs: consistency, (compelling) content, community.

2. Post two to five times a day. There is just so much competition for eyeballs that posting less will hurt your effort.

3. Posts need to be well lit and labeled visually with stickers or red/turquoise labels. Fans need to be able to "get it" while they are doom-scrolling. Use trending hashtags from the discovery page (e.g., #Upcycling).

4. Wait until the artist has organic momentum before verifying their profile or boosting posts to avoid giving the impression that the artist is trying to game the system. The TikTok audience is merciless about that stuff.

5. The TikTok algorithm cares about follows, views, likes, comments (in that order).

6. Be you. Find your tribe. Be authentic.

7. It is about the song and the artist. There are plenty of less-than-stellar songs out there that have blown up on TikTok and then streaming, but it is easier to bank on longevity if you lead with the art.

8. These things take time. If you want something to happen in about two months without existing momentum, that is just a heavy lift. Do not take the "easy" route with third-party playlisting (payola or not) and/or TikTok influencer campaigns.

9. Empowering yourself is always the better road. But it takes time, effort, content, and commitment. Best of luck to all.

Other platforms such as Instagram or Facebook only serve the content to 20%–30% of your owned audience organically. To access a larger percentage of the audience on these as well as other social platforms such as YouTube, a paid campaign on the platforms is necessary. Therefore, it is important to have a core strategy built around TikTok and ad spend marketing on Spotify and other social platforms.

Adoption

Collaborative filtering is measured on the Spotify platform in three ways: followers, listenership (which translates to streams), and saves. These three actions signal to the underlying algorithms that a piece of music or a certain artist has sufficient engagement on the platform. That data has the power and potential to drive editorial decisions from Stockholm to Sao Paolo.

One of the most critical metrics is the saves to listenership ratio.

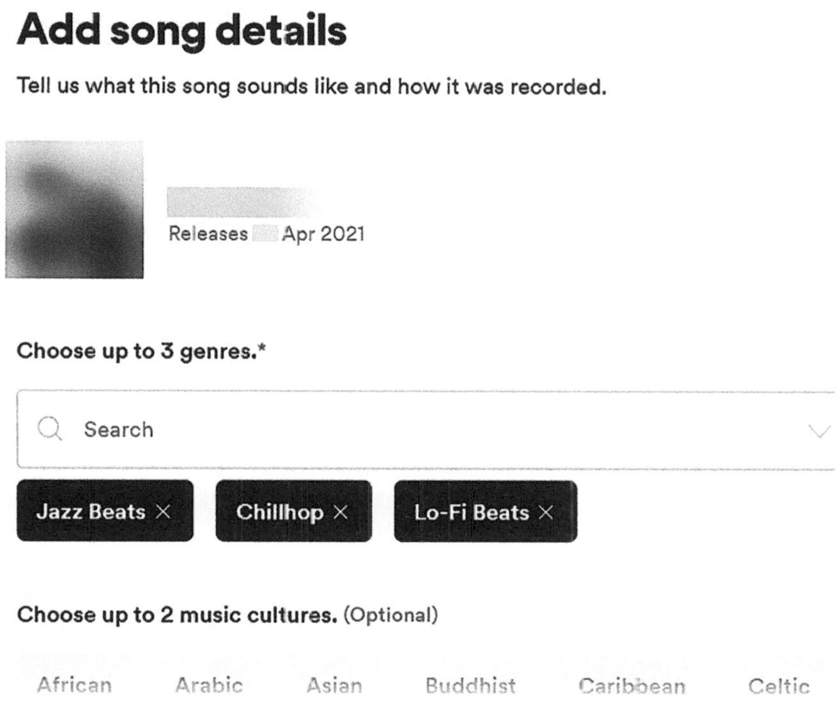

Add song details

Tell us what this song sounds like and how it was recorded.

Releases   Apr 2021

Choose up to 3 genres.*

Search ⌄

Jazz Beats ✕   Chillhop ✕   Lo-Fi Beats ✕

Choose up to 2 music cultures. (Optional)

African   Arabic   Asian   Buddhist   Caribbean   Celtic

FIGURE 10.3   *Spotify's In-Platform Pitch*

One of the most critical metrics is the saves to listenership ratio. If that percentage is around 20% or higher within a 28-day window, the platform serves the music, at a minimum, to the algorithmic and personalized playlists. This can significantly influence growth on the service. Therefore, issuing calls-to-action (CTAs) to fans to save the music to their personal music libraries can really move the needle, especially if more than 50% of streaming consumption comes from the playlists created by individuals who listen to the music repeatedly.

The in-platform pitch process looks like so. Up to two genres may be selected for the release.

Apple Music

Some of the same principles on Spotify apply to Apple Music, but the ability to see some of these data in a front-facing way is limited. To that end, gaining access to the **Apple Music for Artists** platform is critical to execute an effective and efficient streaming marketing strategy. Submitting music priorities to Apple Music for editorial consideration usually happens on the distribution or label side via submissions "grids." These grids require information such as:

- Name of artist
- Identifiers such as International Standard Recording Codes (ISRCs)
- UPCs and Apple IDs
- Previous sales history
- Tours
- Marketing drivers (paid advertising efforts, etc.)
- Upcoming press and media features, as well as any previous editorial coverage

The grid is then submitted to the artist's Apple Music artist and label marketing team representative.

Apple requires a three- to four-week minimum lead time for these submissions to be considered for editorial features. The earlier these submissions occur, the better. As is the case with Spotify, stakeholders in an artist's career try to leverage their own personal relationships with the Apple Music teams, to varying degrees of success.

The effectiveness or success of these efforts are then monitored through Apple Music for Artists or third-party music data analytics platforms such as Chartmetric.

Pandora

As in the previous streaming service platforms, there are similarities to getting the music of your artists in front of Pandora's music curators. Priorities are submitted to your Pandora artist and label marketing team representative either through your distribution, label, or marketing representative.

However, Pandora's real power and differentiator is that it has its own in-platform online marketing system. The system is called Pandora's Artist Marketing Platform or **Pandora AMP**. Additional advertising options on Pandora are covered in depth within the paid media chapter, but we will highlight the tools used most frequently by artists.

First, the artist's profile should be claimed at http://amp.pandora.com. Once an artist has claimed their profile, AMP's system allows for different kinds of audio-based messaging at AMP Playbook. The two main avenues of marketing an artist on the system are:

1. *Artist Audio Messaging or AAM* is for structured and scheduled marketing efforts. These include specific calls to action with respect to the artist's music and content. These tools can result in a 2–8× increase in engagement on Pandora. Artists need to pre-record a 15-second

audio loop that can then be heard prior to or in-between an audio stream on Pandora. Details on how to use this tool are available here:

www.ampplaybook.com/artist-audio-messages

2. *AMPCast* is a real-time, spontaneous marketing channel that the artist can use to do short-form broadcasts on the go. Details on how to use this tool are available here:

www.ampplaybook.com/ampcast-user-guide

Both these tools enable geotargeting of audiences.
The other options available to users of AMP are:

- *Featured Tracks* – lets you "boost" a track for eight weeks. This can result in a significant increase in streams for that track depending on how active the artist's Pandora profile has been.
- *Pandora Stories* – provides the ability to create playlists and mixtapes on the platform from the artist's creator or tastemaker profile. It allows the artist to create a service within the service, but it does require that Pandora approve your Pandora Stories Creator profile.

The progress of all AMP marketing efforts may be found on the AMP dashboard.

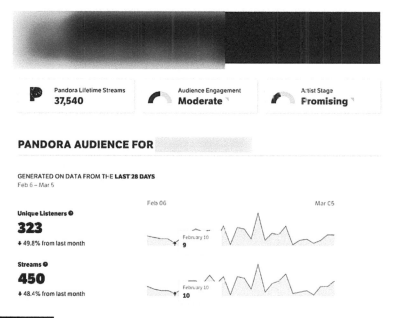

FIGURE 10.4   *Pandora Insights*

PLAYLISTS & DISCOVERY FOR

GENERATED ON DATA FROM THE **LAST 28 DAYS**
Feb 6 – Mar 5

Pandora Streams by Source

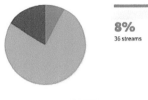

	radio ❓	Other Pandora stations ❓	Interactive plays ❓
	8%	**76%**	**16%**
	36 streams	339 streams	73 streams

Top Pandora Sources for

Top Overall Sources	Top Radio Sources	Top Interactive Play Sources		
TOP SOURCES		STREAMS	STATION OWNERS	LISTEN ON PANDORA
Taylor Swift Artist Radio		274	40.2M	↗
Eva's playlist User Playlist		25	-	↗
Avril Lavigne Artist Radio		16	5.6M	↗

FIGURE 10.5   *Pandora Playlists*

AUDIENCE MAP FOR

Select Your Metric

Pandora Streams ▾

Select Your Date Range

7 days	**28 days**	365 days

GENERATED ON DATA FROM **FEB 8, 2021 - MAR 7, 2021**

The Pandora Streams map shows where _____ tracks have been played on radio or on-demand over the past 28 days. The Pandora Artist Station Adds map shows where new fans have added a _____ station in the past 28 days. Read more about the difference between the maps here.

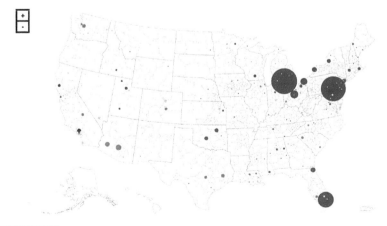

FIGURE 10.6   *Pandora Audience Maps*

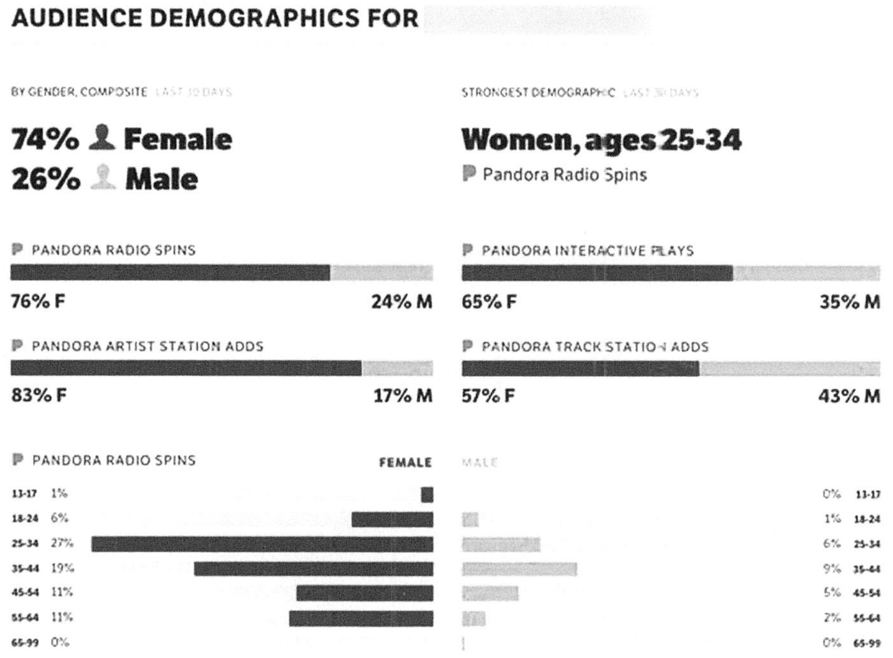

AUDIENCE DEMOGRAPHICS FOR

BY GENDER, COMPOSITE LAST 30 DAYS   STRONGEST DEMOGRAPHIC LAST 30 DAYS

74% ⦿ Female   **Women, ages 25-34**
26% ⦿ Male   ℗ Pandora Radio Spins

℗ PANDORA RADIO SPINS   ℗ PANDORA INTERACTIVE PLAYS

76% F   24% M   65% F   35% M

℗ PANDORA ARTIST STATION ADDS   ℗ PANDORA TRACK STATION ADDS

83% F   17% M   57% F   43% M

℗ PANDORA RADIO SPINS   **FEMALE**   MALE

Age	Female		Male	Age
13-17	1%		0%	13-17
18-24	6%		1%	18-24
25-34	27%		6%	25-34
35-44	19%		9%	35-44
45-54	11%		5%	45-54
55-64	11%		2%	55-64
65-99	0%		0%	65-99

FIGURE 10.7   *Pandora Audience Demographics*

Amazon

As in the previous systems, priorities are submitted primarily to Amazon through the artist's distribution, label, or marketing representative. The progress of the artist's campaigns may be monitored through the **Amazon Music for Artists** profile or through third-party services like Chartmetric.

CONCLUSION

As the competition between the DSPs continues to intensify, it is likely the consumers who stand to benefit the most in the race to determine industry leaders. As in any industry, competitive pressure spurs additional innovation and a drive to better service the needs of the music consumer. Already, Spotify has become not only a liaison for the music but also a central hub for artist information from tour dates to biographical information to lyrics. While the platforms continue their race to the top, it's clear that fans are the true winners.

FIGURE 10.8   *Amazon Music for Artists*

GLOSSARY

Amazon Music for Artists—Amazon's artist platform site that provides tools and other helpful information.

Apple Music for Artists—Apple's artist platform that enables artists to manage their profiles and get detailed insight on their listeners. The platform covers both iTunes and Apple Music.

DSPs—Digital service providers are the music services that deliver streamed tracks to consumers. Examples of DSPs include Spotify and Apple Music.

Freemium model—A marketing model in which a basic account is offered to a consumer for free with the hope that they will upgrade to a premium, or paid, account that offers more features and benefits. Freemium models are a way to encourage trial of products.

Pandora AMP—Pandora's artist marketing platform that provides tools and other resources to help artists maximize their exposure.

Spotify for Artists—Spotify's artist platform site that enables music to be uploaded to the platform. It also provides tools and other resources to help artists maximize their exposure on Spotify.

BIBLIOGRAPHY

Afp. "Spotify Losses Triple in 2020 Despite More Subscribers." *Mint*, 3 Feb. 2021, www.livemint.com/technology/apps/spotify-losses-triple-in-2020-despite-more-subscribers-11612359254100.html.

Farley, Kathleen. "Spotify 101: Understanding the Spotify Playlist Landscape." *Playlist Radar*, 20 Mar. 2021, playlistradar.com/spotify-101-understanding-the-spotify-playlist-landscape.

Herstand, Ari. *How to Make It in the New Music Business: Practical Tips on Building a Loyal Following and Making a Living as a Musician.* 2nd edition, Liveright, 2019.

Houghton, Bruce. "20 Most Popular Playlists on Spotify." *Hypebot*, 12 Nov. 2020, www.hypebot.com/hypebot/2020/11/20-most-popular-playlists-on-spotify.html.

Laker, Benjamin. "The Economics of Music Streaming and the Role of Music Streaming Platforms." *Forbes*, 30 Oct. 2020, www.forbes.com/sites/benjaminlaker/2020/10/28/heres-how-lockdown-has-shown-that-spotify-has-a-sustainability-problem/?sh=3344e03d599b.

Lanham, Tom. "Night Game Gets Boost from John Mayer." *The San Francisco Examiner*, 6 Sept. 2017, www.sfexaminer.com/entertainment/night-game-gets-boost-from-john-mayer.

Porter, Jon. "Spotify Subscribers Surge Past 150 Million." *The Verge*, 3 Feb. 2021, www.theverge.com/2021/2/3/22262508/spotify-q4-2020-earnings-subscriber-numbers-users-podcasts-audiobooks.

Yoo, Noah. "Could Spotify's New Discovery Mode Be Considered Payola?" *Pitchfork*, 10 Nov. 2020, pitchfork.com/thepitch/could-spotifys-new-discovery-mode-be-considered-payola.

CHAPTER 11

Radio

THE STATE OF RADIO

In an era when consumers access media through a myriad of devices and portals, *what* they are consuming and *how* they are consuming it are ever-changing preferences. Whether listening to a podcast on a smartphone, watching a favorite morning show after dinner on a time-shifting TV, or listening to terrestrial radio in the car, consumers have multiple choices by which to be informed and entertained. Looking at media usage of Q1 18 as compared to Q1 20, the increase in daily consumption grew in smartphone applications and websites by nearly 59% in a two-year period, adding to overall media consumption by nearly one hour and a half per day. These 90 minutes, as measured in the smartphone data in hours:minutes of usage, is significant in that consumers are choosing content to engage: be it entertainment through video streams, exercise apps, education blogs, and audio books; accessing new and favorite music through music streaming services; and the list goes on. How to capture the attention of these consumers with so many choices is the ongoing and the overwhelming challenge of content creators and providers universally.

Yet with the abundance of choices that consumers have in being informed and entertained, an interesting phenomena is that terrestrial radio is holding users' attention, even with so much competition. In 2018, monthly listenership averaged 228.5 million users as compared to 2019 monthly listenership averaging 244.5 million users. Listeners increased by 7% even when competition for media attention is fierce (Audio Today, 2018, 2019). The year 2020 and the presence of the COVID-19 virus challenged all norms of

CONTENTS

DOI: 10.4324/9781003153511-11

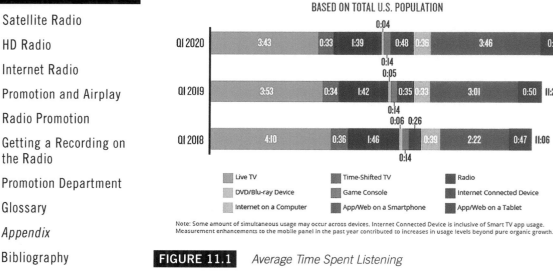

FIGURE 11.1   *Average Time Spent Listening*

(Source: The Nielsen Total Audience Report: Working from Home Edition Aug 2020)

Covid-19 drives 51% increase in broadband traffic in 2020

Editor | 10 February 2021

The ongoing trend towards broadband entertainment and work has been revealed clearly in the Q4 2020 OpenVault Broadband Insights (OVBI) report which has found that average data consumption approached half a terabyte (TByte) at the end of 2020.

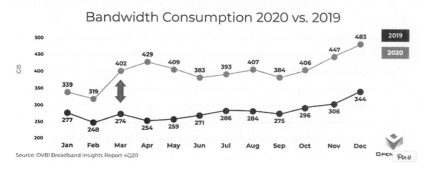

FIGURE 11.2   *Impact of COVID-19 on Broadband Traffic*

(Source: www.rapidtvnews.com/2021021059895/covid-19-drives-51-increase-in-broadband-traffic-in-2020.html?utm_campaign=disney-hits-95mn-subs-as-disney-reveals-tough-q1&utm_medium=email&utm_source=newsletter_2624#axzz6mH8putWi)

business, including the radio industry. As stay-at-home orders shut down active listeners and consumer behavior, broadband traffic increased by 51% as people stayed home and streamed video content and stayed connected to news outlets via television. Commuters were no longer accessing terrestrial radio for their essential traffic report, and the need for national connectivity about the virus was paramount.

Importantly, time spent listening to the radio "in car" dropped as well, by 37.5%. But as shelter-at-home restrictions were lifted, much of the lost ground was gained as listeners rebounded back to the marketplace, as noted by Edison Research. Ad-supported audio has remained relatively the same year-over-year, with AM/FM share in the 18+ age category remaining above 76%.

Nielsen's Audio Today is the clearinghouse of radio data points, conducting exhaustive research with a comprehensive look at pandemic influence on consumer behavior of the year 2020. This fascinating research confirms radio's relevance in today's marketplace as a steady influence of music preference while maintaining a solid foothold as a cultural bedrock in most regions of the United States.

COVID-19 has been disruptive to the majority of lives, and many established work habits were upended. But listening to traditional radio stayed strong, with 75% of the workforce maintaining the "at-work" listening routine and 91% of the U.S. population listening to radio weekly. Average daily listening remained strong at 1 hour 39 minutes of listening per day.

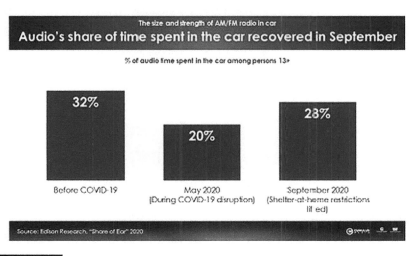

FIGURE 11.3   *COVID-Driven Increase in Broadband Traffic*

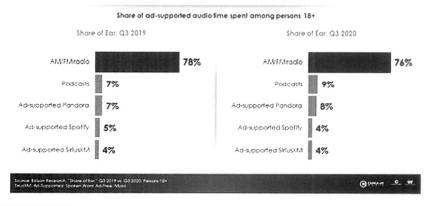

FIGURE 11.4   *Impact of COVID-19 on Radio Audience*

(Source: www.insideradio.com/free/edison-share-of-ear-q3-2020-in-car-audiences-rebound/article_742967a6-232b-11eb-b1ae-4b5231251273.html)

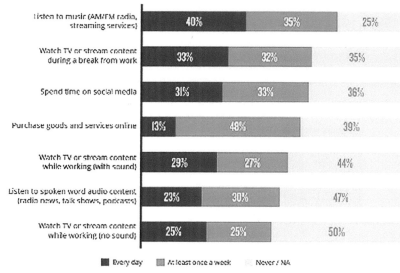

FIGURE 11.5   *How We Entertain Ourselves at Work*

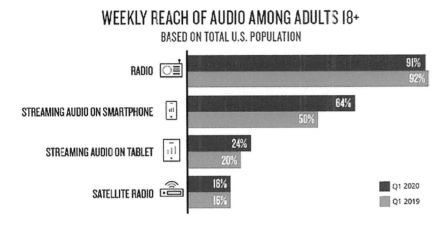

FIGURE 11.6   *Weekly Reach of Audio by Device*

	Total		Black		Hispanic		Asian		White	
Average time spent per adult 18+ per day **Based on total U.S. population**	QI 2019	QI 2020	QI 2019	QI 2020	QI 2019	QI 2020	QI 2019	QI 2020	QI 2019	QI 2020
Live TV	3:53	3:43	5:37	5:17	2:49	2:40	1:55	1:54	3:53	3:43
Time-Shifted TV	0:34	0:33	0:32	0:30	0:19	0:17	0:15	0:16	0:38	0:37
Radio	1:42	1:39	1:48	1:44	1:46	1:40	NA	NA	1:40	1:37
DVD/Blu-ray Device	0:05	0:04	0:05	0:04	0:04	0:03	0:03	0:02	0:05	0:04
Game Console	0:14	0:14	0:17	0:17	0:14	0:14	0:09	0:08	0:14	0:14
Internet-Connected Device	0:35	0:48	0:39	0:59	0:38	0:50	0:42	0:53	0:34	0:46
Internet on a Computer	0:33	0:36	0:25	0:29	0:24	0:25	0:43	0:46	0:34	0:38
App/Web on a Smartphone	3:01	3:46	3:26	4:18	3:09	4:00	3:11	3:42	2:55	3:40
App/Web on a Tablet	0:50	0:58	0:49	0:51	0:45	0:42	0:51	1:05	0:50	0:59
Total	11:27	12:21	13:38	14:29	10:08	10:51	7:50*	8:46*	11:23	12:18

*Radio measurement includes Asian Americans but cannot be separated from the total audience at this time.
Radio measurement for White is inclusive of non-Black and non-Hispanic.

FIGURE 11.7   *Average Daily Time Adults Spent on Entertainment Consumption*

(Source: The Nielsen Total Audience Report: Working from Home Edition Aug 2020)

THE VALUE OF RADIO

Anyone in the business of making money with recorded music is making a mistake by underestimating the reach and impact that radio has with consumers. Labels understand this, which is why they continue to put resources into influencing decisions by radio programmers and other radio gatekeepers to get their music on the airwaves. To make the point, here are weekly audience sizes counted by Broadcast Data Services for the week of February 13, 2021:

Table 11.1	BDSRadioCharts.com Week of 2/13/2021	
Artist	**Song/Chart Position**	**National Audience**
#1 – Ariana Grande	34+35/Top 40	50,787,000
#10 – Tate McRae	You Broke Me First/Top 40	27,016,000
#1 – Luke Combs	Better Together/Country	32,596,000
#10 – Chris Stapleton	Starting Over/Country	17,969,000
#1 – Chris Brown	Go Crazy/Adult R&B	10,781,000
#10 – Giveon	Like I Want You/Adult R&B	3,282,000
#1 – Pop Smoke f. Lil Baby & Dababy	For The Night/Hip Hop	19,316,000
#10 – Internet Money & Gunna f. Don Toliver & Nav	Lemonade/Hip Hop	12,083,000

(Source: BDSRadioCharts.com Week of 2/13/2021)

Social media, largely through our smartphone devices, has become a very important element of all marketing plans, but a competitive international strategy with a goal of selling more than 200,000 equivalent units must include promotion to commercial terrestrial radio as key element of its success. As much as consumers complain about the large number of commercials and repeated playing of the same music, radio is still an important medium the recording industry has to showcase its product to the public. Despite our wired and wireless behavior of consumption, on average, we are listening to the radio one hour and 39 minutes per day (Audio Today 2020).

Given the role radio plays in promoting recordings to consumers, it is important to have an understanding of radio and the people who make programming decisions at those stations. They are the gatekeepers to the radio station's airwaves. When the marketing and promotion staff at the label understand what radio needs, it becomes easier for them to find a way to get their new music programmed.

THE BUSINESS

One of the best adjectives to describe the relationship shared by the recording industry and terrestrial radio is "symbiotic." Though it is a term most often used in science, it means the two industries share a mutual dependence on each other for a mutual benefit. Radio depends on the recording industry to provide elements of its entertainment programming for its listeners, and the recording industry depends on radio to expose its product to consumers.

No two other industries share a unique relationship like this. However, the nature of the businesses of a record company and a radio station are very different. For a record company, it's easy to define the business: to expose new music to listeners. As listeners want more access to specific artists and their recordings, either through purchases, downloads, or more likely streams, revenue is then generated for the record label.

The business of radio is building an audience that it leases to advertisers. Radio uses music to attract listeners in order to attract advertising revenue. The larger the audience the station attracts, the more it can charge for its advertising. Radio, however, is not in the business of building recording careers, nor is it interested in selling recordings. The number of units that a recording is selling might be of interest to a radio programmer, but that information by itself does not necessarily affect programming decisions.

THE RADIO BROADCASTING INDUSTRY

The traditional over-the-air radio broadcast industry in the United States began consolidating when the Telecommunications Act of 1996 was signed into law. Prior to the new law, radio broadcasting companies were limited in ownership to no more than 40 stations (20 AM and 20 FM) nationwide. The current law now allows companies to own unlimited numbers of radio stations nationally but no more than eight stations per market and no more than five in the same service (AM or FM). The numbers decrease on a sliding scale for smaller markets, and in no case may one owner control more than 50% of the stations in a market (Oxenford and Federal Communications Commission).

The relaxing of the radio ownership rules has created some of the largest media companies ever. In the first seven years after passage of the act, the number of radio station owners decreased by 35%, and the largest ownership group swelled to more than 1,200 stations. The landscape continues to shift, and the largest radio companies in America currently are:

Table 11.2 Largest Radio Stations in America 2020

Radio Broadcasting Company	Number of Stations
iHeart Media	850
Cumulus Media	428
Town Square Media	321
Entercom Communications	235

Station counts taken from company websites and reports in 2020. Numbers change often as stations are bought and sold for financial or regulatory reasons. There were a total of 11,264 commercial radio stations in the United States at the end of September 2020. There were an additional 4,169 educational stations.

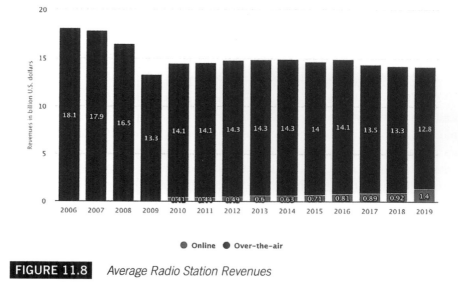

FIGURE 11.8 *Average Radio Station Revenues*

(Source: BIA Advisory Services. "Radio Station Advertising Revenues in The United States from 2006 to 2019, by Source (in Billion U.S. Dollars)." *Statista*, Statista Inc., 24 Jun 2020, www.statista. com/statistics/253185/radio-station-revenues-in-the-us-by-source/)

Revenue for the industry declined steadily from 2005 to 2009, with annual earnings hitting a low of $13.3 billion in 2009. But in 2010, aided by a stronger economy and the additional revenue from advertising online, radio advertising rebounded to $14 billion and maintained this level of income for seven years. In 2019, annual online advertising increased by 56%, but overall revenue dropped by over half a billion dollars to a new low of $12.8 billion.

RADIO STATION STAFFING

Typical Radio Station

In order to see how decisions are made about music choices at a radio station, it is important to understand the organization within the station. The general manager, market manager, or someone with a similar title is responsible for the business success of the station. The sales manager has a staff of people who sell available commercial time to advertisers. Promotions, contests, and other sales-oriented activities that are often the collaborative work of the sales team as well as the programming department. Most stations have an online presence, either a simplistic informational website or a full-blown streaming webcast, both of which are generating ad dollars for

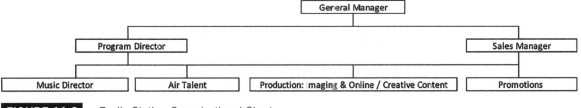

FIGURE 11.9 *Radio Station Organizational Chart*

the station. Graphics as well as streaming content needs to be created for these sites, and an entire department is supporting these commitments for online content. Recognize that within a communication group, there might be more than one station that houses multiple genres of music for that particular market. Most likely, the production department would service *all* the stations for the communication group.

From a record marketing standpoint, the key positions at a radio station are the program director, and to a lesser degree, the music director. The program director (PD) is generally the gatekeeper. Without the "okay" of the programmer, there is no chance that a recording will get on the air at most large radio stations. The PD is directly responsible to the general manager for creating programming that will satisfy the target market and build the existing audience base. The programmer decides what music is played, which announcers are hired, which network services to use, how commercials are produced, and every other aspect of the image the station has within the community it serves.

Critics of radio often say program directors have too much power because they can decide whether a recording is ever exposed to listeners. Large radio chains have group programmers who play an even larger role as a gatekeeper, recommending which music is appropriate for similarly programmed stations owned by the company across the country. As group owners of stations seek economies within their companies, group programmers – rather than a local program director – play a larger role than ever in the decisions regarding which music is played for the station's audience.

As PDs make decisions as to what song will be added to the playlist, in what rotation they will be played, and how many times a song will be played, the music director is asked to generate this playlist weekly, reflecting the number of times a song will be played hourly and daily, along with the inclusion of news, weather, traffic, and the all-important advertising while keeping in mind what are known as quarter hour "sweeps," or keeping listeners on the station for as long as possible. This log of songs takes into consideration: tempo, momentum, song era, energy of song, and if the song is performed by a male or female. Sophisticated software programs generate this type of song list, as reflected by ENCO's MusicMaster playlist examples:

Format Wheel Gold – This wheel represents a Top of Hour (TOH) Time/Station/Legal ID Intro with songs listed by time identifiers. Listeners are "sweeped" through the quarter hours at :15, :30, and :45 minutes of the hour to retain listenership.

Format Wheel – Example of Hot AC or Country – This wheel represents a roll into the (TOH) Time/Station/Legal ID Intro with songs being played right at 0:00. Era songs, station imaging, and artist "liners" that include recording artist IDs add to the mix of station personality. Listeners are "sweeped" through the quarter hours at :15, :30, and :45 minutes of the hour to retain listenership.

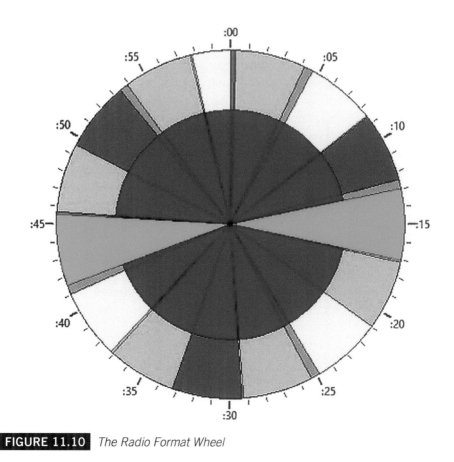

FIGURE 11.10   *The Radio Format Wheel*

Format Clock Report   MusicMaster   GOLD 99 FM

SCHEDULING

Format Clock: A3 - GOLD FM MAIN 1 VT

	Start	Runs	Element	Description	S F
1	0:00	00:19	Fixed	ID - LEGAL ID	
2	0:19	03:52	Fixed	PG - PURE GOLD	
3	4:11	00:30	Fixed	VT - VOICE TRACKS	
4	4:41	03:52	Fixed	GM - GOLD MINE	
5	8:33	00:06	Fixed	LI - LINERS	
6	8:39	03:52	Fixed	GP - GOLD POT	
7	12:31	00:30	Fixed	VT - VOICE TRACKS	
8	13:01	04:00	Traffic	Traffic Merge	S
9	17:01	00:06	Fixed	LI - LINERS	
10	17:07	03:52	Fixed	PG - PURE GOLD	
11	20:59	03:52	Fixed	GM - GOLD MINE	
12	24:51	00:30	Fixed	VT - VOICE TRACKS	
13	25:21	03:52	Fixed	PG - PURE GOLD	
14	29:13	00:06	Fixed	LI - LINERS	
15	29:19	03:52	Fixed	GP - GOLD POT	
16	33:11	03:52	Fixed	PG - PURE GOLD	
17	37:03	00:06	Fixed	LI - LINERS	
18	37:09	03:52	Fixed	GM - GOLD MINE	
19	41:01	00:30	Fixed	VT - VOICE TRACKS	
20	41:31	04:00	Traffic	Traffic Merge	S
21	45:31	00:10	Fixed	AB - AFTER BREAK	
22	45:41	03:52	Fixed	PG - PURE GOLD	
23	49:33	03:52	Fixed	GP - GOLD POT	
24	53:25	00:30	Fixed	VT - VOICE TRACKS	
25	53:55	03:52	Fixed	PG - PURE GOLD	
26	57:47	00:06	Fixed	LI - LINERS	
27	57:53	03:52	Fixed	GM - GOLD MINE	
28	1:01:45	00:06	Fixed	LI - LINERS	
29	1:01:51	03:52	Fixed	PG - PURE GOLD	
30	1:05:43	03:52	Fixed	GP - GOLD POT	

Total: 1:09:35

FIGURE 11.11   *The Radio Format Clock Report*

Format Wheel Twofer Clock – This wheel represents a pre-programmed hour of music with KO kick-offs of voice-overs announcing the next string of "hits." This clock is designed for overnight hours where there might not be any on-air talent available, and the station will string this type of clock back-to-back for three hours straight or more, if needed. The three minutes at the end of the hour could be filled with advertising, PSAs, or other pre-recorded information. (Charts generated by Dave Tyler of MusicMaster. ENCO information from interview with Shane Finch of ENCO.)

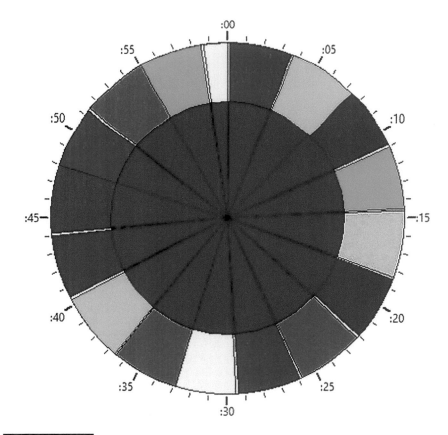

 *Hot Adult Contemporary Radio Format Wheel*

ENCO is the premier global provider of radio and television software solutions, including automation, playout, captioning, visual radio, audio compliance, instant media playout, remote contribution, and cloud-based web-streaming. ENCO delivers world class broadcast solutions to thousands of television and radio broadcasters on six continents. Our solutions range from large multinational radio and television broadcast applications to the needs of small market or community television station. We bring our customers flexible solutions that improve workflow and increase efficiency, while delivering exceptional value, performance, and reliability. Learn why thousands of broadcasters rely on ENCO Solutions every day.

Format Clock Report   **MusicMaster**   The Song Stream

Format Clock: MF · Main Clock with Filters

	Start	Runs	Element	Description	S	F
1	0:00	00:10	Fixed	TOH - Top Of Hour		
2	0:10	03:25	Fixed	A - Currents 1s		
3	3:35	00:06	Fixed	ALB - Album Cut Imaging		
4	3:41	03:45	Fixed	F - Album Cuts		
5	7:28	03:30	Fixed	D - 80/90s Gold		
6	10:58	00:08	Fixed	L - Generic Imaging		
7	11:04	03:28	Fixed	B - Recurrents		
8	14:32	00:07	Fixed	OTP - Other - Classic Imaging		
9	14:39	03:42	Fixed	P - Other - Classic		
10	18:21	00:08	Fixed	L - Generic Imaging		
11	18:29	03:30	Fixed	D - 80/90s Gold		
12	21:59	00:09	Fixed	ART - Artist Liners		
13	22:08	03:39	Fixed	C - New Gold 2000+		1
14	25:47	00:08	Fixed	L - Generic Imaging		
15	25:55	03:25	Fixed	A - Currents 1s		
16	29:20	00:08	Fixed	L - Generic Imaging		
17	29:28	03:21	Fixed	E - Classic Gold		
18	32:49	03:39	Fixed	C - New Gold 2000+		
19	36:28	00:06	Fixed	ALB - Album Cut Imaging		
20	36:34	03:45	Fixed	F - Album Cuts		
21	40:19	00:08	Fixed	L - Generic Imaging		
22	40:27	03:30	Fixed	D - 80/90s Gold		
23	43:57	00:10	Fixed	OTC - Other - Current Imaging		
24	44:07	03:47	Fixed	G - Other - Current		1
25	47:54	03:25	Fixed	A - Currents 1s		
26	51:19	00:08	Fixed	L - Generic Imaging		
27	51:27	03:39	Fixed	C - New Gold 2000+		
28	55:06	03:28	Fixed	B - Recurrents		
29	58:34	00:08	Fixed	L - Generic Imaging		
30	58:42	03:21	Fixed	E - Classic Gold		
31	1:02:03	03:39	Fixed	C - New Gold 2000+		

Total: 1:05:42

FIGURE 11.13   *Monday Through Friday Radio Format Clock*

Unknown to most of the radio audience, outside programming consultants armed with research about both the music and the local audience assist the PD with decisions about music programming. To the record label, the consultant becomes another important gatekeeper. But radio consultants are not used as often as they once were.

RADIO AUDIENCES

Nielsen Audio, the audience measurement company, publishes its annual "Audio Today" in which it provides an analysis of the makeup of audiences who use commercial radio in the United States. The following chart is taken from their 2019 report and shows the percentage of persons by age, ethnicity,

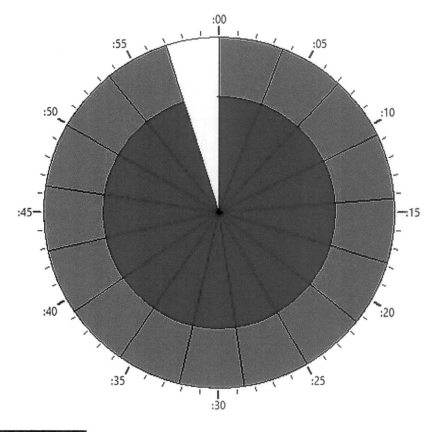

FIGURE 11.14 *Twofer Radio Format Wheel*

and gender who listen to radio during the average week. Quick takeaways include that women out-listen men in both the 18–49 category and 25–54 category, with the country genre dominating the listening genre in overall adult classification.

The size of the audience of a radio station is important to a label because it often determines how much time and other resources are put into promoting a song to that station. Also important to the label is the time of day that the song is scheduled to be played. AQH is a reference to the numerical size of the audience during the *average quarter hour* within the timeframe measured by Nielsen Audio. As you can tell by the Hour-by-Hour Listening chart, audiences are considerably larger during the week and during morning and evening rush hours. Radio refers to rush-hour

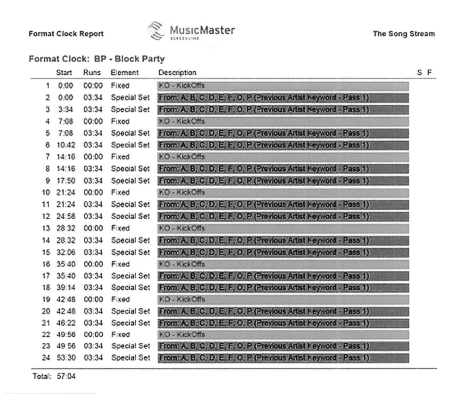

Format Clock Report   MusicMaster   The Song Stream

Format Clock: BP - Block Party

	Start	Runs	Element	Description	S F
1	0:00	00:00	Fixed	KO - KickOffs	
2	0:00	03:34	Special Set	From: A, B, C, D, E, F, O, P (Previous Artist Keyword - Pass 1)	
3	3:34	03:34	Special Set	From: A, B, C, D, E, F, O, P (Previous Artist Keyword - Pass 1)	
4	7:08	00:00	Fixed	KO - KickOffs	
5	7:08	03:34	Special Set	From: A, B, C, D, E, F, O, P (Previous Artist Keyword - Pass 1)	
6	10:42	03:34	Special Set	From: A, B, C, D, E, F, O, P (Previous Artist Keyword - Pass 1)	
7	14:16	00:00	Fixed	KO - KickOffs	
8	14:16	03:34	Special Set	From: A, B, C, D, E, F, O, P (Previous Artist Keyword - Pass 1)	
9	17:50	03:34	Special Set	From: A, B, C, D, E, F, O, P (Previous Artist Keyword - Pass 1)	
10	21:24	00:00	Fixed	KO - KickOffs	
11	21:24	03:34	Special Set	From: A, B, C, D, E, F, O, P (Previous Artist Keyword - Pass 1)	
12	24:58	03:34	Special Set	From: A, B, C, D, E, F, O, P (Previous Artist Keyword - Pass 1)	
13	28:32	00:00	Fixed	KO - KickOffs	
14	28:32	03:34	Special Set	From: A, B, C, D, E, F, O, P (Previous Artist Keyword - Pass 1)	
15	32:06	03:34	Special Set	From: A, B, C, D, E, F, O, P (Previous Artist Keyword - Pass 1)	
16	35:40	00:00	Fixed	KO - KickOffs	
17	35:40	03:34	Special Set	From: A, B, C, D, E, F, O, P (Previous Artist Keyword - Pass 1)	
18	39:14	03:34	Special Set	From: A, B, C, D, E, F, O, P (Previous Artist Keyword - Pass 1)	
19	42:48	00:00	Fixed	KO - KickOffs	
20	42:48	03:34	Special Set	From: A, B, C, D, E, F, O, P (Previous Artist Keyword - Pass 1)	
21	46:22	03:34	Special Set	From: A, B, C, D, E, F, O, P (Previous Artist Keyword - Pass 1)	
22	49:56	00:00	Fixed	KO - KickOffs	
23	49:56	03:34	Special Set	From: A, B, C, D, E, F, O, P (Previous Artist Keyword - Pass 1)	
24	53:30	03:34	Special Set	From: A, B, C, D, E, F, O, P (Previous Artist Keyword - Pass 1)	

Total: 57:04

FIGURE 11.15 *Block Party Radio Format Clock*

programming as "drive time" not only because the time corresponds to our morning and afternoon commutes but because listenership during that time drives the finances of the station. Age of listenership is also affected by time of day and their availability to hear the radio, as noted by teens 12–17, who are busy in school and not able to access the airwaves during instructional hours.

Radio listening peaks in the morning hours, known as morning drive time. Radio listening is divided into day parts of *morning drive, midday, afternoon drive, evening,* and *overnight.* From 6:00 a.m. until 9:00 a.m., listening is greatest as commuters wake up to clock radios and continue to listen as they drive to work. Listening picks up again around noon, declines slightly after lunch but remains relatively strong throughout the afternoon drive time, and then tapers off drastically throughout the evening and into the overnight period.

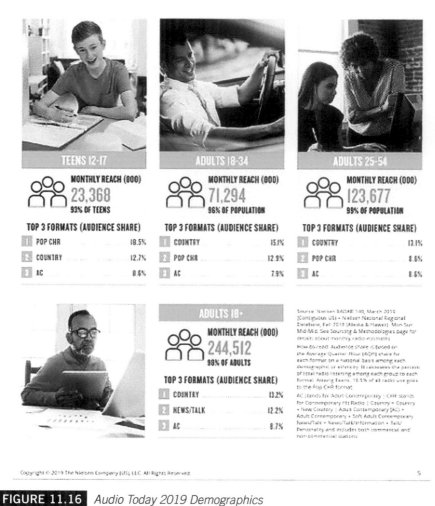

FIGURE 11.16 *Audio Today 2019 Demographics*

(Source: Nielsen Audio Radio Today 2019)

RADIO FORMATS

Station owners choose radio formats by finding an underserved audience that is attractive to advertisers. When the format is chosen and developed, a programmer and staff are hired, and the audience develops. The chart in Figure 11.19 shows the national radio audience sizes by format in 2019. Figure 11.20 breaks down the genre formats by age demographics: teens 12–17, adults 18–34, and adults 25–54, highlighting the variances among

HISPANICS 12+

MONTHLY REACH (000)
45,617
99% OF HISPANICS

TOP 3 FORMATS (AUDIENCE SHARE)

1. MEXICAN REGIONAL 15.9%
2. SPANISH CONTEMPORARY • SPANISH HOT AC 10.5%
3. POP CHR 8.3%

BLACKS 12+

MONTHLY REACH (000)
35,112
98% OF BLACKS

TOP 3 FORMATS (AUDIENCE SHARE)

1. URBAN AC 28.0%
2. URBAN CONTEMPORARY 21.0%
3. RHYTHMIC CHR 6.4%

ADULTS 18-49

MONTHLY REACH (000)
132,414
98% OF POPULATION

TOP 3 FORMATS (AUDIENCE SHARE)

1. COUNTRY 12.5%
2. POP CHR 10.6%
3. AC 8.4%

WOMEN 18-49

MONTHLY REACH (000)
65,908
97% OF POPULATION

TOP 3 FORMATS (AUDIENCE SHARE)

1. COUNTRY 14.6%
2. POP CHR 12.9%
3. AC 10.5%

MEN 18-49

MONTHLY REACH (000)
65,449
98% OF POPULATION

TOP 3 FORMATS (AUDIENCE SHARE)

1. COUNTRY 12.7%
2. POP CHR 8.4%
3. NEWS/TALK 7.8%

WOMEN 25-54

MONTHLY REACH (000)
62,296
98% OF POPULATION

TOP 3 FORMATS (AUDIENCE SHARE)

1. COUNTRY 14.3%
2. AC 11.1%
3. POP CHR 10.8%

MEN 25-54

MONTHLY REACH (000)
61,464
99% OF POPULATION

TOP 3 FORMATS (AUDIENCE SHARE)

1. COUNTRY 12.0%
2. NEWS/TALK 9.8%
3. CLASSIC ROCK 8.4%

Source: Nielsen RADAR 140, March 2019 (Contiguous US) + Nielsen National Regional Database, Fall 2018 (Alaska & Hawaii). Mon-Sun Mid/Mid. Monthly Cume Audience and AQH Share. See Sourcing & Methodologies page for details about monthly radio estimates.

How to read: Audience share is based on the Average Quarter Hour (AQH) share for each format on a national basis among each demographic and ethnicity. It calculates the percent of total radio listening among each group to each format. Among Hispanics 12+, 15.9% of all radio use goes to the Mexican Regional format.

AC stands for Adult Contemporary | CHR stands for Contemporary Hit Radio | Country = Country + New Country
Adult Contemporary (AC) = Adult Contemporary + Soft Adult Contemporary
News/Talk = News/Talk/Information + Talk/Personality and includes both commercial and non-commercial stations

FIGURE 11.16 *(Continued)*

the demographic groups and their preferences musically. Where teens like pop/contemporary hits radio (CHR) best, adults 25–54 prefer country as their favorite genre.

The Nielsen Audio audience measurement service reports the national percentages of radio format shares in the chart. The top radio format is country in the 12+ category, but when looking at just teens 12–17, their top genre is pop CHR and then country.

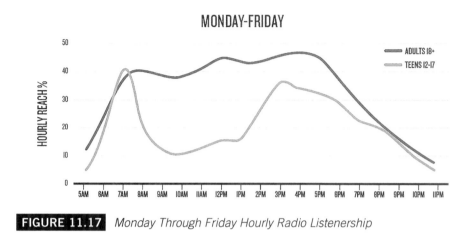

FIGURE 11.17   *Monday Through Friday Hourly Radio Listenership*

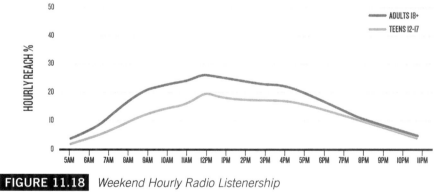

FIGURE 11.18   *Weekend Hourly Radio Listenership*

(Source: Nielsen Audio Today 2020)

Nielsen Audio is a subscription service and is the only major company that measures the size and demographics of radio audiences. While the audience share chart shows the size of the national audience, Nielsen Audio measures the same information radio market by radio market. The share and audience makeup of each individual commercial radio station is measured and reported to subscribing stations and advertising agencies. The size of the station's radio audience is directly related to the amount of money the station can charge for its advertising. The more listeners (or the larger its audience share), the more the station charges companies to access its

TOTAL 12+	
1 COUNTRY*	13.2
2 NEWS/TALK** COMBINED (COMMERCIAL and NON-COMMERCIAL)	12.0
3 ADULT CONTEMPORARY*** (AC)	8.6
4 NEWS/TALK COMMERCIAL (1,629 stations)	8.3
5 POP CHR	7.3
6 CLASSIC ROCK	6.1
7 CLASSIC HITS	5.8
8 HOT AC	4.7
9 URBAN AC	4.1
10 CONTEMPORARY CHRISTIAN	3.9
11 URBAN CONTEMPORARY	3.8
12 ALL SPORTS	3.7
13 RHYTHMIC CHR	2.7
14 MEXICAN REGIONAL	2.5
15 ALTERNATIVE	2.0
16 ADULT HITS + 80s HITS	2.0
17 ACTIVE ROCK	2.0
18 AOR + MAINSTREAM ROCK	1.7
19 SPANISH CONTEMPORARY + SPAN HOT AC	1.6
20 CLASSICAL	1.5

FIGURE 11.19 *Audience Format Trends for 12+*

TEENS 12–17	
1 POP CHR	18.5
2 COUNTRY*	12.7
3 AC***	8.6
4 HOT AC	8.0
5 URBAN CONTEMPORARY	7.5
6 RHYTHMIC CHR	6.4
7 CONTEMPORARY CHRISTIAN	5.8
8 CLASSIC HITS	3.5
9 CLASSIC ROCK	3.5
10 URBAN AC	3.4

FIGURE 11.20 *Audience Format Trends by Age*

(Source: Nielsen 2020)

ADULTS 18–34		
1	COUNTRY*	15.1
2	POP CHR	12.9
3	AC***	7.9
4	URBAN CONTEMPORARY	6.9
5	HOT AC	6.2
6	RHYTHMIC CHR	5.5
7	CLASSIC ROCK	5.2
8	NEWS/TALK** COMBINED	4.5
9	CLASSIC HITS	4.0
10	CONTEMPORARY CHRISTIAN	3.5

ADULTS 25–54		
1	COUNTRY*	13.1
2	POP CHR	8.6
3	AC***	8.6
4	NEWS/TALK** COMBINED	7.9
5	CLASSIC ROCK	6.5
6	HOT AC	5.6
7	CLASSIC HITS	5.2
8	NEWS/TALK COMMERCIAL	4.9
9	URBAN CONTEMPORARY	4.6
10	CONTEMPORARY CHRISTIAN	4.1

FIGURE 11.20   *(Continued)*

audience through advertising. Nielsen Audio charges its clients tens of thousands of dollars for its audience measurement services. Since college and other noncommercial stations do not use traditional advertising, their audience shares are not reported.

With this in mind, a programmer is very careful in choosing music for airplay since the objective is to build their target audience. The program director is not inclined to experiment with an unproven recording that will turn an audience off. This will be discussed in more depth in a later chapter.

TARGETS OF RADIO FORMATS

The ability to obtain airplay can be a major factor in determining whether a recording will be released commercially. To be a commercial product,

Table 11.3 Radio Formats With Target Demographic and Example Artists		
Format Name	**Target Demographic**	**Artists in the Format**
Adult Contemporary	Females 25–54	Harry Stiles, Post Malone, Taylor Swift
Active Rock	Men 18–34	Five Finger Death Punch, Greta Van Fleet, Foo Fighters
Alternative	Persons 18–34	Cage the Elephant, Billie Eilish, Machine Gun Kelly
CHR*/Top 40	Persons 18–34	Ariana Grande, Justin Bieber, The Weeknd
CHR/Rhythmic	Persons 12–24	Megan Thee Stallion, Pop Smoke, Drake
Country	Persons 25–54	Kelsea Ballerini, Luke Combs, Florida Georgia Line
Urban	Persons 18–34	H.E.R., Chris Brown, Lil Baby

*CHR is for contemporary hit radio. Artists listed in this chart are as they appear in the February 2020 charts for the BDS Radio Charts.

recorded music must find a target that is able and willing to buy it. Finding that target is the first step in the marketing process, followed by the development of a strategy to reach the target. This table provides some broad definitions of music formats and their targets.

RADIO FORMATS

80s Hits

Popular music and hits from the 1980s; can include a variety of genres
   Adults ages 35+

Active Rock

New and recent hard rock and heavy metal, with some alternative rock songs
   Men ages 25 to 44

Adult Contemporary

An adult-oriented pop/rock station with no hard rock, often with a greater emphasis on non-current music and softer hits from the 1980s and 1990s
   Women ages 25 to 54

Adult Hits

Does not adhere to a specific music genre; many adult hits stations play a mix of rock, pop, adult contemporary, and select oldies hits, predominantly drawing on music from the 1970s through the 2000s, including classic hits
    Adults ages 25+

Adult Standards/Middle of the Road

Eclectic rock, often with wide variations in musical style
    Adults ages 25 to 44

Album Adult Alternative

Broad, diverse playlist; music tends to include numerous types of formats, often avoiding hard rock and rap music
    Adults ages 35+

Album-Oriented Rock

Mainstream rock & roll, which can include guitar-oriented "heavy metal" and classic rock
    Men ages 25 to 44

All News

All-news, either local or national in origin; stations having this description must broadcast all-news programming during a majority of a normal broadcast day
    Adults ages 35+

All Sports

Listed only if all or a substantial block of a broadcast day is devoted to play-by-play, sports news, interviews, or telephone talk
    Men ages 25+

Alternative

Eclectic rock, often with wide variations in musical style
    Adults ages 25 to 44

Children's Radio

Programming geared towards children, often including music and spoken stories. The family hits format is similar but likely contains more all-ages music geared towards older children and all ages

Christian Adult Contemporary

Current, popular music, geared to a Christian adult audience; contemporary inspirational is a format with a similar style
Adults ages 25+

Classic Country

An oldies format, focusing on the 1960s, 1970s, and 1980s, often paired with another type of programming
Adults ages 25+

Classical

Classical music; often includes opera, theater, and/or culture-oriented news and talk
Adults ages 35+

Comedy

Programming is humorous and comedic in nature; primarily talk programming
Adults ages 25+

Country

Country music, including contemporary and traditional styles
Adults ages 25+

Easy Listening

Primarily instrumental cover versions of popular songs, with more up-tempo varieties of this format including soft rock originals, which may be mixed with "smooth jazz" or adult standards
Adults ages 35+

Educational

Primarily spoken-word, focused on offering educational programming
   Variety of ages

Gospel

Current gospel songs and sermons geared toward African-Americans
   Adults ages 35+

Hot Adult Contemporary

A variety of classic and contemporary mainstream music geared towards adults; some concentrate slightly more on mainstream pop music and alternative rock while often excluding the more youth-oriented teen music
   Adults ages 25+

Jazz

Mostly instrumental; also includes music from the Blues genre
   Adults ages 25+

Mainstream Rock

Current rock, mainstream "alternative," and heavier guitar-oriented hits
   Teens and adults ages 20 to 25

Modern Adult Contemporary

An adult-oriented softer modern rock format with less heavy, guitar-oriented music than the younger new rock
   Mostly women ages 25 to 44

New Country

Modern, current country music; popular hits
   Adults ages 25+

News Talk Information

A format devoted to informative news and talk programming; either local or national in scope
   Adults ages 35+

Nostalgia

A genre encompassing stations whose emphasis is on playing music from decades past, mainly focusing on the 1940s and prior
    Adults ages 35+

Oldies

Popular music, usually rock-oriented, with 80% or more non-current music
    Adults ages 25 to 55

Pop Contemporary Hit Radio

Focuses on playing current popular music, as determined by the top 40 music charts; includes contemporary hits dance
    Teens and adults ages 18+

Religious

Local or syndicated religious programming, often spoken-word, sometimes mixed with music
    Adults ages 25+

Rhythmic

Includes rhythmic adult contemporary and rhythmic contemporary hit radio, which are often a cross between mainstream, Top 40, and urban contemporary formats; also includes rhythmic oldies concentrating on disco and dance genres
    Adults ages 25+

Smooth Adult Contemporary

Incorporating popular new Adult Contemporary (AC)/smooth jazz artists while including more mainstream and urban adult contemporary material in an effort to reach a more diverse age demographic
    Adults ages 25+

Soft Adult Contemporary

A cross between adult contemporary and easy listening, primarily non-current, soft rock originals
    Mostly women ages 25+

Southern Gospel

Country-flavored gospel music, also includes the "Christian country" or "positive country" format
> Adults ages 25+

Spanish

Spanish-language programming, often paired with another type of programming, including Spanish adult hits, Spanish contemporary, Spanish contemporary Christian, Spanish hot adult contemporary, Spanish news/talk, Spanish oldies, Spanish religious, Spanish sports, Spanish tropical, Spanish variety, Mexican regional, Tejano, Latino urban
> Variety of Ages

Talk/Personality

Talk, either local or network in origin, which can be telephone talk, interviews, information, or a mix
> Adults ages 25+

Urban Adult Contemporary

Mix of contemporary R&B and traditional R&B, usually not including rap music
> Adults ages 25+

Urban Contemporary

Playlists made up mainly of hip-hop, R&B, electronic dance music, and Caribbean music
> Teens and adults ages 15+

Urban Oldies

Primarily playing R&B music dating back to the late 1950s and the early 1960s through the early 1990s
> Adults ages 35+

Variety

Incorporating four or more distinct formats, either block-programmed or airing simultaneously
> Variety of ages

World Ethnic

Programs geared to various ethnicities, primarily in languages other than English

Variety of Ages

(https://newsgeneration.com/broadcast-resources/guide-to-radio-station-formats/)

The target market of a particular radio format is the logical consumer target for commercial recordings. The 2020 New Generation Guide to Radio Station Formats lists over 35 musical radio formats ranging from adult album alternative (AAA) to CHR to variety. Music marketers carefully study the listeners of each format so that they can pitch their artist to the stations that reach their target market.

Radio station group owners further refine their format audience by gender and age. For example, many country radio stations specifically target females 35–44, while stations with CHR/rhythmic formats target males 12–24.

One of the key components of most of these radio markets is the 18–34-year-old. Young adults are big consumers, setting up new households and making substantial purchases like appliances, furniture, or that first new car. Women in this age group heavily influence or actually make the purchase decisions for households, and advertisers highly value this demographic as a target for their messages.

WHAT IS IMPORTANT TO PROGRAMMERS?

Convincing radio to play new music is "selling" in every sense of the word. And in order to sell someone anything, you must know what is important to them and what their needs are. High on that list of important things to radio is Nielsen Audio's measurement of radio audiences because it directly impacts the earnings of the station for its owners. Understanding concepts like this and their importance to programmers will help marketers of recorded music better relate to the needs of radio and its programming gatekeepers.

RATINGS RESEARCH AND TERMINOLOGY

The term **P-1** listeners represents one of the prized numbers of radio programming. "These first preference listeners . . . [are] referred to in radio station boardrooms, focus groups, and inside the headquarters of Arbitron doing diary reviews." Experts say the term P-1 describes "the most loyal of radio listeners," and every radio programmer courts this primary core of their radio audiences (www.mcvaymedia.com). The terms P-2 and P-3 refer to listeners with a lower degree of connection and loyalty to a particular radio station.

Cume is the total number of unduplicated persons included in the audience of a station over a specified time period. This programming term comes from the word cumulative, and it refers to the total of all different listeners who tune into a particular radio station, measured by Nielsen Audio in quarter-hour segments.

AQH, or average quarter-hour, refers to the number of people listening to a radio station for at least 5 minutes during a 15-minute period.

TSL means "time spent listening." It is an estimate of the amount of time the average listener spent listening to a station (or all stations) during a particular daypart. TSL is calculated by the following formula.

TSL = [Quarter-hours in a time period × AQH persons]/Cume persons

A radio station's **share** refers to the percentage of persons tuned to a station out of all the people listening to radio at the time.

Share = AQH persons tuned to a specific station/AQH persons in market currently listening to radio × 100

Rating refers to the percentage of persons tuned to a station out of the total market population.

Rating = AQH persons tuned to a specific station/Persons in market × 100

A complete list of radio rating terms and their definitions can be downloaded from www.arbitron.com/downloads/terms_brochure.pdf

Nielsen Audio manually measures and rates radio listener habits in about 290 markets in the United States. Ratings are measured using the diary method in all but 52 markets. These latter markets are measured using electronic devices called the Portable People Meter, shown in Figure 11.20 (https://tenwatts.blogspot.com/2019/04/nielsen-radio-survey-part-5.html? m=1).

For the diary method of tracking listening, Nielsen Audio selects households at random and asks members aged 12 and above to carry the diary for

FIGURE 11.21 *A Portable People Meter (PPM)*

(Source: Nielsen Audio)

one week and record their radio listening. Potential diary keepers are first contacted by telephone, and then diaries are sent to the household. Completed diaries are returned to Nielsen Audio, and the data are entered into computers and analyzed on the following characteristics:

- Geographic survey area (metro or total survey area)
- Demographic group
- Daypart
- Each station's AQH: the estimated number of persons listening
- Each station's rating: the percent of listeners in the area of study during the daypart

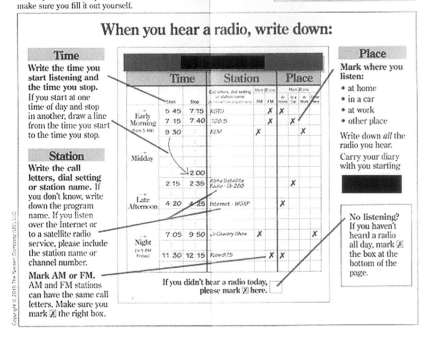

You count in the radio ratings!

No matter if you listen a lot, a little or not at all, you're important!

You're one of the few people picked in your area to have the chance to tell radio stations and other businesses what you listen to.

This is *your* radio diary. Please make sure you fill it out yourself.

Here's what we mean by "listening":

Listening is any time you can hear a radio – whether you choose the station or not. You may be listening to radio on AM, FM, the Internet or satellite. Be sure to include all your listening.

Any time you hear radio from ▮▮▮▮▮▮ write it down – whether you're at home, in a car, at work or someplace else.

When you hear a radio, write down:

Time

Write the time you start listening and the time you stop. If you start at one time of day and stop in another, draw a line from the time you start to the time you stop.

Station

Write the call letters, dial setting or station name. If you don't know, write down the program name. If you listen over the Internet or to a satellite radio service, please include the station name or channel number.

Mark AM or FM. AM and FM stations can have the same call letters. Make sure you mark ☒ the right box.

Place

Mark where you listen:
• at home
• in a car
• at work
• other place

Write down *all* the radio you hear.

Carry your diary with you starting ▮▮▮▮▮

No listening? If you haven't heard a radio all day, mark ☒ the box at the bottom of the page.

If you didn't hear a radio today, please mark ☒ here.

FIGURE 11.22 *Nielsen Survey Instructions*

- Each station's share: the percent of one station's total daypart estimated listening audience
- Cume: the total unduplicated audience during the daypart for an average week

For those using the Portable People Meter, it is a unique audience measurement system that tracks what consumers listen to on the radio, and what consumers watch on broadcast television, cable and satellite TV. The portable People Meter is a pager-sized device that consumers wear throughout the day. It works by detecting identification codes that can be embedded in the audio portion of any transmission.

(Arbitron)

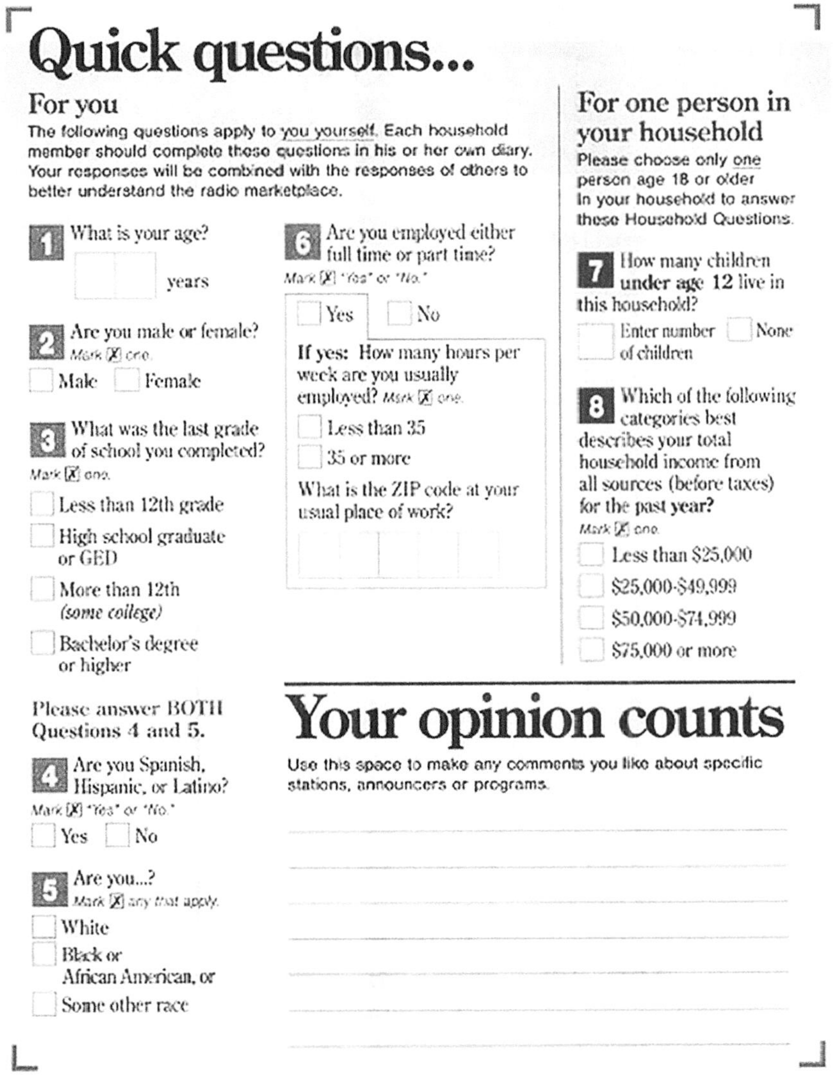

Quick questions...

For you

The following questions apply to you yourself. Each household member should complete these questions in his or her own diary. Your responses will be combined with the responses of others to better understand the radio marketplace.

1 What is your age?

_____ years

2 Are you male or female?
Mark [X] one.

☐ Male   ☐ Female

3 What was the last grade of school you completed?
Mark [X] one.

☐ Less than 12th grade

☐ High school graduate or GED

☐ More than 12th (some college)

☐ Bachelor's degree or higher

Please answer BOTH Questions 4 and 5.

4 Are you Spanish, Hispanic, or Latino?
Mark [X] "Yes" or "No."
☐ Yes ☐ No

5 Are you...?
Mark [X] any that apply.
☐ White
☐ Black or African American, or
☐ Some other race

6 Are you employed either full time or part time?
Mark [X] "Yes" or "No."

☐ Yes   ☐ No

If yes: How many hours per week are you usually employed? Mark [X] one.

☐ Less than 35

☐ 35 or more

What is the ZIP code at your usual place of work?

_____

For one person in your household

Please choose only one person age 18 or older in your household to answer these Household Questions.

7 How many children under age 12 live in this household?

☐ Enter number of children   ☐ None

8 Which of the following categories best describes your total household income from all sources (before taxes) for the past year?
Mark [X] one.

☐ Less than $25,000

☐ $25,000-$49,999

☐ $50,000-$74,999

☐ $75,000 or more

Your opinion counts

Use this space to make any comments you like about specific stations, announcers or programs.

_____
_____
_____
_____
_____
_____

FIGURE 11.23 *Nielsen Survey*

The meter is about the size of a pager and is worn at all times during the day. At night, the People Meter is placed into a base unit, and information is uploaded to a central database. Whereas the diary method of tracking radio listenership uses diaries targeted to a balanced yet random population sample, the PPM enlists families of people, pays them a fee, and seeks a two-year commitment for them to be part of the program.

RADIO MARKET SURVEY POPULATION, RANKINGS & INFORMATION—SPRING 2020

MARKET SURVEY RANKINGS, FREQUENCY, TYPE AND POPULATION

BY RANKING

MKT CODE	RANK	TYPE	FREQ	MARKET	DST	METRO 12+ POPULATION	HISPANIC 12+ POPULATION*	BLACK 12+ POPULATION**
001	1	PPM	13	New York	BH	16,110,500	4,092,600	2,682,700
003	2	PPM	13	Los Angeles	BH	11,469,700	4,993,500	794,000
005	3	PPM	13	Chicago	BH	7,952,400	1,698,900	1,322,300
009	4	PPM	13	San Francisco	BH	6,764,400	1,507,600	438,600
024	5	PPM	13	Dallas-Ft. Worth	BH	6,339,800	1,742,200	1,052,000
033	6	PPM	13	Houston-Galveston	BH	5,979,700	2,137,700	1,041,000
015	7	PPM	13	Washington, DC	BH	5,019,400	815,800	1,346,600
047	8	PPM	13	Atlanta	BH	4,971,100	510,100	1,760,500
007	9	PPM	13	Philadelphia	BH	4,627,200	432,000	959,200
013	10	PPM	13	Boston	BH	4,376,900	507,100	356,600
429	11	PPM	13	Miami-Ft. Lauderdale-Hollywood	BH	4,159,800	2,296,000	840,000
039	12	PPM	13	Seattle-Tacoma	BH	4,006,500	388,200	268,900
011	13	PPM	13	Detroit	B	3,848,800	159,800	844,000
057	14	PPM	13	Phoenix	BH	3,815,900	1,113,200	240,600
027	15	PPM	13	Minneapolis-St. Paul	BH	3,032,400	164,100	278,900
063	16	PPM	13	San Diego	BH	2,873,100	935,300	156,500
087	17	PPM	13	Tampa-St. Petersburg-Clearwater	BH	2,797,700	544,600	336,500
035	18	PPM	13	Denver-Boulder	BH	2,796,400	582,500	161,400
540	19	12S	12	Puerto Rico		2,740,700	*	*
321	20	PPM	13	Nassau-Suffolk (Long Island)	BH	2,458,400	455,300	244,800
051	21	PPM	13	Portland, OR	H	2,413,400	307,400	85,600
021	22	PPM	13	Baltimore	BH	2,408,800	142,300	710,800
093	23	PPM	13	Charlotte-Gastonia-Rock Hill	BH	2,391,900	221,100	553,300
017	24	PPM	13	St. Louis	B	2,342,200	71,100	439,300
059	25	PPM	13	San Antonio	BH	2,151,300	1,175,900	154,500
379	26	PPM	13	Riverside-San Bernardino	BH	2,145,300	1,173,600	184,100
065	27	PPM	13	Sacramento	BH	2,052,500	411,200	164,300
101	28	PPM	13	Salt Lake City-Ogden-Provo	H	2,012,700	297,900	33,400
023	29	PPM	13	Pittsburgh, PA	B	1,972,100	36,100	176,800
131	30	PPM	13	Orlando	BH	1,956,800	670,800	331,400
257	31	PPM	13	Las Vegas	BH	1,945,100	579,100	255,500
135	32	PPM	13	Austin	BH	1,877,300	580,000	143,200
031	33	PPM	13	Cincinnati	B	1,856,600	57,300	242,100
041	34	PPM	13	Kansas City	BH	1,777,000	153,000	235,300
019	35	PPM	13	Cleveland	BH	1,774,400	99,800	355,700
045	36	PPM	13	Columbus, OH	B	1,722,000	69,900	297,400
215	37	PPM	13	San Jose	H	1,674,400	401,000	47,200
115	38	PPM	13	Raleigh-Durham	BH	1,651,400	185,900	365,400
049	39	PPM	13	Indianapolis	BH	1,585,200	104,000	266,100
073	40	PPM	13	Nashville	BH	1,512,100	105,600	249,800

FIGURE 11.24   *Radio Markets by Population (41–83)*

Considerable pushback from radio has included the complaint that just a few PPM wearers within a specific market determine station ratings. As an example: Nashville has a 1.5 million 12+ population with 975 PPM wearers. Of these PPM wearers, only 30 might be dedicated AAA genre

RADIO MARKET SURVEY POPULATION, RANKINGS & INFORMATION—SPRING 2020

MKT CODE	RANK	TYPE	FREQ	MARKET	COT	METRO 12+ POPULATION	HISPANIC 12+ POPULATION	BLACK 12+ POPULATION
393	41	2S	2	Hudson Valley	BH	1,508,800	337,700	195,100
043	42	PPM	13	Milwaukee-Racine	BH	1,508,500	153,900	235,800
413	43	PPM	13	Middlesex-Somerset-Union	BH	1,483,000	354,000	203,000
077	44	PPM	13	Providence-Warwick-Pawtucket	BH	1,419,800	177,800	85,300
109	45	PPM	13	Norfolk-Virginia Beach-Newport News	BH	1,411,600	98,600	448,100
107	46	PPM	13	Jacksonville	BH	1,345,400	121,500	285,400
299	47	PPM	13	West Palm Beach-Boca Raton	BH	1,321,300	297,600	243,900
166	48	PPM	13	Greensboro-Winston-Salem-High Point	BH	1,315,100	116,800	308,200
083	49	12S	12	Oklahoma City	BH	1,279,500	152,200	141,700
063	50	12S	12	New Orleans	BH	1,258,400	103,800	399,100
075	51	PPM	13	Memphis	B	1,127,600	59,100	532,400
061	52	PPM	13	Hartford-New Britain-Middletown	BH	1,078,200	169,000	127,000
105	53	12S	12	Richmond	D	1,069,300	67,000	323,300
065	54	12S	12	Louisville	B	1,053,000	49,200	167,100
516	55	2S	2	Monmouth-Ocean	H	1,049,400	103,600	56,500
209	56	2S	2	McAllen-Brownsville-Harlingen	H	1,036,900	966,800	4,300
515	57	12S	12	Ft. Myers-Naples	BH	1,027,200	228,900	77,300
191	58	12S	12	Greenville-Spartanburg	B	1,003,700	68,200	176,000
037	59	12S	12	Buffalo-Niagara Falls	D	960,800	46,300	119,800
079	60	12S	12	Rochester, NY	B	955,600	67,000	106,000
005	61	12S	12	Birmingham	B	925,500	37,300	277,700
207	62	12S	12	Tucson	H	906,300	324,400	36,200
067	63	12S	12	Dayton	B	842,100	22,400	124,500
009	64	12S	12	Honolulu		832,800	77,600	35,100
103	65	12S	12	Tulsa	H	827,500	76,000	73,500
089	66	12S	12	Fresno	H	817,300	424,600	41,300
060	67	12S	12	Albany-Schenectady-Troy		814,900	44,400	68,000
127	68	12S	12	Grand Rapids	H	804,900	75,100	62,500
141	69	12S	12	Albuquerque	H	775,400	375,700	21,300
373	70	2S	2	Sarasota-Bradenton	H	755,300	87,100	47,500
071	71	12S	12	Des Moines		751,900	44,100	38,400
189	72	12S	12	Metro Fairfield County	BH	744,900	148,000	92,400
121	73	12S	12	Knoxville		739,500	31,100	46,900
145	74	12S	12	Allentown-Bethlehem	H	736,400	123,700	43,300
085	75	12S	12	Omaha-Council Bluffs	H	726,800	75,200	63,500
161	76	12S	12	El Paso	H	696,500	580,800	23,300
223	77	12S	12	Baton Rouge	B	695,200	29,500	241,700
175	78	12S	12	Wilkes Barre-Scranton	H	692,500	71,500	41,500
231	79	12S	12	Charleston, SC	D	690,400	35,700	173,300
143	80	12S	12	Bakersfield	H	661,300	364,700	35,300
291	81	2S	2	Stockton	H	632,400	254,900	50,700
139	82	2S	2	Wilmington, DE	D	625,900	54,000	138,200
311	83	2S	2	Lakeland-Winter Haven	BH	620,900	137,400	92,500

FIGURE 11.25   *Radio Markets by Population (1–42)*

(Source: https://worldradiohistory.com/Archive-Arbitron/Sample-Targets/Radio%20PPM%20 Markets_Daily%20In-Tab%20Targets%20Oct18.pdf)

NIELSEN RADIO PPM® MARKETS WITH DAILY IN-TAB TARGETS

SORTED BY P12+ RANK

OCTOBER 2018 P12+ RANK	MARKET	DAILY IN-TAB TARGET	PPM CURRENCY MONTH	OCTOBER 2018 P12+ RANK	MARKET	DAILY IN-TAB TARGET	PPM CURRENCY MONTH
1	New York[2]	4,494	Sept. 08	27	Sacramento	1,065	Dec. 09
2	Los Angeles[1]	2,702	Sept. 08	28	Pittsburgh, PA[1]	1,197	Sept. 09
3	Chicago[1]	2,141	Sept. 08	29	Salt Lake City-Ogden-Provo[1]	975	Dec. 09
4	San Francisco[1]	2,366	Sept. 08	30	Las Vegas	975	Dec. 09
5	Dallas-Ft. Worth[1]	1,600	Dec. 08	31	Orlando	975	Sept. 10
6	Houston-Galveston[1]	1,600	Jun. 07	32	Cincinnati[1]	1,040	Dec. 09
7	Washington, DC	1,600	Dec. 08	33	Austin	975	Sept. 10
8	Atlanta[1]	1,600	Dec. 08	34	Cleveland[1]	1,118	Dec. 09
9	Philadelphia[1]	1,683	Mar. 07	35	Kansas City[1]	975	Dec. 09
10	Boston	1,671	Mar. 09	36	Columbus, OH[1]	975	Sept. 10
11	Miami-Ft. Lauderdale-Hollywood[1]	1,703	Jun. 09	37	San Jose[1]	975	Sept. 08
12	Seattle-Tacoma	1,341	Jun. 09	38	Raleigh-Durham	975	Sept. 10
13	Detroit[1]	1,584	Dec. 08	39	Indianapolis[1]	1,019	Sept. 10
14	Phoenix[1]	1,200	Jun. 09	41	Milwaukee-Racine[1]	975	Sept. 10
15	Minneapolis-St. Paul[1]	1,250	Jun. 09	42	Middlesex-Somerset-Union[3]	975	Sept. 08
17	San Diego[1]	1,200	Jun. 09	43	Nashville	975	Sept. 10
18	Tampa-St. Petersburg-Clearwater[1]	1,200	Sept. 09	44	Providence-Warwick-Pawtucket	975	Sept. 10
19	Denver-Boulder[1]	1,200	Sept. 09	45	Norfolk-Virginia Beach-Newport News	975	Sept. 10
20	Nassau-Suffolk (Long Island)[1]	1,200	Sept. 08	46	Jacksonville[1]	975	Dec. 10
21	Baltimore[1]	1,188	Sept. 09	47	West Palm Beach-Boca Raton	975	Dec. 10
22	Portland, OR[1]	978	Dec. 09	48	Greensboro-Winston Salem-High Point[3]	975	Dec. 10
23	Charlotte-Gastonia-Rock Hill[1]	1,011	Sept. 10				
24	St. Louis[1]	1,214	Sept. 09	51	Memphis	975	Dec. 10
25	Riverside-San Bernardino[1]	975	Sept. 08	52	Hartford-New Britain-Middletown	975	Dec. 10
25	San Antonio[1]	975	Dec. 09				

1 FM Radio Ratings Data accredited by Media Rating Council[*]
2 Includes embedded markets: Nassau-Suffolk and Middlesex-Somerset-Union[+]

3 Embedded market
4 Includes embedded market: San Jose, which is not accredited by the MRC

PPM ratings are based on audience estimates and are the opinion of Nielsen and should not be relied on for precise accuracy or precise representativeness of a demographic or radio market.

FIGURE 11.26   *Portable People Meter Markets*

station listeners that will determine the P-1 and P-2 station in the market. Statistically speaking, this small sample size may not be significantly large enough to validate the findings of the data according to the stations affected.

Additionally, the PPM is "listening" to what the wearer is hearing externally. Earphones cannot be picked up by the PPM, and extraneous "sounds" such as background music of a restaurant or store can also be "registered" by the device, skewing the data. Nielsen has tried to adjust for the earphone exception by asking PPM wearers to "log" their earphone listening and submit to Nielsen to complete their profiles. But, as one can surmise, collecting listening data is both a science and an art and is not always spot-on accurate.

Each Nielsen Audio Radio Market Report covers a 12-week period for the specified market and contains numerous pages like the example in Figure 11.27. At the top of each page, the target demographic is listed. Beneath that, the dayparts are laid out in columns. Then for each daypart, the AQH, the Cume, the AQH rating and AQH share are listed for each radio station in the area (listed in the left-hand column).

RADIO PROGRAMMING RESEARCH

Knowing how radio researches its audience can be helpful to marketers to understand how programmers define benchmarks for their decisions to add

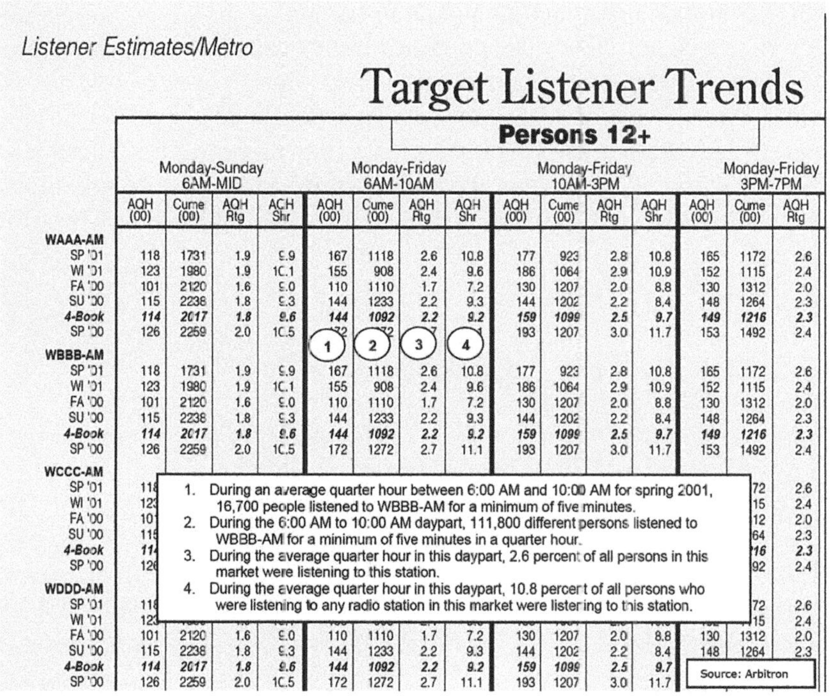

FIGURE 11.27 *How to Read a Nielsen Audio Book Report*

(Source: Nielsen Audio)

or remove music from their playlists or increase or decrease the frequency songs are played. The key research tools used by programmers are discussed in the next chapter in the section on charts.

Many stations use panels of listeners for programming research. The panel method is a research technique in which the same people are studied at different points in time. Members of the panel are selected to reflect a representative sample of a station or format's listenership and are periodically surveyed on their opinions of music and programming. The panel members are contacted either by telephone or email and asked to respond to song hooks played either over the phone or through the Internet. The programmer then analyzes the listener response to make decisions about the continued use of the song on the air and how frequently it will be played.

A second method for collecting listener responses to music is the auditorium test. As the name implies, listeners are brought together in a single place, most often a hotel ballroom, and asked to listen to and rate the music at the same time. The biggest advantage of the auditorium test is control, but improved sound quality is another major plus. Invited listeners record their responses either on paper or using a Wi-Fi monitored dial, and the researchers compile the data for the radio station.

Another tool being used by programmers is called MScore, which uses the audience preference tracking of the Portable People Meter to judge whether listeners switch stations while a song is playing. For programmers, it is useful feedback about the music they program; for record labels, the information is useful to continuously monitor a song in the marketplace and modify promotional strategy (Albright 2009; mediamonitors.com).

THE CHANGING FACE OF RADIO

The first commercial radio stations, all 30 of them, were licensed by the U.S. government in 1922, 12 years before the FCC was established. The following year, in 1923, there were 556 commercial radio stations, all of them broadcasting on AM frequencies. The technology didn't change much until the 1960s when the FCC approved a new band of radio frequencies for commercial broadcasters – FM. FM promised better sound quality and fewer commercials (6 minutes per hour instead of the 18 to 20 minutes typical at the time) to gain market share.

SATELLITE RADIO

XM Satellite Radio was founded in 1992 and launched on September 25, 2001. The next year, Sirius Satellite Radio was launched. In 2008, the two companies merged. Sirius XM offers an array of music and other entertainment channels, which are fed to proprietary radio receivers, meaning one must own a special receiver in order to access programming. The subscriber to Sirius XM pays a monthly fee for basic services and a higher fee for additional services, much like a satellite television service. The opportunities for marketers of recordings with satellite radio is that the company has longer playlists within each genre of music, there are more opportunities for new music to be played, and every song that is played displays the artist and song title to the listener on the faceplate of the receiver. Since the purchase of Pandora in 2018, the collective subscription base is now over 38 million strong, with the company having been profitable since 2009.

HD RADIO

Opportunities for marketers of recorded music have improved with the addition of new technology for the radio broadcast industry. Commercial stations have mostly converted their AM and FM stations to HD radio, greatly improving the quality of the signals for both. HD does not mean "high definition." It is trademarked term for the in-band on-channel (IBOC) digital radio technology used by AM and FM radio stations in the United States. This technology allows for more than one station to broadcast on a single frequency. What this means to music marketers is that songs played on the radio will deliver near-CD quality audio and have the ability to display the artist's name and the song title on the radio receiver, just like satellite radio. Radio announcers rarely provide artist or song information to listeners, and this new technology will help consumers of recorded music to identify artists and songs. As radio stations convert to digital broadcasting, they are also given up to two additional signals that are adjacent to their primary broadcast frequency. These new "stations" give the station owners opportunities to explore experimental programming and to broaden their listener bases.

INTERNET RADIO

Not to be confused with subscription streaming services, internet radio could also be termed "webcasting" or "streaming radio' where radio stations are simulcasting their live radio shows on their internet stream. This also includes services such as free Pandora, free Spotify, and Last.fm, including advertising, which pays for the service to listeners while still delivering a tailored listening experience as prescribed by the consumer. (More on streaming services as well as the social media impact in Chapter 10.) With expansive Wi-Fi availability and the power of 5G networks, streaming radio-like music services over wireless devices bring music closer to specific consumer interests and pose a competitive challenge for traditional commercial radio. An additional advantage of Internet radio is that the song information is displayed on the device's screen as the song is played, often with a link to an online retail store to facilitate the impulse purchase.

An early contender and lasting example is Pandora Radio. Pandora has been a market leader in streaming services. As one of the first to offer consumers a tailored listening experienced based on their proprietary Music Genome Project technology, Pandora uses sophisticated algorithms to create playlists that match musical tastes. Although Pandora now offers subscriptions services that eliminate streaming advertising, the service has thrived with its free service offering. With approximately 70 million users monthly, Pandora is the largest streaming music provider in the United States, with an industry-leading digital audio advertising platform, and is a subsidiary of Sirius XM Holding Inc. (www.pandora.com/aboutGetting airplay).

It is the job of the record promoter to get airplay on commercial radio stations. This has become more difficult with the consolidation of radio because there are fewer music programmers, and competition for getting added to the playlist is fierce. The process of record promotion is outlined in the following section.

PROMOTION AND AIRPLAY

From the earliest days of product marketing, salespeople have constantly sought as many ways as possible to say their wares are the best available and that theirs are the number one products in the eyes of consumers. In music especially, bragging rights of having a "number one" provide leverage for promoters at the label to ask the chart makers to "join the crowd" and move their single or album higher on their charts, ultimately impacting revenue through increased streams and sales. Chart success also ushers in other opportunities that benefit the artist, label, and songwriter, as they could bring more television synchronization licenses, a higher booking rate for concerts, and an increase in royalty payments for the songwriter. In this section, we will explore how radio promotion leads to airplay.

We have all been to the doctor's office and seen drug company marketers drag their wheeled bag of samples and a box of donuts into the office area "behind the door," a practice that continues today even with the shift to conducting business online. They are touting benefits of medicines and building relationships with doctors so they can earn bonuses based on how many prescriptions are written in their territories. Doctors do not need medicines, but their patients do. In the music business, promoters seek to convince program directors and playlist curators to "prescribe" the label's music to the station's or DSP's listeners. And like the promoter of medicine, the music promoter is compensated based on their level of success in convincing programmers to choose the label's music to play. This section provides a deeper dive into airplay and promotion surrounding the **terrestrial**, satellite, and online radio platforms.

Airplay, in the traditional sense of music being played on terrestrial radio, is viewed by some as a dying medium and a waste of promotion money; however, radio continues to draw huge audiences, exposing recorded music to millions of consumers every day. During any given week of the year, the top single on *Billboard*'s Mainstream Top 40 National Airplay chart may be heard by a U.S. audience of 50 million. Any medium that connects a product with its target market with such impact is far from dead and must be an important component in the marketing mix.

RADIO PROMOTION

Lobbying and lobbyists have been around as long as any one person has been responsible for a decision or a vote, and people have always wanted to influence that decision or vote in their favor. Every day, lawmakers at the national and state levels meet with representatives of special interest groups who ask them to vote on matters that are in the best interests of their groups or their clients.

The same thing happens between a radio programmer and a music promoter. Music promoters are lobbyists in the purest form. They are either on the staff of a record company or they are part of a company specifically hired by the label to promote new music.

The people lobbied are usually the music directors and/or program directors of radio stations that report their station's **playlist** to the major trade and data organizations that publish the charts. Some stations have a music director who is responsible for the day-to-day aspects of the music programmed on the station, but in most cases it is the program director who is the primary influencer in terms of deciding what songs can added to the playlist and moved up in **rotation**. Program directors have the ultimate responsibility for everything that a radio station broadcasts – banter by personalities, advertising, information such as news and traffic reports, and all music played by the station. Simply said, the radio programmer can decide whether a recording ever gets on the air at their station.

Decisions by radio programmers are one of the keys to the life of a recording and have become the basis for savvy, smart, and creative record promotion. Programming decisions about music determine:

- Whether a new recording is added to the playlist of a radio station that reports its chart and airplay to trade and data organizations.
- Whether the recording receives at least light rotation on the playlist.
- Whether it eventually receives heavy rotation on the station's playlist.

Importantly, however, consolidation has altered the influence some PDs have, as many programming decisions for large conglomerates such as iHeart, Entercom, and Cumulus are now made by a single person within those organizations and then communicated to the individual stations. At the very least, a "suggested" or "preferred" playlist is distributed to the stations in the group.

The Relationship Between Promoter and Radio

The promotion department at a record label has the responsibility of securing radio airplay for the company's artists. Quite simply, they lobby the station decision makers – the gatekeepers – to add recordings by the label's artists to the station's playlist. Once added, the radio promoter with the label continues to push for increased **spins** with the station to maximize the artist's exposure with the station's listener base and to continue to move the single up the charts.

The effectiveness of a record promotion person hinges on the strength of the relationships he or she has with radio programmers and other music gatekeepers. These relationships are built much like anyone in business that has a client base requiring regular service. The promoter makes frequent contact with the programmer, whether that be in-person visits, phone calls, or email or text. The promoter arranges lunch or dinner meetings; provides the station access to the label's artists; and helps the station promote and market itself with contests, performances by artists, and giveaways. Even in today's digitally focused world, programmers spend a good amount of their time traveling to the radio stations for which they are responsible to have face-to-face meetings with key influencers. A promoter who has effectively

FIGURE 11.28 *In the Radio Studio*

developed a good relationship with a programmer can make the request for the programmer to treat his current record project favorably, which is difficult for smaller labels to do because of their limited promotional resources.

If the promoter has no relationship with the radio programmer from a reporting station, it is highly unlikely that telephone calls will ever be returned. Programmers today have too many things to do and little time to listen to promotional pitches from people and companies they do not know. And because of centralized programming by some radio groups, there is more pressure on fewer programming personnel to find time to take a call from a promoter.

History and the Payola Laws

Record promotion and its regulation by the federal government began not long after the advent of commercial radio broadcasting. In 1934, Congress created and passed the Communications Act, which restricted radio licensees – the stations themselves – from taking money in exchange for airing certain content unless the broadcast was commercial in nature. However, this early act contained nothing that prohibited disc jockeys (DJs) from taking payments in exchange for airplay. During the big band era of the 1940s, and the rock 'n' roll days of the 50s, DJs were routinely taking money from record promoters in exchange for the promise to play a record on the air.

Disc jockeys during this time often made their own decisions about which records would be included on their programs, and promoters would approach them directly to influence their record choices.

Lawmakers railed against the rampant bribes being given to DJs to play records. In 1960, Congress amended the 1934 act to include a provision that was intended to eliminate illegal bribes to play music, so-called **payola**. Under the revised law, disc jockeys and radio stations were permitted to receive money and gifts to play certain songs, but the amendment placed a requirement that these inducements must be disclosed to the public on the air. If this disclosure was not made, it exposed the DJ and management to possible fines and imprisonment (Freed 2005). The change in the law also created the requirement that record labels must report their cash payments and major gifts for airplay to the station. This 1960 amendment continues to guide the radio and records industries today.

Despite the stronger laws against payola, federal investigators were called upon to investigate scandals within the record promotion business in the 1970s and 1980s. There were no major convictions despite the appearance that money, drugs, and prostitution were being used as leverage by promoters to get radio airplay for recordings (Katunich 2002).

Entanglement by the payola laws began when radio stations asked the label promoters for something of value, whatever it was. The first question to the record company then becomes, "How does this help me sell copies of the artist's record?" One label says that everything they do today with radio stations to promote an artist "has to really pass the smell test." They require proof that promotional prizes and free concerts by artists are acknowledged on the air as being given by the label. They expect to receive affidavits from the station showing when the announcements were aired and at how much money those announcements were valued, underscoring the disclosure requirement for radio stations taking payments as well as the record labels providing payment. In the late 1990s and early 2000, stations and labels began to break the payola law and in 2005 came under investigation by the New York attorney general's office. The examination into these practices resulted in Sony BMG, EMI, Warner Bros., and Universal Music paying over $30 million in penalties for paying radio stations to play their music. The Federal Communications Commission followed the New York inquiry, which resulted in $12.5 million in fines levied against Citadel Broadcasting, CBS, Entercom, and iHeart Media Communications (Barbington 2007). Following the investigation, labels of all sizes spent considerable time and fees on lawyers to ensure that promotion stayed within the guidelines of the law. The following email from a promotion employee who was unhappy with the times he was being given for spins of a song is of one of hundreds acquired by and used by the New York Attorney General's office as evidence of pay-for-play violation of Payola laws. (Source: NY Attorney General's website.) "OK, HERE IT IS IN BLACK AND WHITE AND IT'S SERIOUS: IF A RADIO STATION GOT A FLYAWAY TO A CELINE [DION] SHOW IN LAS VEGAS FOR THE ADD, AND THEY'RE PLAYING THE SONG ALL IN OVERNIGHTS, THEY ARE NOT GETTING THE FLYAWAY. PLEASE FIX THE OVERNIGHT ROTATIONS IMMEDIATELY."

GETTING A RECORDING ON THE RADIO

As noted previously, the consolidation of terrestrial radio has concentrated some of the music programming decisions into the hands of a few programmers who provide consulting and guidance from the corporate level to programmers at their local stations. In many cases, local programmers can add songs to their playlists based upon the preferences of the local audiences. The following steps detail how songs are typically added to station's playlists for those stations that program new music:

1. A record label promoter or an independent promoter hired by the label calls the corporate programmer, the station music director (MD), and/or program director announcing an upcoming release (it is the job of the promoter to know the best person to contact at each station). Radio music directors have "call times." These are designated times of the week that they will take calls from record promoters. The call times vary by station and broadcasting company and are subject to change. For example, a music director may take calls only on Tuesdays and Thursdays, 2:00–4:00 p.m.

2. Leading up to the **add date**, meaning the day the label is asking that the single be added to the station's playlist, the promoter will call again, touting the positives of the recording, and ask that the recording be added.

3. The music director and/or program director will consider the selling points by the promoter, study the streams the song is generating on Spotify and Pandora in the market, review their preferred charts for performance of the recording in other cities, consider current research

on the local audience and its preferences, look at any guidance provided by their corporate programmers or consultants, and then decide whether to add the song.

4. The PD will look for reaction or response to adding the song.

Strategic Decisions

It is important to understand that promotion by a label is focused on radio stations that program current, new music. Stations airing older music are using record company catalogues as the basis for their entertainment programming, and many fans of the artists they program already have the music in their digital or physical libraries. Older albums and their related singles have been around for years, sometimes decades, and there is limited interest on the part of a label to promote them to radio. Most energy and money invested by record companies will be used in promoting the newest singles and album projects.

Record promotion to radio by labels typically needs to answer four questions:

1. What is the return on investment (ROI)? How does it help the label generate streams or sell recordings (digital or physical)?

2. What does it do for the artist in terms of improving relationships or building the artist's brand?

3. What does it do to further the label's agenda and help them market the artist?

4. Does it make sense for the radio station?

The goal of the promotions team is to try to time the album release date with the momentum created by significant airplay at radio as well as increased streaming and playlist activity on the music streaming platforms. This objective is particularly challenging because the length of time it takes for a single to reach peak position on the charts differs by many variables, including the genre, the level of the artist (singles from superstars tend to move more quickly than those from new artists), and, of course, how strongly the music resonates with the audience. It is now not uncommon for labels to roll out two to three singles from a project on streaming services before the full album is dropped to keep the music coming while the singles at radio are taking longer to climb the chart.

Radio Promotion Tour

When an artist from a major label is preparing to release a new album, the radio promotions team will generally organize a **radio promotion tour**. In normal (pre-COVID) times, the radio promotion tour would send the artist and their radio promotion rep from the label to each targeted station over a period of several weeks. At each station, the artist would talk about the project on-air with the radio station's personalities or perhaps be featured in a listener appreciation event.

As a radio promotional tour is developed for an artist with new music, emphasis is generally placed on the most popular radio stations in targeted cities, especially those "reporting" stations that report their airplay charts to major trade magazines such as *Billboard*. At stationratings.com as well as through other online sources, data is available to enable labels to determine which stations are rated the highest in audience shares by market. Specific information about the station programmer's name, address, and telephone number is available through several media databases, including Cision, Meltwater, and Propel, which all include radio. One handy website for radio coverage maps by location is radiolocator.com. Another consideration is how much time should be spent targeting commercial stations (those that sell advertising) vs. non-commercial stations (known as "non-comms"). Most commercial stations have shorter playlists because their on-air time is also filled with the paid ads, while non-comm stations like college radio and National Public Radio (NPR) often have a broader set list that features a wider range of music.

During the coronavirus pandemic, these in-person activities were forced to change to an entirely digital presentation format where live streaming platforms were utilized to allow the artist to connect in a meaningful way with each station.

Some artists are taking to platforms like Twitch to conduct their radio tour in a virtual environment. Under this scenario, a tour might take place over a five- or six-week period, with several slots available every day. Each station that confirmed participation would be sent a link to access the live event, where the band or artist would be introduced by the VP of promotion with the label. The act would then perform multiple tracks on the project live for the radio executives, followed by an informal, online chat with the on-air personalities.

After a single has been released to radio, nearly everyone at the label tracks its progress. For the promotion staff, the monitoring tools and charts discussed in this chapter help guide their work. Some labels also have staff members who monitor the "buzz" on their new music on social media and in blogs. All this information helps the label determine whether they have a hit on their hands, whether they should step up the promotional effort, or whether it is time to cut their losses and move on to the next project. Importantly, streaming success is now a part of the decision-making process in the label determining the potential trajectory of a track. Many labels now use streaming data as an early indicator of potential success at radio. Conversely, many radio station executives will monitor stream activity for a single to inform their decision to add it.

PROMOTION DEPARTMENT

An important component of promoting a recording to radio is the effectiveness of the record company promotion department or the independent promoters hired to get radio airplay. Again, the promotions team is tasked with convincing the key personnel at stations to add the record and, later, to increase the spins. This would appear to be a simple process, but the competition for space on playlists is fierce. Thousands of recordings are sent to radio stations every year, and the rejection rate is high because of the limited number of songs a station can program for its audience. The promotion team must convince radio to add the song and, once added, to increase the number of times the song is played during the day.

Record promotion has been a big-dollar investment, which made it a key marketing element necessary to stimulate consumers' interest in buying new music. Costs for a major label to market and promote new music can easily reach $1 million or more, not including the production of the recording or any advances to the artist or producer.

Most large labels have a promotion department whose sole purpose is to achieve the highest airplay chart position possible. While most consumers assume a number one song is the biggest seller at retail, the number one song on many *Billboard* charts is the song that has the most airplay on radio, or it may be a combination of sales, airplay, streams, and/or video views. Even with the rise in popularity of streaming, radio airplay is still a very important component to a marketing plan, and this is particularly true in some radio-focused genres such as country.

Labels often have a senior vice president of promotion who usually reports directly to the label president (major labels may have an executive VP of promotion to whom the senior VP reports) and oversees all radio promotion across the label imprints. At pop music labels, there is often one VP who handles each genre, whereas in country, there is often a national director of promotion or VP, with 4–6 "regionals" reporting to that position (see Appendix). These regional promotion people are viewed by the label as field representatives of the promotion department. They are liaisons to key radio stations in their region, and they, along with the VP, are the front line for the label attaining airplay. It is their responsibility to create and nurture relationships with programmers and influencers for the purpose of convincing them to add the company's recordings to their playlists. Some major labels have floating vice presidents, with each imprint having a VP or senior VP who oversees the regionals.

It cannot be overstated that the success of a record promotion person and, in turn, the success of a recording hinges on the strength of the relationship he or she has with key radio programmers. For that reason, the business of a radio rep at a label is knowing everything they can about who makes the decisions at every local station in their region or in the corporate office. The promoter must know who the final decision maker for new music is, as well as who the influencers are at the station. The promoter will understand what type of sound each of those influencers like and what motivates them.

The availability of real-time airplay data has made the work of the promoter a constant process, as they are continually analyzing successes and failures of singles and planning their next move with the stations. Every week, custom reports are pulled by the label's marketing and promotions teams from all the major services and charts, including Music Connect, Mediabase, BDS, Spotify, Amazon, Apple Music, and more. The promotions team will review airplay, sales, stream activity, and chart movement for the singles it is pushing. When available, this data is reviewed for the specific markets the promotions person handles. Promoters will often use data to show how well the song is streaming in the local market or to demonstrate how the song is performing in similar markets.

Making a Data-Driven Case

The promotions team for Olivia Rodrigo's "Drivers License" would have likely used charts such as the following to convince radio stations to add or increase spins on the song. Note that the single was the most streamed song during the week in February, while it ranked only 27th in radio airplay. That discrepancy could have been shared with the radio station's PD and MD as the basis for an argument that they should "get on board" with the single, as it was clearly resonating with the audience.

The following trending data also demonstrates the momentum the single had in weeks one through five.

Using BDS, the label team can quickly see the stations in the largest markets that are not playing the single, providing a quick snapshot of the stations that the label might want to prioritize for a potential add.

Chart with Six Weeks for Adult Contemporary
Fri Feb 12 '21 - Thu Feb 18 '21 / 87 Stations

Select a format
AC: Adult Contemporary

															Current	Recurrent	
														Songs%	4.8	1.5	
														Spins%	9.6	6.2	

Title	Artist	TW Airplay Rank	LW Airplay Rank	Peak Chart Rank	Aud Rank	Aud +/-	US av Stream Rank	Nsh a Stream Rank	CRG	Play	Add	download song	Label	Stns 5	+/-	TW Airplay Spins
drivers license	Olivia Rodrigo	㉗	31	30	28	26k	①	1	c	◄	+	↓	Geffen/Intersc...	8	+22	60
Blinding Lights	The Weeknd	1	1	1	1	3.1m	2	2	c	◄	+	↓	XO/Republic	86	-39	2377
Love Story (Taylo...	Taylor Swift	㉜		19		292k	③	20	c	◄	+		Republic	8	+29	29
Levitating	Dua Lipa Feat. D...	16	16	15	18	-33k	4	5	c	◄	+		Warner Records	12	+13	182
willow	Taylor Swift	8	8	8	8	1.4m	5	6	c	◄	+	↓	Republic	57	+70	768
HOLY	Justin Bieber Fea...	10	11	10	10	834k	6	4	c	◄	+	↓	Raymond Brau...	26	+46	496
Bang!	AJR	㉕	24	24	21	2.7k	⑦	7	c	◄	+		AJR/BMG/S-Cu...	5	+10	68
I Hope	Gabby Barrett	6	6	3	4	1.8m	8	3	c	◄	+	↓	Warner Music ...	72	-28	1542
Circles	Post Malone	4	4	1	6	76k	9	12	c	◄	+	↓	Republic	82	-138	1726
Before You Go	Lewis Capaldi	2	3	2	2	2.5m	10	10	c	◄	+	↓	Vertigo/Capitol	79	+23	2070

FIGURE 11.29   *Adult Contemporary Rankings*

Drivers License – Olivia Rodrigo							
	% Chg	This Wk	Last Wk	3 Wks	4 Wks	5 Wks	6 Wks
Rank		1	1	1	1	1	1
Streams	-16.8	26m	32m	38m	51m	78m	54m
On-Demand Streaming Rank and Streams for this Report							

FIGURE 11.30   *Olivia Rodrigo's "Drivers License" Trend*

They may also wish to share any key promotional efforts that are planned to increase awareness of the song, such as any tours scheduled, the song's inclusion in advertising campaigns, or other paid marketing efforts. Armed with this data, the promotions team develops a plan for targets to add the music or conversion targets (those stations they might be able to convince to jump from light to moderate rotation, for example).

FROM THE RADIO STATION'S PERSPECTIVE

According to radio veteran and Nashville-area VP/Market Manager for Cromwell Media, Dennis Gwiazdon, "most new music, unless it is from a superstar artist, will either start in overnights (12–5 AM) or at night (7PM–12M) to determine its legitimacy and acceptance by the audience. If the song receives a positive reaction (such as requests, listener comments on the station's website, or through social media, M-scores, research, BDS streaming, etc.) then it could move up to a light or even medium rotation. Songs that eventually make it to the 'power category' (the highest number of spins) have typically checked every box as a possible hit and have become part of the station's core library. After a song has peaked in power, most get put into **recurrent** status and become part of the station's regular library. Only a few songs each year – maybe 10–15 – make it to that level."

drivers license - Olivia Rodrigo   🖨 ✖

		Spinning				**Not**	
Spins	Calls	Metro	Market		Calls	Metro	Market
10	WKJY	20	Nassau-Suffolk		WLTW	1	New York
6	WHUD	41	Hudson Valley		KOST	2	Los Angeles
4	WRVE	67	Albany/Schenectad...		WSHE	3	Chicago
1	WLEV	74	Allentown/Bethlehem		KOIT	4	San Francisco
1	WDEF	89	Chattanooga		KODA	6	Houston
14	WAJI	116	Ft. Wayne		WASH	7	Washington, DC
1	WCRZ	140	Flint		WSB	8	Atlanta
23	WGFB	161	Rockford		WBEB	9	Philadelphia
					WMJX	10	Boston
					KRWM	12	Seattle
					WNIC	13	Detroit
					KESZ	14	Phoenix
					KYXY	16	San Diego
					WMTX	17	Tampa/St. Petersburg
					WWRM	17	Tampa/St. Petersburg
					KOSI	18	Denver
					KKCW	21	Portland, OR
					WLIF	22	Baltimore
					KEZK	24	St. Louis
					KYMX	27	Sacramento
					KBEE	28	Salt Lake City
					WSHH	29	Pittsburgh
					KSNE	31	Las Vegas
					KKMJ	32	Austin
					WREW	33	Cincinnati

FIGURE 11.31 *Olivia Rodrigo's "Drivers License" Spin Report*

Although every station is different, the general goal of the promotions team is to convince the station decision makers to move a single up the different levels of rotation. For example, an overnight position might give the single one spin per night. Once the single is bumped up to light rotation, it might receive 15–20 spins, media rotation might provide 20–35 spins, then a power rotation to provide anywhere from 35–72 spins.

Throughout all the negotiations, radio and the label each must be very careful not to engage in **payola**, the illegal practice of "pay for play" by providing payment or other valuable goods like trips and cars to a radio station or to the personnel who work there in exchange for increasing the number of spins on the station when that music is being featured as part of the station's normal day's broadcast. The goal of the promoter is to provide marketing support for the music while staying within the parameters of the law, and there are certainly legitimate marketing efforts around the project that are expected. These might include in-market appearances and free concerts for the local audience, private acoustic performances for a winner and ten friends, flyaways to see the artist perform at an exotic location, meet-and-greet opportunities at a local concert, autographed merchandise, and other items that provide value to the station and/or the station's listeners. Radio groups and conglomerates generally seek the type of unique experience that is truly priceless.

Independent Promoters

Labels sometimes hire independent promoters, or "**indies**," to augment their own promotional efforts. Since promoting songs for airplay relies on well-developed relationships, indies may have developed stronger relationships with some key stations than the label has, and the record company is willing to pay indies for the value of those relationships. Some successful promoters are very selective in what projects they accept to push to radio, so there is a level of trust that develops between the promoter and station that the work they are representing is worthy of the station's consideration.

In the 1980s, a practice of labels hiring independent promoters to attain airplay at radio came under very tight scrutiny by federal law enforcement. Fred Dannen, in his book *Hit Men*, found that CBS Records was paying $8–10 million per year to indies to secure airplay for their acts. By the mid-80s, he says, that amount was $60–80 million for all labels combined. The resurgence of independent promotion in the 1990s caused the New York attorney general in 2004 to look closely at these lucrative agreements, with the result that many indies shuttered their doors or went into other businesses.

Independent promoters traditionally made their money this way: they would sign an agreement with a radio station to be the stations' exclusive consultant on new music for a year and then pay the radio station for that right. The indie promoter did not require the station to play specific songs, according to the typical agreement, but the station did promise to give its playlist for the following week to the indie promoter before anyone else. Then, the promoter would send an invoice back to the record company for $3,000 to $4,000 for each single that was charted on stations that they represented. Because of this "exclusive" arrangement with the radio station, record label promotion people had no dealings with the radio stations represented by indies. One label promotion vice president said the $4,000 can easily turn into $30,000 per single if the indie had to provide contest prizes to help promote the single at radio and to pay for other promotional expenses. Many arrangements with indie promoters included provisions for bonuses based on their success at charting records with individual stations (Phillips 2002).

The use of independent promoters provided the record company a layer of insulation between themselves and radio. The record companies did not deal directly with certain radio stations in matters of adds and spins; rather, they dealt only with independent promoters who promote to these stations. As one executive said, "The use of independent promoters creates a clearinghouse by removing the label one step away from . . . making any compromises that some might make to get a song on the air." Another says, "This way the money doesn't go directly from the label to the radio station" (Personal interviews, fall 2004).

Critics of the independent promotion system claimed it had tended to shut off access to radio airplay by independent labels and artists. Alfred Liggins is the CEO of Urban One, Inc. (formerly known as Radio One), one of the largest urban market radio companies in the country. Liggins acknowledged in a 2002 interview that their exclusive relationships with independent promoters meant that labels without an indie promoter were less likely to get a record played on his stations (ABC Television, 20/20, 2002). Nearly every record label has discontinued its relationship with independent promoters who pay radio stations for access.

Satellite Radio Promotion

Satellite radio has expanded the horizon for promoting recordings to radio. The services provided by Sirius XM have a relatively low monthly subscription fee and offer more than 100 music channels. Though the service requires that a vehicle or home be equipped with a special receiver or subscription with online computer service, most newer cars and trucks have satellite receivers integrated into their standard radios. Satellite radio delivers an audience of approximately 35 million subscribers and offers exclusive talk, live news, major sports events, and commercial-free and curated music channels with sub-genres and niche formats, which gives labels great opportunity for airplay. Record company promoters actively work with satellite music channel programmers seeking adds to their playlists of current music, and the company's merger with Pandora has made them the largest audio entertainment company in North America. Although most of the music-oriented stations are prohibited from accepting any type of paid promotional effort, there are many creative ways to gain exposure on those stations, from having the artist record liners for the stations to use to having them guest host in a program prior to a new music release.

A remote broadcast is often organized by producers of live events (such as televised award shows) to generate a significant amount of promotion within the few days leading up to the airing of the special. Multiple broadcasters such as radio stations, syndicators, radio groups, online media outlets, and broadcast television stations are each given dedicated space within a larger room so they may capture liners and interviews from many of the artists participating in the event. For example, an artist may visit the booth of a top radio station for a five-minute chat before they are whisked away by their escorts to visit a Spotify booth, five minutes after which they may be moved to a booth hosted by CMT talent, as pictured in the following.

FIGURE 11.32   *CMA Awards Broadcast*

FIGURE 11.33   *CMT host Cody Alan with Old Dominion at the CMA Awards*

GLOSSARY

Add date—This is the day the label is asking that the record be added to the station's playlist.

Average quarter-hour (AQH)— The number of people listening to a radio station during a 15-minute period as measured by Nielsen Audio is called the AQH.

Cume—The total of all different listeners who tune into a particular radio station is its cume.

Format—The kind of programming used by a radio station to entertain its audience is the format.

Heavy rotation—These recordings are among the most popular songs played on a radio station.

Indie—A shorthand term meaning an independent record promoter who works for radio stations and record labels under contract.

Light rotation—These are recordings that are played fewer times on a radio station than songs in heavy rotation.

Payola—The illegal practice of giving and receiving money in exchange for the promise to play certain recordings on the radio without disclosing the arrangement on the air.

P-1—The primary core of listeners to a specific radio station are P-1s.

Playlist—The weekly listing of songs that are currently being played by a radio station.

Program director—This is an employee of a radio station or a group of radio stations who has authority over everything that goes over the air.

Radio promotion tour—An effort by the artist to visit (either in person or virtually) radio station executives and on-air personalities to showcase their new music.

Recurrents—Songs that used to be in high rotation at a station but are now on the way down, reduced to limited spins.

Rotation—Mix or order of music played on a radio station. This may also refer to the range of spins a single receives (the level of rotation).

Terrestrial radio—This is traditional commercial radio broadcasting using a transmitter and tower consisting of AM, FM, and HD radio

Share—A share is the radio audience of a specific station measured as a percentage of the total available audience in the market.

Spin—This is a reference to the airing of a recording on a radio station one time. "Spins" refers to multiple airings of a recording.

TSL—"Time spent listening" by radio station listeners at particular times of the day.

Trades—This is a reference to the major music business trade magazines.

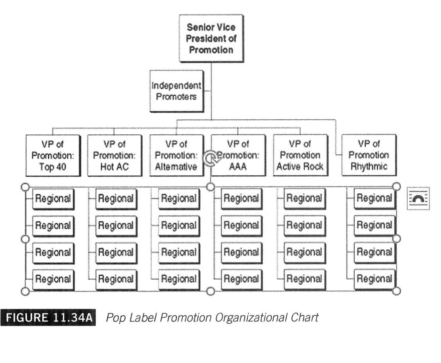

FIGURE 11.34A *Pop Label Promotion Organizational Chart*

FIGURE 11.34B *Country Label Promotion Organization Chart*

APPENDIX

In Figure 11.34a is an example of an organizational chart for a typical pop label record promotion department. Note that the VPs of the various formats are specific to the music. However, the individual regional promoters typically promote all current music types marketed by the label.

BIBLIOGRAPHY

ABC Television, 20 Nov. 2002, www.abcnewsstore.com/store/index.cfm?fuseaction¼customer.product&product_codeT020524%2002.

Albright, J. "The PPM Music Test 'MScore Switch' Is On." Presentation at A&O Pre-CRS Seminar, 3 Mar. 2009, Nashville, TN, Oxenford and Federal Communications Commission.

"Arbitron Terminology." http://www.arbitron.com/Downloads/terms_brochure.pdf, Nielsen Audio, 2013, www.arbitron.com/downloads/terms_brochure.pdf.

Barbington, C. "Big Radio Settles Payola Charges." *Washington Post*, 6 Mar. 2007, p. D1.

"BDSradio." *BDS Radio*, www.bdsradio.com. Accessed 23 Mar. 2021.

BDSRadioCharts.com Week of 2/13/2021.

BIA Advisory Services. "Radio Station Advertising Revenues in the United States from 2006 to 2019, by Source (in Billion U.S. Dollars)." *Statista*, 24 June 2020, www.statista.com/statistics/253185/radio-station-revenues-in-the-us-by-source/.

Charts generated by Dave Tyler of ENCO. ENCO information from interview with Shane Finch of ENCO.

"Edison Share of Ear 2020 in Car Audiences Rebound." *Insider Radio*, 10 Nov. 2020, www.insideradio.com/free/edison-share-of-ear-q3-2020-in-car-audiences-rebound/article_742967a6–2.32b-11eb-b1ae-4b5231251273.html.

Federal Communications Commission. FCC, n.d., www.fcc.gov/guides/review-broadcast-ownership-rules. Accessed 30 July 2014.

Freed, A. 2005, www.historychannel.com/speeches/archive/speech_106.html.

Fritz, Jose. "Nielsen Radio Survey (Part 5)." *ARCANE RADIO TRIVIA*, 12 Apr. 2019, tenwatts.blogspot.com/2019/04/nielsen-radio-survey-part-5.html?m=1.

Hoovers, Inc. "Company Reports." *CC Media Holdings, Inc.*, 17 Feb. 2009.

Hull, G. *The Recording Industry*, London: Routledge, 2014, pp. 201–202.

Katunich, L. J. "Time to Quit Paying the Payola Piper." 22 *Loyola of Los Angeles Entertainment Law Review*, REV. 643, 654, 29 Apr. 2002

Marketing Charts. 2014, www.marketingcharts.com/wp/radio/radio-ad-revenue-growth-forecast-downgraded-28169/attachment/biakelsey-radio-indus-advertising-revenue-2006–2017-mar2013/.

McVay, Mike. "McVay Media." www.mcvaymedia.com.

"MRC Data – Sign In." *Music Connect*, www.musicconnect.mrc-data.com. Accessed 23 Mar. 2021.

Nielsen Audio Today 2018, 2019, 2020.

"Nielsen Radio PPM Markets." *Nielsen Audio*, 2020, worldradiohistory.com/Archive-Arbitron/Sample-Targets/Radio%20PPM%20Markets_Daily%20In-Tab%20Targets%20Oct18.pdf.

Nielsen Total Audience Report: Working from Home Edition Aug 2020, www.pandora.com/about.

Oxenford, David. "On the 15th Anniversary of the Telecommunication Act of 1996, the Effect on Broadcasters Is Still Debated." *Broadcast Law Blog*. Ed. David Oxenford, 9 Feb. 2011. Path www.broadcastinglawblog.com. Accessed 31 July 2014.

Phillips, C. "Logs Link Payments with Radio Airplay." *LA Times*, 29 May 2001, Radio Advertising Bureau, and rab.org.

Phillips, C. "Congress Members Urge Investigation of Radio Payola." *LA Times*, 24 May 2002.

"Radio Genre Formats." *About/Radio-Genre-Formats*, 2020, https//newsgeneration.com/broadcast-resources/guide-to-radio-station-formats/.

Radio Today: How America Listens to Radio. "Arbitron Report. Recording Industry Association of America." 2008.

"Sony Settles Payola Investigation | New York State Attorney General." *Attorney General of New York*, ag.ny.gov/press-release/2005/sony-settles-payola-investigation. Accessed 24 Mar. 2021.

Williams, G. *Review of the Radio Industry, 2007*, Federal Communications Commission, 2007, http://hraunfoss.fcc.gov/edocs_public/attachmatch/DA-07-3470A11.pdf.

CHAPTER 12

The Charts

This chapter looks at charts of all types from the viewpoint of their value to marketers in promoting music to the gatekeepers. From a historical perspective, music charts have been among the most valued information provided by **trade** magazines, which is why our discussion of charts begins with the trades.

THE HISTORY OF TRADE MAGAZINES

Billboard Magazine has, for more than a century, been the leading music business publication, providing comprehensive weekly views of the recording industry, the music business, and commercial radio.

Billboard Magazine was first introduced in 1894 as a publication supplying its readership with information about advertising and, a few years later, about the carnival industry. Its reporters then began writing stories about sales of sheet music, and early in the last century, it added regular features about silent films and commercial radio. In 1936, it published its first "hit parade," which was a term used at the time to rank popular songs and then became a term used by radio to denote its most popular music. In 1940, *Billboard* compiled and published its first music popularity chart, called "Best Selling Records Chart." The first national number one recording reported by *Billboard* was Tommy Dorsey's "I'll Never Smile Again" with vocals by Frank Sinatra. It was the decade of the 1940s when *Billboard* became a leader and ultimately the icon for music charts by publishing numerous charts for various genres of music. The *Billboard* Hot 100, which ranked single releases based on both sales and airplay, was launched in 1958. That chart remains one of the continuing staples of the magazine (Billboard.com).

Other national trade magazines have competed with *Billboard* over the years, including the *Gavin Report*, which closed in 2002; *Cashbox*, which closed in 1995; and *Radio & Records*, which closed in 2009. *Gavin*, *Cashbox*, and *R&R* magazines were all key industry trade publications for many years until they were retired for economic reasons. *Gavin* and *Cashbox* relied upon "reported" airplay by radio stations, which gave the publications their airplay charts for the coming week, whereas *Billboard* (currently owned by MRC, a media company that also owns Dick Clark Productions and *The Hollywood Reporter*), now relies on electronically monitored airplay and other data by Broadcast Data Systems to generate the charts.

Another data provider for charts is MediaBase, and although it does not publish a hard copy trade magazine for distribution, its data is published weekly in *USA Today* and is used by many entertainment outlets, such as the nationally syndicated show *America's Top 40 with Ryan Seacrest*.

THE IMPORTANCE OF CHARTS

Music charts are a measurement of the momentum of a music track, album, or video. It is a way to determine the overall popularity of songs, and a high chart position is perhaps the most important piece of real estate a record label can own on a chart because of its influence on gatekeepers. For example, a top-streaming song on a prominent digital chart may influence a radio programmer's decision to add music to its airplay list or a decision by a respected blogger to mention it online. Credible, influential charts of all types are key tools used by record labels to promote music to people who can provide access to their audiences. The chart position can have the effect of demonstrating "word of mouth" interest between and among audiences and give a label's promotion department the information they need to encourage industry decision-makers to engage the music. The higher the music moves up in the charts, the greater the opportunity to expand the artist's fan base, which may result in an even higher number of streams, increased video views, and improved live ticket sales. A high chart position helps catch the eye of many interested parties, all of which generates additional exposure for the artist and the music, thereby building the artist's brand value.

Charts are especially important for use by radio programmers to indicate the music that is gaining traction in their market, as evidenced by increased streams, or to give them a basis to compare their audience offerings with those audiences of similar cities. If radio programmers see a song quickly climbing the charts, they may decide to add it to their station's playlist. In fact, promoters at the label will often present this type of data to radio as a way to convince them to add a single or increase **spins** on tracks that are seeing increased streams in the marketplace.

CREATING THE CHARTS

Charts are compiled using a variety of criteria. While there are some charts that still focus exclusively on radio airplay, most charts today are inclusive of all the ways in which fans consume music, including digital and physical sales, digital downloads and streams, radio airplay, and video views.

Nearly every major music genre has an airplay chart in *Billboard*. The airplay charts compile the national airplay of singles on radio stations as detected by a proprietary system now owned by MRC, called Broadcast Data Systems. Most genres are also represented in MediaBase airplay charts, but first let us look at BDS. Although labels are concerned with getting airplay on all stations within the format, their focus centers around the stations that report to BDS/*Billboard* and Mediabase because that is how the charts are derived.

Before we dive into chart specifics, let us first recap the changing data analytics and ownership landscape, as there have been significant moves within the space in the last few years. In 2019, Nielsen's music data and analytics division (which included BDS and the database formerly known as Soundscan) was sold to Valence Media, the parent company of *Billboard*, *Hollywood Reporter*, Dick Clark Productions, *Spin*, *Vibe*, and *Stereogum*. In 2020, Valence Media merged with MRC, and the joined companies took on the MRC name in a rebranding effort. This merger now puts all the previous products under the MRC umbrella.

UNDERSTANDING THE *BILLBOARD* CHARTS

Billboard is a leading music entertainment brand owned by Billboard-Hollywood Reporter Media Group, a division of MRC. *Billboard* is known for its published pieces, including *Billboard Magazine*, as well as its music charts that cover virtually every area or format of the business. As the industry has changed over the years, so has the way *Billboard* charting is determined. While the *Billboard* charts were initially focused on airplay only, the media company began including digital downloads in 2005, digital streams in 2007, and YouTube video data in 2013. *Billboard* publishes numerous weekly charts as well as reporting yearly and decade-focused charts for the

Hot 100 and Hot 200 charts. Many weekly charts today represent a variety of consumer access methods. For the most part, charts reflect sales and airplay between Friday and Thursday of any given week. An exception to that rule may be found in mixed-data charts, such as the *Billboard* Hot 100, which also uses an airplay cycle of Monday through Saturday. Updated charts are published every Tuesday on Billboard.com and reflect the date of the *Billboard* issue in which they appear.

Billboard Top 200 Chart

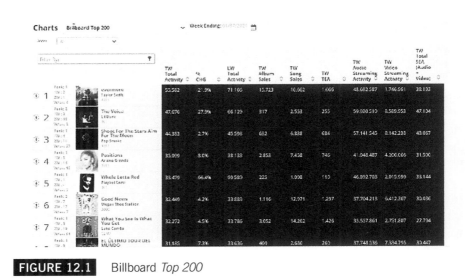

| FIGURE 12.1 | Billboard *Top 200* |

BROADCAST DATA SYSTEMS AND *BILLBOARD* CHARTS

The engine behind *Billboard* is Broadcast Data Systems, also owned by *Billboard*'s parent company, MRC. Simply put, BDS is a system that digitally "recognizes" tracks played on radio, television, and online, and it then records the number of spins. To be recognized by the system, a song first must be input into the system's computer by the label, which in turn creates a digital fingerprint of that song. The fingerprint is downloaded to BDS monitors, and as the computerized airplay monitor "listens" to a song being played on the radio, it compares its digital fingerprint to that on file and then logs it as a detected play of the song.

BDS monitors more than 1,200 stations in 130 radio markets in the United States and 22 in Canada to detect more than 100 million songs annually. BDS data is used by *Billboard*, *Airplay Monitor*, and *Canadian Music Network* magazines to compile the music charts they publish. Additionally, the service compiles detections of airplay and streaming on satellite and cable music channels. The information is tabulated and sold to subscribing customers like labels, management companies, and radio groups.

Billboard Alternative National Airplay Chart

This is an example of the *Billboard* Alternative National Airplay Chart for a recent week as measured by BDS. It provides the ranking for this week (a circle around the number means it either stayed the same or moved up in the rankings), followed by last week's ranking or LW. Also displayed are the

Monitored: Chart / New & Active / Most Added / Most Increased / Recurrent Chart
Indicator: Chart / New & Active / Most Added / Most Increased / Recurrent Chart

billboard Powered By BDS
ALTERNATIVE NATIONAL AIRPLAY

Issue Date 2/13/2021

TW	LW	WKS	ARTIST TITLE IMPRINT / PROMOTIONAL LABEL	PLAYS TW	+/-	AUDIENCE MILLIONS	RANK
			*** NO. 1 ***				
1	3	17	CAGE THE ELEPHANT Skin And Bones (Mix 2020) RCA *1 week(s) at number 1*	2064	+186	4.261	4
2	1	13	BILLIE EILISH Therefore I Am DARKROOM/INTERSCOPE	1375	-19	4.556	2
3	5	16	MACHINE GUN KELLY X BLACKBEAR My Ex's Best Friend EST19XX/BAD BOY/INTERSCOPE	1362	+157	4.805	1
4	4	38	ALL TIME LOW FEAT. BLACKBEAR Monsters FUELED BY RAMEN/EMG	1765	-49	4.367	3
5	2	24	CANNONS (ELECTRONIC) Fire For You COLUMBIA	1757	-162	3.825	5
6	7	14	GLASS ANIMALS Heat Waves WOLF TONE/REPUBLIC	1503	+43	2.725	10
7	6	21	ROYAL & THE SERPENT Overwhelmed ATLANTIC	1436	-157	2.864	8
8	8	19	ROYAL BLOOD Trouble's Coming WARNER RECORDS	1274	-67	1.936	13
9	9	36	MACHINE GUN KELLY Bloody Valentine EST19XX/BAD BOY/INTERSCOPE	1157	89	3.207	6
10	12	44	TWENTY ONE PILOTS Level Of Concern FUELED BY RAMEN/EMG	1089	-9	2.958	7
11	10	27	PEACH TREE RASCALS Mariposa HOMEMADE PROJECTS/TENTHOUSAND PROJECTS/CAPITAL	1076	165	2.780	9
12	13	17	GRANDSON Dirty FUELED BY RAMEN/EMG	1058	-11	1.573	18
13	16	16	CLAIRO Sofia FADER LABEL	1047	+122	2.562	11
14	11	25	I DONT KNOW HOW BUT THEY FOUND ME Leave Me Alone FEARLESS/CONCORD	1012	-190	1.647	16
15	15	5	KINGS OF LEON The Bandit RCA	1005	+61	1.531	19
16	17	10	FOUSHEE Deep End RCA	958	+114	2.422	12
17	20	3	WEEZER All My Favorite Songs ATLANTIC/EMG/CRUSH MUSIC	922	+229	1.581	17
			◄ MOST INCREASED PLAYS ►				
18	23	2	FOO FIGHTERS Waiting On A War RCA	900	+351	1.824	14

FIGURE 12.2   *Alternative Airplay Chart*

number of weeks the song has been on the chart, the artist, title, label, the total number of plays this week, the increase or decrease in the number of plays for the week, the audience (in millions), and the ranking by audience.

BDS gives the label's marketing department considerable information about how singles are performing and where. Combining this information with Nielsen data, label marketers have continuing feedback on the performance of their recorded music projects. And, most importantly, this feedback gives marketers the information needed to modify marketing plans to draw as much commercial activity out of the marketplace as possible.

BILLBOARD CHARTS ARE LARGELY BROKEN INTO SEVERAL CATEGORIES

Sales data charts are compiled by Music Connect/MRC Data from a sample of a wide variety of retailers across the United States including music stores, music departments at electronics retailers and mass merchandisers like Wal-Mart and Target, and direct-to-consumer transactions of both physical albums, as well as digital downloads. The data used in creating these charts represents more than 90% of the U.S. music retail market. The system uses the same point-of-sale information that music merchants use to maintain their sales and inventory, so an itemized receipt serves this dual purpose. The Digital Song Sales chart is an example of this type of chart.

Airplay data charts are compiled using Music Connect/MRC Data that electronically monitors the digital fingerprints of each track across more than 140 markets in the United States. Charts may be based on spin count – the number of times a song is played weekly on monitored radio stations within a specified period of time (such as the Top 40 Charts for each genre) – or may incorporate the size of the audience when the spin occurs. For example, a song that is played on a top-ranked station in a large market like Chicago is "worth" more in the tabulation of those charts than a spin that occurs on a station in a smaller city because it is reaching a larger number of listeners. Also, a song played during morning drive (when there are a lot of people listening) is worth more than a song being played on the same station at 3 a.m. The charts using an audience method are Radio Songs, Rock Airplay, Country Airplay, R&B/Hip-Hop Airplay, Rap Airplay, and Latin Airplay.

Streaming data charts utilize radio streams, on-demand songs (from listener-controlled channels), and video plays. The Streaming Songs chart, for example, ranks the top streamed radio and on-demand songs and videos.

More on the *Billboard* charts at the end of this chapter.

The *Billboard* Hot 100 is the U.S. music industry standard record chart in the United States. It has been a part of the *Billboard* chart offerings since 1956. The weekly chart is developed using a formula that combines online streams, radio airplay, and sales (physical and digital), regardless of the genre of music or the radio format in which the song is programmed. The Hot 100 chart has spawned numerous other genre-specific charts.

Mixed data charts combine airplay, streaming, and sales information to give a comprehensive view of the popularity of singles. Charts that use this

FIGURE 12.3   *Hot Bubbling Under Chart*

mingled data include the Hot 100, Hot R&B/Hip-Hop Songs, Hot Rap songs, Hot Country, Hot 100 Bubbling Under chart, and more.

The *Billboard* Hot Bubbling Under Chart identifies songs that are gaining traction but have not yet charted on the Hot 100. The ranks are compiled using radio audience, streaming activity, and sales.

Other *Billboard* Charts

There are a few other important *Billboard* charts that are based on data from other sources. For example, the *Billboard* Social 50 chart (temporarily

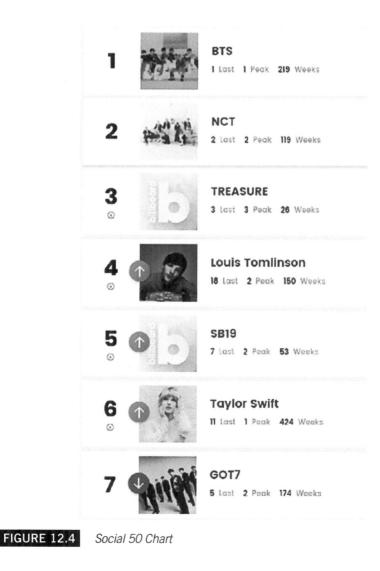

1   BTS
1 Last   1 Peak   **219** Weeks

2   NCT
2 Last   2 Peak   **119** Weeks

3   TREASURE
3 Last   3 Peak   **26** Weeks

4   Louis Tomlinson
18 Last   2 Peak   **150** Weeks

5   SB19
7 Last   2 Peak   **53** Weeks

6   Taylor Swift
11 Last   1 Peak   **424** Weeks

7   GOT7
5 Last   2 Peak   **174** Weeks

FIGURE 12.4   *Social 50 Chart*

paused) ranks the most popular artists on Instagram, Facebook, Twitter, YouTube, and Wikipedia, blending weekly additions of friends, fans, and followers with artist page views and engagement. Another example is the Dance Club Songs chart, which is compiled using data submitted from a nationwide panel of club DJs.

Award Designations

Within the charts, *Billboard* indicates recognition for accomplishments each week with the use of special awards. Designations are as follows (*Billboard*, 2021).

Table 12.1 *Billboard* Chart Designations	
Designation	**Description**
Bullet	Greatest weekly gains.
Hot Shot Debut	A chart's highest-ranking new entry.
Greatest Gainer (GG)	Includes the title with the chart's largest unit increase (for album charts) or the largest increase in plays or audience (for airplay charts).
Airplay Gainer (AG)	On mixed charts, indicates the largest increase in radio audience.
Sales Gainer	On mixed charts, indicates the largest sales unit increase.
Streaming Gainer (SG)	On mixed charts, indicates largest increase in streaming.
Airpower	On airplay charts, indicates titles appearing in the top 20 for the first time with increased plays and audience.
PaceSetter (PS)	On album charts, indicates biggest percentage growth.

MEDIABASE

Mediabase, a division of iHeart Media, is a competitor to BDS. The music industry service monitors the airplay of recordings on more than 1,800 U.S. and Canadian radio stations in approximately 160 markets, and it publishes its data weekly in *USA Today*. Mediabase also incorporates listening behavior and reactions through Mscore from Media Monitor and music influencer HitPredictor. In addition to being a valuable resource to labels, management, and other industry executives, this data is also used by television music programs and radio countdown shows.

These stations that provide data to BDS and Mediabase are referred to as "reporting stations." Generally, reporters are the top stations in the largest markets.

SPOTIFY CHARTS

This powerhouse music player makes numerous charts available to give an almost instantaneous read on music popularity around the world:

> *Stats may be pulled via the Viral Chart either Globally or by market. This chart measures the popularity of songs by day or by week. The Top 200 chart has the same options in that the rankings may be pulled globally or by market, as well as by day or by week.*

The Weekly Top 50 Charts highlight the biggest tracks and albums each week. The charts are published each Monday within the app and through the @SpotifyCharts Twitter account. Options include weekly album or song rankings either globally or in 47 countries and territories, including the United States.

The U.S. Top 10 Debuts and Global Top 10 Debuts gauges immediate momentum by ranking the popularity of songs and albums within the first 72 hours after release. Spotify for Artists also provides additional charts, including the list of Top 100 Songs (globally or by market).

FIGURE 12.5   *Spotify Viral 50*

YOUTUBE

YouTube's charts of the most popular music videos and artists create yet another opportunity for the record label to promote its music to gatekeepers.

As most users of the website have discovered, it provides an efficient way to find the most popular music videos in a variety of genres as well as videos of those who are rising stars, whether they are on independent labels or the majors.

Chart availability on YouTube includes:

Market-specific rankings of Top Tracks, Top Artists, Top Videos played most the week prior, as well as the Trending Chart highlighting new videos that immediately gained traction upon release. Each of these charts is available for 44 countries. Global charts available include Global Top Songs, Global Top Artists, and Global Top Music Videos throughout the world.

ITUNES/APPLE MUSIC CHARTS

In recent years, the music industry has applied a more favorable weight to paid streaming over free. Thus, the charts provided by Apple are important to labels, artists, managers, and other gatekeepers, given the company's 50 million paid worldwide subscribers to Apple Music. To that end, Apple Music has released playlists of the Top 100 Songs on Apple Music by country (115 market-specific charts) or the Top 100 Global Songs. These Top 100 playlist charts are updated daily at 12:00 a.m. PT and based on Apple Music streams. Other charts include the Top 200 Songs on iTunes, Top 200 Albums, Top Music Videos, and the iTunes The Voice Rankings.

AMAZON CHARTS

One of the oldest online retailers, Amazon, creates Best Sellers in Digital Music Top 100 lists that are updated hourly charts.

Digital Albums: Top 100 Paid and Top 100 Free (by genre)
Digital Songs: Top 100 Paid and Top 100 Free (by genre)

Their overall charts, Best Sellers in Albums and Best Sellers in Songs, are not genre specific like their others are. Unlike iTunes, Amazon sells digital versions of recorded music as well as physical CDs, vinyl albums, and used CDs. To the record label, the near real-time charts provided by Amazon can give the promotion and publicity departments another tool to use in acquiring more exposure for the music.

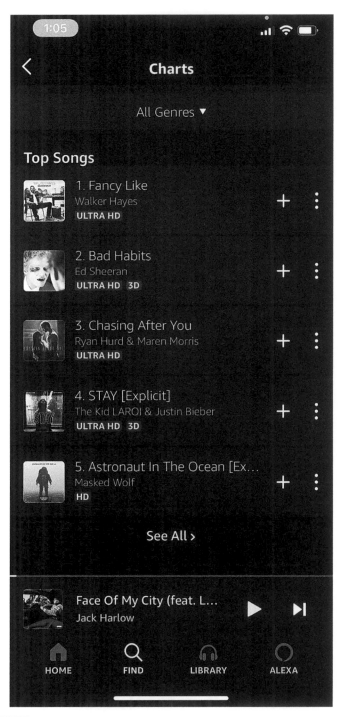

FIGURE 12.6   *Amazon Top Songs*

Amazon also charts top songs through its app. A dropdown feature allows for segmenting by genre.

OTHER CHARTS THAT MEASURE POPULARITY

TRACtion, a tool by Nashville-based CDX, uses song-recognition technology to monitor the online stream of more than 600 terrestrial radio stations in both major and secondary radio markets. The data is compiled is real time and is used by record labels, radio promoters, and other industry executives who need radio airplay measurement of areas that are not serviced by the traditional radio station monitoring services such as BDS and Mediabase. TRACtion powers the Americana chart, among others.

For years, CMJ charts were the industry standard for college radio airplay. After CMJ's dissolution in 2017, several other chart organizations emerged to fill the gap left by CMJ's departure:

- North American College & Community Radio Charts (NACC) publishes the Weekly NACC 200 and the Top 30 Genre charts. The charts are developed using airplay data from more than 200 participating college radio stations across North America.
- Spinitron is an online playlist management tool used by non-commercial ("non-comm") and educational radio stations. It publishes charts based on the activity of the stations using its services.
- Muzooka offers both weighted (larger stations are assigned higher value) and non-weighted (all stations are treated equally regardless of size) radio charts for college stations.

Big Champagne is a technology-driven service that was purchased by Live Nation in 2011. The company publishes a comprehensive chart that is based on a combination of sales, radio, streaming, and social media called the Ultimate Chart.

Rate the Music (RTM) is an internet-based music research effort developed and published by All Access Media Group that acts like Call Out Research performed by radio. The chart tests the Top 30 songs in each radio format as determined by Mediabase and uses internet listeners to rate song familiarity, familiarity percentage, and **burn** (a term that is used industrywide to indicate when an audience is getting tired of a particular song).

ADDITIONAL SOURCES TO TRACK MUSIC

Pollstar magazine is a primary source of concert and livestream information. *Pollstar* tracks the popularity of recording artists based on their ticket sales at performance venues and paid livestream views. It is a subscription-based publication that lists artists, venues, ticket sales percentages, and gross revenue from single appearances, but some of the basic data and charts are available for free on Pollstar.com. Continued sell-outs by an artist or an impressive number of livestream views can make an important statement and talking point about an artist's growing fan base and popularity. For example, a label promotion person

FIGURE 12.7   *Pollstar Live Chart*

(Source: Pollstar.com)

might use the information from *Pollstar* to tell a programmer that the group sold out its three-day run at a local venue. LIVE75 is *Pollstar's* main chart that ranks the average ticket sales for active tours over the previous 30 days.

The number of sources available to the label to help promote music past the gatekeepers continues to grow, some of which include a fee to subscribe for their services. Social networking sites track views of videos and streams of songs with data that is available to the label at no cost. Counting "Likes" and "Followers" on Facebook, Twitter, and Instagram and streaming activity on YouTube, TikTok, and VEVO contributes to the overall social media action of an artist or their project. Chartmetric and other analytics companies have created interfaces to help leverage this data in a comprehensive, easy-to-read manner. A more detailed deep dive on these tools may be found in the DSP chapter of the book.

A DEEPER DIVE INTO NIELSEN, MRC DATA, AND IMPORTANT TERMINOLOGY

The Nielsen Company has been providing important data to all kinds of products in the United States and abroad since the 1920s. In the mid-1930s, the company began to monitor and created an index of radio stations for the United States and replicated this concept for television in the 1950s as the format began to penetrate the U.S. marketplace. In doing so, the company established several key measurement devices that have aided in evaluating consumer behavior quarter over quarter, year after year. In doing so, Nielsen invented its own vernacular of terms that apply to the entertainment business that are important in evaluating the marketplace:

Designated Market Area – A term used by Nielsen Media Research to identify an exclusive geographic area of counties in which the home market television stations hold a dominance of total hours viewed. There are 210 DMAs in the United States These markets are listed in order of population density – meaning they are stratified largest to smallest by size of city. These markets apply to the analysis of music as well.

County Size – The classification of counties according to Census household counts and metropolitan proximity. There are four county size classes "A," "B," "C," and "D". In general, **"A" counties** are highly urbanized, **"B" counties** relatively urbanized, "C" counties relatively rural, and **"D" counties** very rural. Why is this important? These counties establish the zones

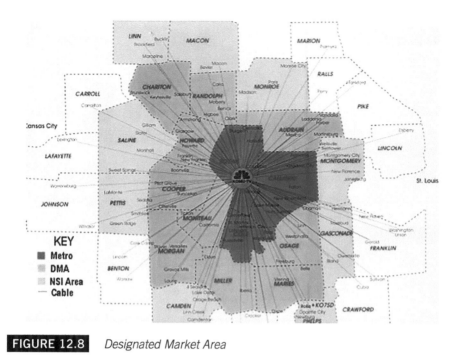

FIGURE 12.8 *Designated Market Area*

known as City, Suburb, or Rural and can help marketers determine strategic plans. As you look at this map of Columbia, Missouri, the DMA is outlined in pink, with urbanized counties A and B in hot and pale pink, relatively rural counties C in blue and very rural counties D that are serviced just outlined.

Other industry terms:

Geographic Regions – states are grouped by physical location, which can lead to sales and marketing analysis based on geographic data.

Store Types – As an industry standard, stores have been identified by types to aid the industry in identifying source of sales. These store types include Mass Merchants; Traditional Retailers, which include chain stores and electronic superstores; Independent Retailers; Non-traditional Outlets that include online and venue sales; and Digital portals.

Formats – The type of music format that consumer purchase continues to evolve – be it physical or digital. Physical formats include Compact

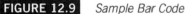 **FIGURE 12.9**   *Sample Bar Code*

Discs (CD), Vinyl (LP), Cassette, and DVD with Digital formats, including Downloads and Streaming.

Other technologies have aided in the data collection of the entertainment business and include the application of inventory management tools as well as digital management devices that allow for the important tracking and processing of royalty payments.

UPC – The UPC contains a unique sequence of numbers that identifies a product. GS1US standards recommend a GS1 company prefix number that identifies a unique code for a specific business. The Global Trade Item

Number (GTIN) initial numbers, which can vary in size depending on the variety of products the company sells, identify the business, with the last digits identifying the products. The small digit is the "check" digit, which validates the barcode equation.

Many record companies designate a five- to six-digit number that identifies the label. The record company then assigns a four- to five-digit product code that identifies the release, including artist and title of the album. The 11th digit can designate the configuration of the product. The last digit is known as the "check" digit. When scanned, a mathematical equation occurs determining if the product has been correctly scanned. The "check" digit is the "answer" to that equation, verifying an accurate scan. In the United States, the standard UPC contains 12 digits.

ISRC – For digital product, the ISRC is the international identification system for sound recordings and music video recordings. Just like the UPC, each ISRC has a unique and permanent sequence of numbers that identifies each specific recording that can be permanently encoded into a product as its digital fingerprint. The encoded ISRC provides the means to automatically identify recordings for royalty payments, key to publishers and songwriters alike. Because these numbers are embedded into the product, this coding system is perfect for the electronic distribution of music, with an ease of adoption into the international music community that has been reliable and cost effective.

Unlike the sophisticated algorithm of the UPC, the ISRC is an internal tracking code. The codes are simple, with country of origin, label or artist identifier, year of creation, and finally song designation number.

A sample code could look like: US ASM 15 00001

To capture sales data, the company that has rights to the digital track gives tracking agencies the ISRCs and should be embedded during production. ISRCs can also be provided by a digital retail portal such as CD Baby.

To learn more about UPCs, go to the GS1US website:
www.gs1us.org/resources/education-and-training/gs1-company-prefix-course
To learn more about ISRCs and the application process, check out the International Standard Recording Code website:
www.usisrc.org/

THE *BILLBOARD* CHARTS

Prior to the invention of barcodes, an archaic reporting system based on undocumented sales information was used by *Billboard* to produce the sales charts. Prior to 1991, *Billboard* had a panel of "reporting" retail stores that identified the best-selling record in their store, based on genre. Oftentimes, this information was not supported through actual sales data but was based on what store managers "thought" was their best seller. Record labels employed retail promotion teams to help influence these reports; hence, the sales charts were not always valid depictions of true sales throughout the nation.

Remember, this reporting structure was what was used prior to the creation of barcode scanning systems and the use of point-of-sales data. So capturing accurate sales data was difficult, even for the retailer. The use of barcodes or UPCs has greatly improved product management.

With the introduction of barcodes and efficient computer management of inventory, a new idea was introduced. Mike Shallet, an ex-record label promotion guy, along with Mike Fine, a statistician who had previously worked with major newspapers and magazines with a focus on surveys, conceived a revolutionary concept that would use this newfound technology to derive the top-selling records of the week. And in 1991, SoundScan was born.

Since its inception, this data source has been the foundation of the *Billboard* charts, based on validated sales and now streaming data that certifies the most "accessed" record in the nation. Now owned by MRC Data, SoundScan has been renamed Music Connect and provides a comprehensive view of streaming, radio airplay and sales data for artists, albums and songs. Music Connect helps clients dive deep into datasets to uncover insightful metrics, identify trends and track and compare performance across markets to help make informed decisions.

Using UPCs from point-of-sale cash registers as well as ISRCs from digital files, "sales" data is collected with a rolling daily accounting of what has generated revenue and sold.

Currently, the album that sells the most earns the #1 position for the week on the *Billboard* Top 200 Chart.

To register music with MRC's Music Connect, labels and artists must submit an online form to be included in sales data. Forms can be found at https://support.mrc-data.com/portal/en/kb/articles/register-music

ABOUT MRC DATA

MRC Data is the most comprehensive global provider of data and analytics to the entertainment and music industry and consumers. Established in 2019, MRC Data services all digital service providers, labels, airplay, and music retailers. MRC Data includes the industry's definitive *Billboard* charts, encompassing the most complete and well-respected database of charts across all music genres. MRC Data products include Music Connect, Broadcast Data Systems, and Music360, which collectively capture and represent the most robust dataset related to music sales, performance, artist activity, and consumer engagement. MRC Data is part of PMRC Holdings, a joint venture between MRC and Penske Media Company (PMC). Source: MRCEntertainment.com

A LOOK AT MUSIC CONNECT DATA

The Charts

By compiling and organizing sales data, Music Connect derives many charts that help describe the marketplace. These charts include the *Billboard* Top 200, artist/album/song reports with sales (with and without TEA and SEA), streaming, radio airplay, and social data, with the capability to create unlimited favorite lists and share any report, as well as industry-level stats. This is the data that drives the *Billboard* charts. Music Connect can also filter information and create data queries that can answer questions that are specific to artists, their markets, and help with strategic decisions.

Figure 12.10 is a screen shot of the *Billboard* Top 200 chart for the week ending June 4, 2020.

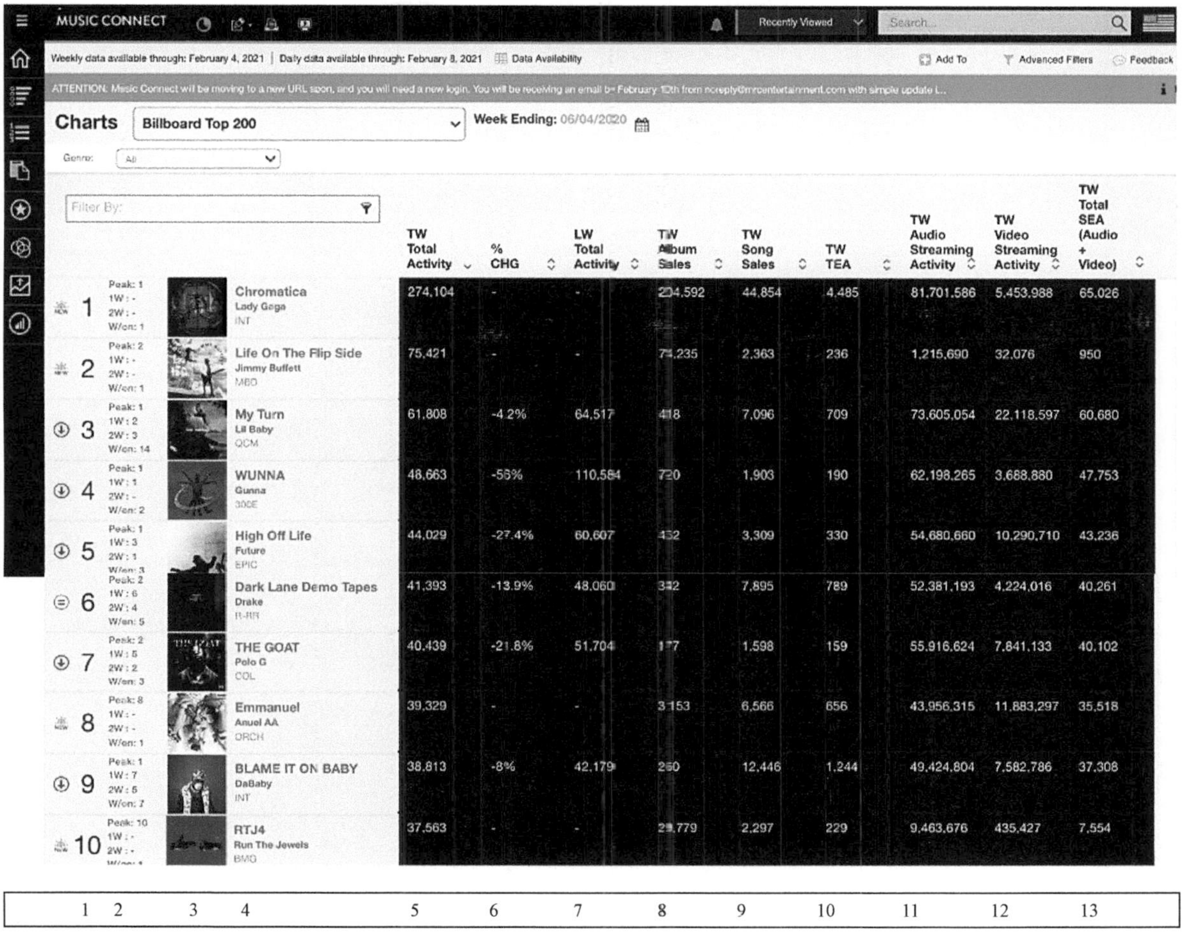

Weekly data available through: February 4, 2021 | Daily data available through: February 8, 2021 ⊞ Data Availability   ⊡ Add To   ⊤ Advanced Filters   ⊙ Feedback

ATTENTION: Music Connect will be moving to a new URL soon, and you will need a new login. You will be receiving an email b– February 12th from noreply@mrcentertainment.com with simple update L...

Charts Billboard Top 200   Week Ending: 06/04/2020

Genre: All

					TW Total Activity	% CHG	LW Total Activity	TW Album Sales	TW Song Sales	TW TEA	TW Audio Streaming Activity	TW Video Streaming Activity	TW Total SEA (Audio + Video)
1	Peak: 1 1W: - 2W: - W/on: 1		Chromatica Lady Gaga INT		274,104	-	-	204,592	44,854	4,485	81,701,586	5,453,988	65,026
2	Peak: 2 1W: - 2W: - W/on: 1		Life On The Flip Side Jimmy Buffett MBO		75,421	-	-	74,235	2,363	236	1,215,690	32,076	950
3	Peak: 1 1W: 2 2W: 3 W/on: 14		My Turn Lil Baby QCM		61,808	-4.2%	64,517	418	7,096	709	73,605,054	22,118,597	60,680
4	Peak: 1 1W: 1 2W: - W/on: 2		WUNNA Gunna 300E		48,663	-56%	110,584	720	1,903	190	62,198,265	3,688,880	47,753
5	Peak: 1 1W: 3 2W: 1 W/on: 3		High Off Life Future EPIC		44,029	-27.4%	60,607	432	3,309	330	54,680,660	10,290,710	43,236
6	Peak: 2 1W: 6 2W: 4 W/on: 5		Dark Lane Demo Tapes Drake R-RN		41,393	-13.9%	48,060	332	7,895	789	52,381,193	4,224,016	40,261
7	Peak: 2 1W: 5 2W: 2 W/on: 3		THE GOAT Polo G COL		40,439	-21.8%	51,704	177	1,598	159	55,916,624	7,841,133	40,102
8	Peak: 8 1W: - 2W: - W/on: 1		Emmanuel Anuel AA ORCH		39,329	-	-	3,153	6,566	656	43,956,315	11,883,297	35,518
9	Peak: 1 1W: 7 2W: 5 W/on: 7		BLAME IT ON BABY DaBaby INT		38,813	-8%	42,179	260	12,446	1,244	49,424,804	7,582,786	37,308
10	Peak: 10 1W: - 2W: - W/on: 1		RTJ4 Run The Jewels BMG		37,563	-	-	29,779	2,297	229	9,463,676	435,427	7,554

1	2	3	4	5	6	7	8	9	10	11	12	13

This is a sample of the Top 200 Album Chart. Each column is identified by number:

1—Current chart position
2—Identifies peak position, position one week ago, two weeks ago, and number of weeks on the chart
3—Image of album art
4—Album title/artist/label
5—This week's total activity including album sales with TEA and SEA
6—Percentage of change from this week to last week
7—Last week's total activity including album sales with TEA and SEA
8—This week's album sales
9—This week's song sales
10—This week's track equivalent albums
11—This week's audio streaming activity
12—This week's video streaming activity
13—This week's streaming equivalent albums = audio and video streaming combined (weighted)

FIGURE 12.10 Billboard *Top 200*

Figure 12.11 is a screen shot of the *Billboard* Top 200 Song Consumption chart for the week ending July 2, 2020.

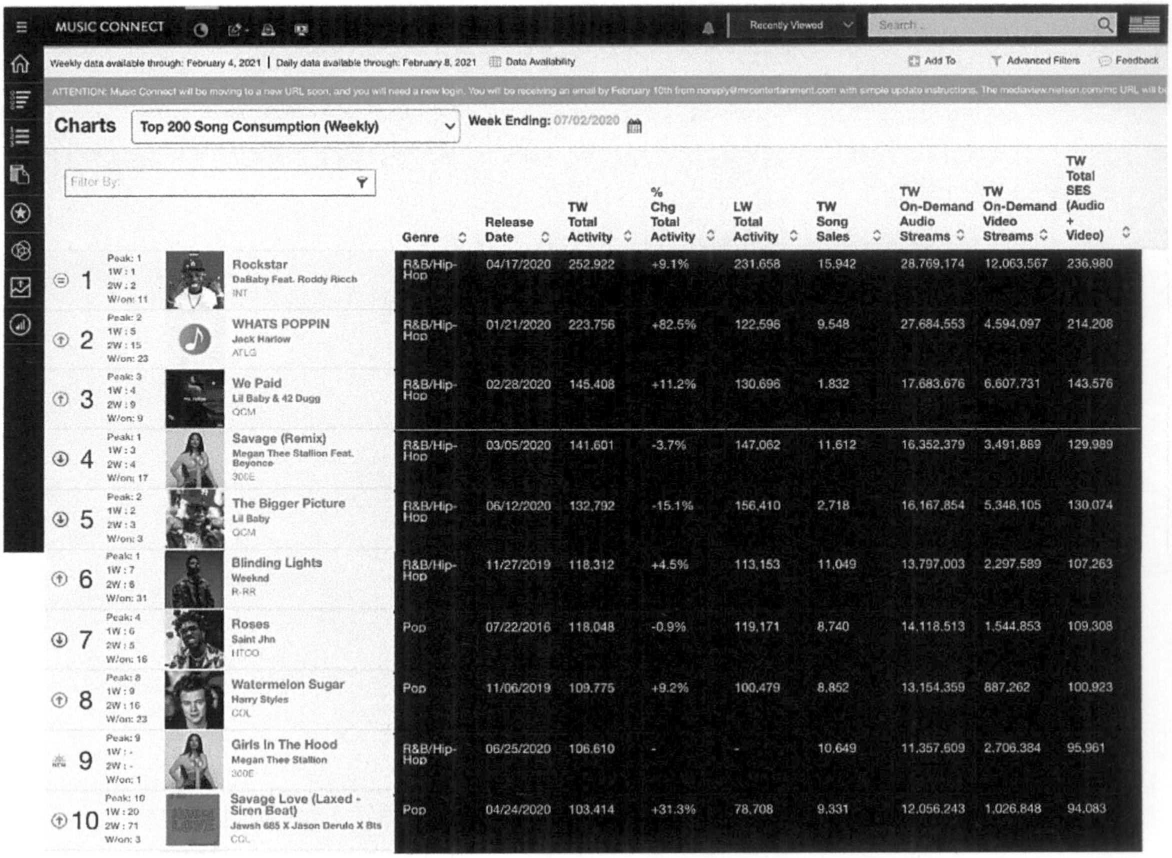

		Genre	Release Date	TW Total Activity	% Chg Total Activity	LW Total Activity	TW Song Sales	TW On-Demand Audio Streams	TW On-Demand Video Streams	TW Total SES (Audio + Video)
1	Peak: 1 / 1W : 1 / 2W : 2 / W/on: 11 — Rockstar — DaBaby Feat. Roddy Ricch — INT	R&B/Hip-Hop	04/17/2020	252.922	+9.1%	231.658	15.942	28,769,174	12,063,567	236.980
2	Peak: 2 / 1W : 5 / 2W : 15 / W/on: 23 — WHATS POPPIN — Jack Harlow — ATLG	R&B/Hip-Hop	01/21/2020	223.756	+82.5%	122.596	9.548	27,684,553	4,594.097	214.206
3	Peak: 3 / 1W : 4 / 2W : 9 / W/on: 9 — We Paid — Lil Baby & 42 Dugg — QCM	R&B/Hip-Hop	02/28/2020	145.408	+11.2%	130.696	1.832	17,683,676	6,607,731	143.576
4	Peak: 1 / 1W : 3 / 2W : 4 / W/on: 17 — Savage (Remix) — Megan Thee Stallion Feat. Beyonce — 300E	R&B/Hip-Hop	03/05/2020	141.601	-3.7%	147.062	11.612	16,352,379	3,491.889	129.989
5	Peak: 2 / 1W : 2 / 2W : 3 / W/on: 3 — The Bigger Picture — Lil Baby — QCM	R&B/Hip-Hop	06/12/2020	132.792	-15.1%	156.410	2.718	16,167,854	5,348,105	130.074
6	Peak: 1 / 1W : 7 / 2W : 6 / W/on: 31 — Blinding Lights — Weeknd — R-RR	R&B/Hip-Hop	11/27/2019	118.312	+4.5%	113.153	11.049	13,797,003	2,297.589	107.263
7	Peak: 4 / 1W : 6 / 2W : 5 / W/on: 16 — Roses — Saint Jhn — IITCO	Pop	07/22/2016	118.048	-0.9%	119.171	8.740	14,118,513	1,544.853	109.308
8	Peak: 8 / 1W : 9 / 2W : 16 / W/on: 23 — Watermelon Sugar — Harry Styles — COL	Pop	11/06/2019	109.775	+9.2%	100.479	8.852	13,154,359	887,262	100.923
9	Peak: 9 / 1W : - / 2W : - / W/on: 1 — Girls In The Hood — Megan Thee Stallion — 300E	R&B/Hip-Hop	06/25/2020	106.610	-	-	10.649	11,357,609	2,706,384	95.961
10	Peak: 10 / 1W : 20 / 2W : 71 / W/on: 3 — Savage Love (Laxed - Siren Beat) — Jawsh 685 X Jason Derulo X Bts — COL	Pop	04/24/2020	103.414	+31.3%	78.708	9.331	12,056,243	1,026.848	94.083

1	2	3	4	5	6	7	8	9	10	11	12	13

Each Column is identified by number:

1—Current chart position
2—Identifies peak position, position one week ago, two weeks ago, and number of weeks on the chart
3—Image of artist
4—Song title/artist/label
5—Genre
6—Date released
7—This week's total activity
8—Percentage of change from this week to last week
9—Last week's total activity
10—This week's songs sales
11—This week's on-demand audio streaming activity
12—This week's on-demand video streaming activity
13—This week's streaming equivalent albums = audio and video streaming combined (weighted)

FIGURE 12.11 Billboard *Top 200 Song Consumption*

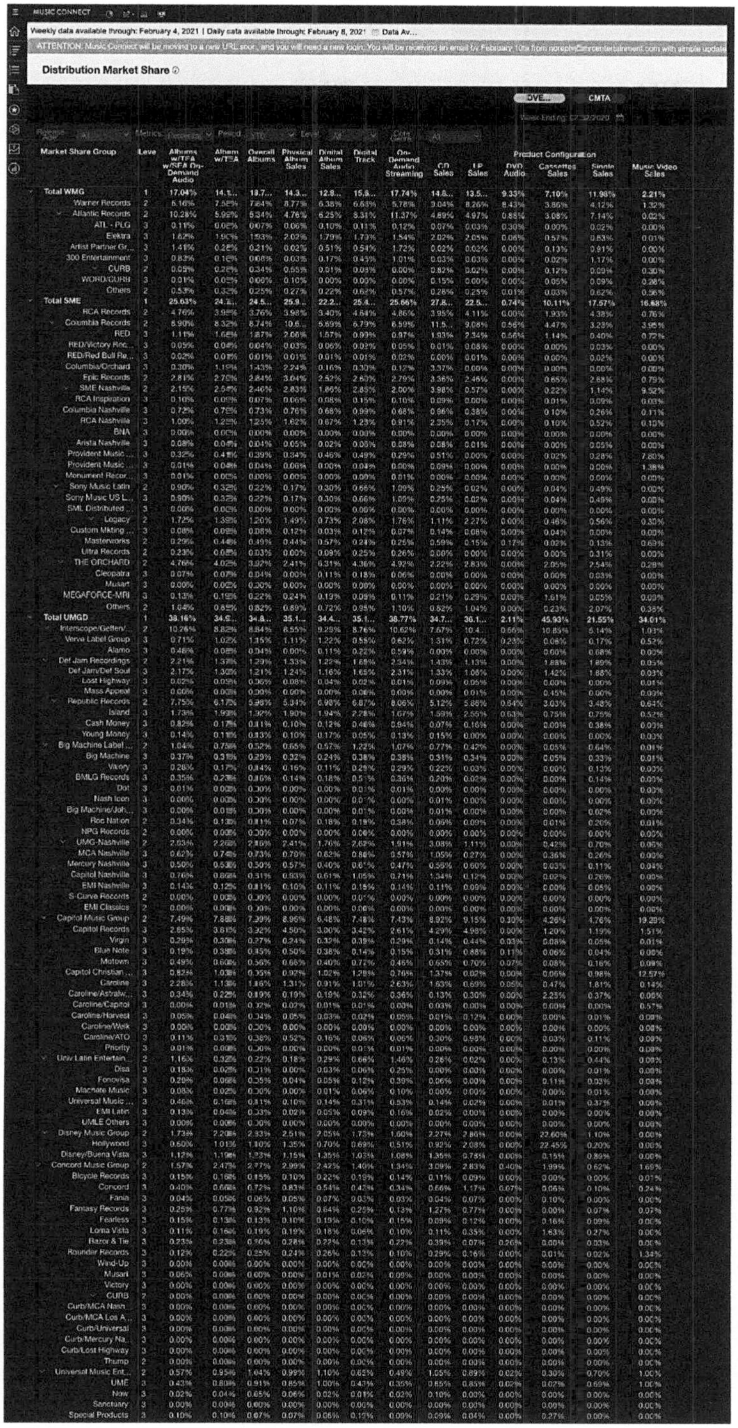

FIGURE 12.12A *Distribution Market Share by Label*

Ingrooves	2	1.24%	0.94%	0.89%	0.48%	1.53%	1.12%	1.30%	0.41%	0.62%	0.00%	0.17%	1.57%	0.59%
Ark 21	2	0.00%	0.00%	0.00%	0.00%	0.00%	0.00%	0.00%	0.00%	0.00%	0.00%	0.00%	0.00%	0.00%
ABKCO	2	0.08%	0.17%	0.19%	0.23%	0.12%	0.10%	0.06%	0.22%	0.26%	0.00%	0.00%	0.01%	0.06%
Bungalo Records	2	0.00%	0.00%	0.00%	0.00%	0.00%	0.00%	0.00%	0.00%	0.00%	0.00%	0.00%	0.00%	0.00%
VI Music	2	0.00%	0.00%	0.00%	0.00%	0.00%	0.00%	0.00%	0.00%	0.00%	0.00%	0.00%	0.00%	0.00%
VP Records	2	0.00%	0.00%	0.00%	0.00%	0.00%	0.00%	0.00%	0.00%	0.00%	0.00%	0.00%	0.00%	0.00%
Eagle Rock Ent.	2	0.01%	0.04%	0.05%	0.05%	0.05%	0.02%	0.00%	0.06%	0.03%	0.06%	0.00%	0.02%	8.50%
Glassnote	2	0.01%	0.04%	0.05%	0.07%	0.01%	0.00%	0.00%	0.01%	0.20%	0.00%	0.00%	0.01%	0.01%
Shout Factory!	2	0.02%	0.03%	0.03%	0.03%	0.03%	0.03%	0.01%	0.05%	0.00%	0.00%	0.00%	0.00%	0.95%
Mass Appeal	2	0.01%	0.01%	0.01%	0.02%	0.00%	0.01%	0.01%	0.01%	0.04%	0.00%	0.00%	0.00%	0.00%
Others	2	0.53%	0.27%	0.20%	0.18%	0.25%	0.51%	0.59%	0.16%	0.21%	0.00%	0.17%	0.19%	0.03%
Total Under Review	1	**0.55%**	**0.24%**	**0.06%**	**0.07%**	**0.04%**	**0.90%**	**0.60%**	**0.04%**	**0.13%**	**0.00%**	**0.30%**	**0.02%**	**0.00%**
Others	2	0.55%	0.24%	0.06%	0.07%	0.04%	0.90%	0.60%	0.04%	0.13%	0.00%	0.30%	0.02%	0.00%
Total Others	1	19.16%	26.1...	26.8...	24.5...	30.4...	23.7...	17.80%	22.5...	27.7...	87.80%	36.85%	48.88%	47.08%
Madacy	2	0.00%	0.00%	0.00%	0.00%	0.00%	0.00%	0.00%	0.01%	0.00%	0.00%	0.17%	0.00%	0.00%
Daptone Records	2	0.00%	0.01%	0.01%	0.02%	0.00%	0.00%	0.00%	0.00%	0.06%	0.00%	0.00%	0.00%	0.00%
Laserlight/Delta	2	0.00%	0.00%	0.00%	0.00%	0.00%	0.00%	0.00%	0.01%	0.00%	0.00%	0.06%	0.00%	0.00%
Bad Taste	2	0.00%	0.00%	0.00%	0.00%	0.00%	0.00%	0.00%	0.00%	0.00%	0.00%	0.00%	0.00%	0.00%
Roots Music	2	0.00%	0.00%	0.00%	0.00%	0.00%	0.00%	0.00%	0.00%	0.00%	0.00%	0.00%	0.00%	0.00%
TVT	2	0.00%	0.00%	0.00%	0.00%	0.00%	0.00%	0.00%	0.00%	0.01%	0.00%	0.00%	0.00%	0.00%
Rounder	2	0.00%	0.00%	0.00%	0.00%	0.00%	0.00%	0.00%	0.00%	0.00%	0.00%	0.00%	0.00%	0.00%
Balboa	2	0.00%	0.00%	0.00%	0.00%	0.00%	0.00%	0.00%	0.00%	0.00%	0.00%	0.00%	0.00%	0.00%
Starbucks	2	0.00%	0.00%	0.00%	0.00%	0.00%	0.00%	0.00%	0.00%	0.00%	0.00%	0.00%	0.00%	0.00%
Malaco/Muscle Sh...	2	0.00%	0.00%	0.00%	0.00%	0.00%	0.00%	0.00%	0.00%	0.00%	0.00%	0.01%	0.00%	0.03%
Freddie Records	2	0.00%	0.00%	0.00%	0.00%	0.00%	0.00%	0.00%	0.00%	0.00%	0.00%	0.00%	0.00%	0.00%
Image Entertainment	2	0.00%	0.00%	0.00%	0.00%	0.00%	0.00%	0.00%	0.00%	0.00%	0.05%	0.00%	0.00%	0.07%
Oh boy	2	0.00%	0.00%	0.00%	0.00%	0.00%	0.00%	0.00%	0.00%	0.00%	0.00%	0.00%	0.00%	0.00%
Kill Rock Stars	2	0.00%	0.01%	0.02%	0.02%	0.01%	0.00%	0.00%	0.00%	0.06%	0.00%	0.07%	0.00%	0.00%
Bloodshot	2	0.00%	0.00%	0.01%	0.01%	0.01%	0.00%	0.00%	0.00%	0.01%	0.00%	0.00%	0.00%	0.00%
CMH	2	0.00%	0.00%	0.00%	0.00%	0.00%	0.00%	0.00%	0.00%	0.00%	0.00%	0.00%	0.00%	0.00%
Ultra	2	0.00%	0.00%	0.00%	0.00%	0.00%	0.00%	0.00%	0.00%	0.00%	0.00%	0.03%	0.00%	0.00%
Mailboat	2	0.00%	0.00%	0.00%	0.00%	0.00%	0.00%	0.00%	0.00%	0.00%	0.00%	0.00%	0.00%	0.01%
NAXOS	2	0.00%	0.02%	0.03%	0.04%	0.02%	0.00%	0.00%	0.06%	0.00%	0.08%	0.00%	0.00%	0.02%
Red House Records	2	0.00%	0.00%	0.00%	0.00%	0.01%	0.00%	0.00%	0.01%	0.00%	0.00%	0.00%	0.00%	0.00%
Sundazed Records	2	0.00%	0.02%	0.03%	0.05%	0.00%	0.00%	0.00%	0.02%	0.13%	0.00%	0.00%	0.00%	0.00%
Collectables Records	2	0.00%	0.01%	0.02%	0.03%	0.00%	0.00%	0.00%	0.05%	0.00%	0.00%	0.00%	0.00%	0.00%
Mountain Apple	2	0.00%	0.00%	0.00%	0.00%	0.00%	0.00%	0.00%	0.00%	0.00%	0.00%	0.00%	0.00%	0.00%
Yep Roc	2	0.00%	0.03%	0.04%	0.05%	0.03%	0.00%	0.00%	0.03%	0.08%	0.00%	0.03%	0.01%	0.00%
Barsuk Records	2	0.01%	0.03%	0.04%	0.05%	0.01%	0.00%	0.00%	0.02%	0.12%	0.00%	0.22%	0.02%	0.00%
Warp Records	2	0.00%	0.03%	0.04%	0.05%	0.02%	0.00%	0.00%	0.01%	0.14%	0.00%	0.00%	0.04%	0.00%
Genius Products	2	0.00%	0.00%	0.00%	0.00%	0.00%	0.00%	0.00%	0.00%	0.00%	0.00%	0.00%	0.00%	0.00%
X5 Music	2	0.00%	0.01%	0.01%	0.00%	0.03%	0.00%	0.00%	0.00%	0.00%	0.00%	0.00%	0.00%	0.00%
Select-O-Hits	2	0.00%	0.00%	0.00%	0.00%	0.00%	0.00%	0.00%	0.00%	0.00%	0.00%	0.00%	0.00%	0.00%
E1 Entertainment	2	0.48%	0.58%	0.61%	0.60%	0.63%	0.46%	0.44%	0.69%	0.43%	0.43%	0.05%	0.44%	0.67%
E1 Music	3	0.26%	0.12%	0.07%	0.00%	0.19%	0.29%	0.28%	0.00%	0.00%	0.00%	0.00%	0.16%	0.07%
Shanachie	3	0.00%	0.02%	0.03%	0.02%	0.04%	0.00%	0.00%	0.04%	0.00%	0.00%	0.00%	0.00%	0.08%
Navarre	3	0.01%	0.07%	0.09%	0.13%	0.01%	0.00%	0.00%	0.19%	0.03%	0.03%	0.02%	0.00%	0.14%
Cleopatra	3	0.00%	0.01%	0.02%	0.03%	0.00%	0.00%	0.00%	0.03%	0.04%	0.00%	0.00%	0.02%	0.01%
E1 Other	3	0.19%	0.34%	0.39%	0.39%	0.38%	0.16%	0.16%	0.41%	0.34%	0.40%	0.02%	0.24%	0.19%
Third Party PLG	2	0.00%	0.00%	0.00%	0.00%	0.00%	0.00%	0.00%	0.00%	0.00%	0.00%	0.00%	0.00%	0.00%
Epitaph	2	0.21%	0.42%	0.46%	0.49%	0.39%	0.30%	0.16%	0.27%	0.95%	0.00%	0.40%	0.47%	0.12%
Sub Pop	2	0.17%	0.30%	0.36%	0.46%	0.20%	0.08%	0.13%	0.16%	1.02%	0.00%	6.17%	0.12%	0.28%
Secretly Canadian	2	0.00%	0.02%	0.02%	0.03%	0.01%	0.00%	0.00%	0.00%	0.09%	0.00%	0.01%	0.00%	0.00%
Surfdog	2	0.00%	0.00%	0.00%	0.00%	0.00%	0.00%	0.00%	0.00%	0.00%	0.00%	0.00%	0.00%	0.00%
Alligator Records	2	0.00%	0.02%	0.02%	0.03%	0.02%	0.00%	0.00%	0.04%	0.00%	0.36%	0.04%	0.00%	0.00%
Saddle Creek	2	0.00%	0.01%	0.01%	0.02%	0.01%	0.00%	0.00%	0.00%	0.06%	0.00%	0.30%	0.00%	0.00%
Temporary Residence	2	0.00%	0.01%	0.01%	0.01%	0.01%	0.00%	0.00%	0.00%	0.03%	0.00%	0.00%	0.00%	0.00%
Beggars Group	2	0.40%	0.66%	0.78%	1.01%	0.42%	0.23%	0.33%	0.30%	2.45%	0.00%	0.68%	0.35%	0.00%
Side One Dummy	2	0.00%	0.01%	0.01%	0.00%	0.02%	0.00%	0.00%	0.00%	0.01%	0.00%	0.00%	0.00%	0.00%
Jagjaguwar Records	2	0.02%	0.06%	0.08%	0.11%	0.03%	0.00%	0.01%	0.01%	0.32%	0.00%	0.34%	0.03%	0.00%
Merge	2	0.01%	0.07%	0.09%	0.12%	0.04%	0.00%	0.00%	0.02%	0.33%	0.00%	0.02%	0.03%	0.00%
Domino	2	0.05%	0.11%	0.14%	0.20%	0.06%	0.01%	0.04%	0.05%	0.48%	0.00%	0.34%	0.13%	0.00%
Comedy Central	2	0.05%	0.02%	0.02%	0.00%	0.06%	0.00%	0.05%	0.00%	0.00%	0.00%	0.00%	0.00%	0.00%
Touch & Go	2	0.00%	0.01%	0.02%	0.03%	0.00%	0.00%	0.00%	0.00%	0.07%	0.00%	0.00%	0.03%	0.00%
New West Records	2	0.00%	0.03%	0.05%	0.07%	0.00%	0.00%	0.00%	0.04%	0.13%	0.12%	0.07%	0.00%	0.14%
VP Records	2	0.04%	0.04%	0.03%	0.02%	0.04%	0.07%	0.04%	0.02%	0.03%	0.01%	0.00%	0.01%	0.00%
Bad Son Recordings	2	0.00%	0.00%	0.00%	0.00%	0.00%	0.00%	0.00%	0.00%	0.00%	0.00%	0.00%	0.00%	0.00%
BMG	2	1.07%	1.35%	1.44%	1.37%	1.55%	1.06%	0.99%	1.63%	0.83%	2.38%	0.25%	1.32%	0.09%
12Tone	2	0.15%	0.15%	0.15%	0.15%	0.16%	0.17%	0.15%	0.17%	0.10%	0.00%	0.00%	0.09%	0.00%
Curb	2	0.44%	0.33%	0.20%	0.07%	0.40%	0.81%	0.46%	0.07%	0.08%	0.00%	0.00%	0.25%	0.00%
Big Loud Records	2	0.28%	0.12%	0.06%	0.03%	0.10%	0.35%	0.30%	0.04%	0.00%	0.00%	0.00%	0.03%	0.00%
Vydia	2	0.00%	0.00%	0.00%	0.00%	0.00%	0.00%	0.00%	0.00%	0.00%	0.00%	0.00%	0.00%	0.00%
SMH Records	2	0.00%	0.00%	0.00%	0.00%	0.00%	0.00%	0.00%	0.00%	0.00%	0.00%	0.00%	0.00%	0.00%
EONE Christian	2	0.00%	0.00%	0.00%	0.00%	0.00%	0.00%	0.00%	0.00%	0.00%	0.00%	0.00%	0.00%	0.16%
Malaco Records	2	0.00%	0.00%	0.00%	0.00%	0.00%	0.00%	0.00%	0.00%	0.00%	0.00%	0.00%	0.00%	0.00%
New Day Distribution	2	0.00%	0.01%	0.02%	0.03%	0.00%	0.00%	0.00%	0.05%	0.00%	0.00%	0.00%	2.29%	1.23%
Crossroads Distribu...	2	0.00%	0.00%	0.00%	0.00%	0.00%	0.00%	0.00%	0.00%	0.00%	0.00%	0.00%	0.06%	0.00%
Infinty Music Dist	2	0.00%	0.00%	0.00%	0.00%	0.00%	0.00%	0.00%	0.00%	0.00%	0.00%	0.00%	0.00%	0.00%
Hymns of Worship	2	0.00%	0.00%	0.00%	0.00%	0.00%	0.00%	0.00%	0.00%	0.00%	0.00%	0.00%	0.00%	0.00%

FIGURE 12.12B *Distribution Market Share by Label*

Figures 12.12A and 12.12B identify market share of distribution companies by label, week ending 7/2/2020 – year to date. This information breaks down the data by configuration across each label represented and summarizes a total distribution value by conglomerate: WMG, SME, UMGD, Others.

GENRE PERCENTAGES

Figures 12.13A–B show genre percentages by time periods and values, including their percentage of change over time as of 7/2/20.

TP = This period, LP = Last period, TP TY = This period this year, TP LY = This period last year, YTD = Cumulative for the year

Figure 12.14 reveals DMA percentages by time periods and values, including their percentage of change over time. Screen shot was of 7/2/20

TP = This period, LP = Last period, TP TY = This period this year, TP LY = This period last year, YTD = Cumulative for the year

Figure 12.15 highlights overall sales by format, store strata, and product configuration by time period.

Figures 12.16A–F represent a dashboard look at the single "In My Feelings" from Drake's album *Scorpion*.

The dashboard displays this period sales/on-demand streaming/airplay audience for the week as well as activity to date (cumulative data from release). The graph displays a visual image of airplay audience, on-demand total, songs w SES on-demand, and digital track sales.

Deeper data reveals streaming sources as well as radio formats playing the single this period as well as cumulatively.

Similar to the singles dashboard, Figure 12.17 represents a dashboard look at Drake's album *Scorpion*.

The dashboard displays this period sales/streaming/airplay audience for the week as well as activity to date (cumulative data from release). The graph displays a visual image of airplay audience, albums w/TEA w/SEA on-demand, and on-demand total.

Deeper data reveals streaming and sales sources, product configurations, and source of streaming by song by this period, YTD, and ATD.

Genres	TP	TP vs LP % Chg	LP	TP TY	TP TY vs TP LY % Chg	TP LY	YTD TY	YTD TY vs YTD LY % Chg	YTD LY
Blues									
Albums w/TEA w/SEA On-Demand	39,747	-2.0%	40,545	39,747	-18.8%	48,936	1,038,...	-13.2%	1,196,870
Albums w/TEA w/SEA On-Demand Audio	38,760	-2.0%	39,546	38,760	-16.4%	46,369	1,011,...	-10.9%	1,135,503
Albums w/TEA	18,394	-3.9%	19,142	18,394	-34.5%	28,078	526,856	-23.7%	690,517
Overall Albums	16,418	-4.5%	17,189	16,418	-36.2%	25,749	475,286	-24.2%	627,244
Physical Album Sales	10,881	-6.6%	11,642	10,881	-45.0%	19,780	337,877	-28.1%	470,305
Digital Album Sales	5,537	-0.2%	5,547	5,537	-7.2%	5,969	137,409	-12.4%	156,909
Digital Tracks	19,760	1.2%	19,525	19,760	-15.2%	23,290	515,658	-18.5%	632,728
On Demand Streaming	30,754...	-0.1%	33,782...	30,754...	-9.0%	33,787...	745,87...	-9.2%	821,002...
On Demand Audio Streaming	27,056...	0.1%	27,037...	27,055...	12.0%	24,158...	644,04...	9.0%	591,064...
On Demand Video Streaming	3,698...	-1.2%	3,744...	3,698...	-61.6%	9,628...	101,83...	-55.7%	229,937...
Children									
Albums w/TEA w/SEA On-Demand	192,336	1.8%	188,973	192,336	16.8%	164,661	5,188,...	25.8%	4,124,976
Albums w/TEA w/SEA On-Demand Audio	182,907	1.8%	179,758	182,907	30.1%	140,557	4,878,...	36.7%	3,568,262
Albums w/TEA	30,551	6.5%	28,672	30,551	-20.3%	38,316	936,758	-11.2%	1,054,914
Overall Albums	27,071	7.1%	25,281	27,071	-20.3%	33,949	818,397	-11.7%	926,438
Physical Album Sales	16,247	3.8%	15,645	16,247	-23.5%	21,247	499,124	-16.7%	599,414
Digital Album Sales	10,824	12.3%	9,636	10,824	-14.8%	12,702	319,273	-2.4%	327,024
Digital Tracks	34,799	2.6%	33,906	34,799	-20.3%	43,673	1,183,...	-7.9%	1,284,757
On Demand Streaming	233,29...	1.1%	230,81...	233,29...	4.4%	223,37...	6,296,...	17.7%	5,351,12...
On Demand Audio Streaming	197,93...	0.9%	196,25...	197,93...	48.8%	132,98...	5,135,...	57.4%	3,263,46...
On Demand Video Streaming	35,358...	2.3%	34,562...	35,358...	-60.9%	90,391...	1,161,...	-44.4%	2,087,07...
Christian/Gospel									
Albums w/TEA w/SEA On-Demand	278,960	0.8%	276,614	278,960	-7.6%	302,011	7,749,...	-16.5%	9,285,235
Albums w/TEA w/SEA On-Demand Audio	265,673	0.8%	263,653	265,673	-2.2%	271,765	7,342,...	-12.8%	8,419,515
Albums w/TEA	55,474	1.2%	54,789	55,474	-39.4%	91,614	1,726,...	-48.0%	3,318,261
Overall Albums	39,544	-0.3%	39,678	39,544	-44.6%	71,361	1,252,...	-53.4%	2,684,741
Physical Album Sales	21,312	-7.3%	23,002	21,312	-57.5%	50,180	686,739	-65.1%	1,966,520
Digital Album Sales	18,232	9.3%	16,676	18,232	-13.9%	21,181	565,468	-21.3%	718,221
Digital Tracks	159,294	5.4%	151,108	159,294	-21.3%	202,529	4,740,...	-25.2%	6,335,199
On Demand Streaming	334,24...	0.8%	331,56...	334,24...	-6.7%	358,38...	9,153,...	-10.3%	10,206,6...
On Demand Audio Streaming	284,41...	0.5%	282,95...	284,41...	16.1%	244,95...	7,626,...	9.6%	6,960,18...
On Demand Video Streaming	49,827...	2.5%	48,600...	49,827...	-56.1%	113,42...	1,526,...	-53.0%	3,246,44...
Classical									
Albums w/TEA w/SEA On-Demand	137,721	-1.8%	140,285	137,721	2.0%	135,040	3,897,...	-6.3%	4,159,576
Albums w/TEA w/SEA On-Demand Audio	135,592	-1.9%	138,189	135,592	7.1%	126,625	3,806,...	-3.1%	3,926,808
Albums w/TEA	38,072	-8.1%	41,414	38,072	-17.1%	45,945	1,099,...	-18.8%	1,353,196
Overall Albums	34,874	-9.2%	38,406	34,874	-16.1%	41,580	1,007,...	-18.8%	1,239,713
Physical Album Sales	19,449	-12.9%	22,323	19,449	-16.0%	23,167	554,504	-21.1%	702,671
Digital Album Sales	15,425	-4.1%	16,083	15,425	-16.2%	18,413	452,506	-15.7%	537,041
Digital Tracks	31,981	6.3%	30,081	31,981	-26.7%	43,646	923,581	-18.6%	1,134,823
On Demand Streaming	134,84...	0.8%	133,77...	134,84...	-1.7%	137,13...	3,872,...	-8.9%	4,249,79...
On Demand Audio Streaming	126,86...	0.8%	125,91...	126,86...	20.2%	105,57...	3,529,...	4.5%	3,376,91...
On Demand Video Streaming	7,983...	1.5%	7,860...	7,983...	-74.7%	31,558...	343,11...	-60.7%	872,880...
Comedy									
Albums w/TEA w/SEA On-Demand	37,139	27.5%	29,138	37,139	3.4%	35,921	748,067	-23.6%	979,682
Albums w/TEA w/SEA On-Demand Audio	35,493	27.8%	27,779	35,493	9.4%	32,437	697,103	-20.5%	876,568
Albums w/TEA	10,209	87.7%	5,437	10,209	21.5%	8,400	148,174	-44.7%	267,976
Overall Albums	7,772	67.7%	4,634	7,772	10.8%	7,011	126,357	-45.5%	231,834
Physical Album Sales	1,535	-0.8%	1,547	1,535	-43.6%	2,724	42,349	-58.7%	102,604
Digital Album Sales	6,237	102.0%	3,087	6,237	45.5%	4,287	84,007	-35.0%	129,229
Digital Tracks	24,372	203.2%	8,038	24,372	75.5%	13,890	218,171	-39.6%	361,425
On Demand Streaming	40,677...	13.5%	35,837...	40,677...	-11.7%	46,074...	945,84...	-22.9%	1,226,19...
On Demand Audio Streaming	34,504...	12.2%	30,742...	34,504...	4.5%	33,006...	754,73...	-10.1%	839,480...
On Demand Video Streaming	6,173...	21.2%	5,095...	6,173...	-52.8%	13,067...	191,11...	-50.6%	386,716...
Country									
Albums w/TEA w/SEA On-Demand	1,304...	1.0%	1,291...	1,304...	5.1%	1,240...	30,701...	7.2%	28,644,675
Albums w/TEA w/SEA On-Demand Audio	1,268...	1.0%	1,255...	1,268...	9.3%	1,160...	29,737...	11.0%	26,795,503
Albums w/TEA	221,521	1.6%	218,026	221,521	-17.5%	268,577	5,782,...	-18.7%	7,110,690
Overall Albums	157,143	-0.1%	157,325	157,143	-16.4%	188,016	4,295,...	-17.1%	5,183,091
Physical Album Sales	100,915	-6.9%	108,349	100,915	-20.2%	126,525	2,943,...	-16.3%	3,516,005
Digital Album Sales	56,228	14.8%	48,976	56,228	-8.6%	61,491	1,351,...	-18.9%	1,667,065
Digital Tracks	643,783	6.0%	607,005	643,783	-20.1%	805,606	14,872,...	-22.8%	19,275,989
On Demand Streaming	1,523...	0.9%	1,509,...	1,523...	2.7%	1,483,...	35,566...	6.8%	33,280,1...
On Demand Audio Streaming	1,387,...	0.7%	1,377,...	1,387...	17.2%	1,183,...	31,945...	21.2%	26,356,9...
On Demand Video Streaming	136,23...	2.6%	132,75...	136,23...	-54.7%	300,66...	3,611,...	-47.8%	6,923,14...

FIGURE 12.13A *YTD Genre Percentage*

Markets	TP	TP vs LP %Chg	LP	TP TY	TP TY vs TP LY %Chg	TP LY	YTD TY	YTD TY vs YTD LY %Chg	YTD LY
New York, NY									
Albums w/TEA w/SEA-On-Demand Audio	260,441	-1.8%	265,210	260,441			12,901...		
Albums w/TEA	131,438	-3.2%	135,846	131,438	-17.6%	159,607	3,600...	-17.5%	4,371,446
Overall Albums	98,295	-5.7%	104,257	98,295	-15.5%	116,325	2,732...	-15.8%	3,244,875
Physical Album Sales	50,678	-5.9%	53,881	50,678	-15.1%	59,678	1,485...	-18.3%	1,816,758
Digital Album Sales	47,617	-5.5%	50,376	47,617	-15.9%	56,647	1,247...	-12.7%	1,428,117
Digital Tracks	331,430	4.9%	315,889	331,430	-23.4%	432,822	8,747...	-22.4%	11,265,709
On Demand Audio Streaming	161,25...	-0.3%	161,*0...	161,25...			13,140...		
Airplay Audience	7,192...	14.7%	6,2*3	7,192...	-37.7%	11,551...	259,11...	-12.3%	295,619...
Los Angeles, CA									
Albums w/TEA w/SEA-On-Demand Audio	244,970	-3.0%	252,651	244,970			15,031...		
Albums w/TEA	113,371	-8.4%	123,796	113,371	-27.6%	156,570	3,693...	-17.3%	4,465,093
Overall Albums	85,245	-12.0%	96,879	85,245	-30.3%	122,379	2,977...	-16.2%	3,552,785
Physical Album Sales	44,056	-19.4%	54,637	44,056	-43.2%	77,621	1,926...	-19.0%	2,377,455
Digital Album Sales	41,189	-2.5%	42,242	41,189	-8.0%	44,758	1,051...	-10.5%	1,175,330
Digital Tracks	281,256	4.5%	269,175	281,256	-17.7%	341,913	7,159...	-21.5%	9,123,081
On Demand Audio Streaming	164,49...	2.1%	161,36...	164,49...			16,284...		
Airplay Audience	6,658...	14.3%	5,834...	6,658...	-37.9%	10,724...	233,22...	-13.5%	269,757...
Chicago, IL									
Albums w/TEA w/SEA-On-Demand Audio	132,649	-1.6%	134,837	132,649			7,841...		
Albums w/TEA	63,028	-7.4%	68,043	63,028	-21.0%	79,764	1,771...	-18.8%	2,182,224
Overall Albums	49,501	-10.2%	55,136	49,501	-20.5%	62,281	1,427...	-17.5%	1,731,528
Physical Album Sales	29,833	-10.9%	33,498	29,833	-21.9%	38,206	915,691	-20.5%	1,152,291
Digital Album Sales	19,668	-9.1%	21,638	19,668	-18.3%	24,075	512,058	-11.6%	579,237
Digital Tracks	135,269	4.8%	129,073	135,269	-22.6%	174,826	3,442...	-23.6%	4,506,961
On Demand Audio Streaming	87,026...	4.2%	83,492...	87,026...			8,605...		
Airplay Audience	3,939...	12.2%	3,5*2...	3,939...	-27.6%	5,438...	135,52...	-6.4%	144,805,...
Philadelphia, PA									
Albums w/TEA w/SEA-On-Demand Audio	114,968	-4.7%	120,617	114,968			4,740...		
Albums w/TEA	54,359	-10.3%	60,582	54,359	-19.1%	67,201	1,480...	-20.2%	1,855,597
Overall Albums	42,088	-13.7%	48,787	42,088	-17.8%	51,199	1,152...	-19.6%	1,432,679
Physical Album Sales	23,589	-19.9%	29,440	23,589	-22.0%	30,231	877,192	-23.6%	886,629
Digital Album Sales	18,499	-4.4%	19,347	18,499	-11.8%	20,968	475,370	-12.9%	546,049
Digital Tracks	122,716	4.0%	117,958	122,716	-23.3%	160,020	3,280...	-22.4%	4,229,182
On Demand Audio Streaming	75,760...	0.9%	75,043...	75,760...			4,607...		
Airplay Audience	2,757...	5.3%	2,61 7...	2,757...	-35.6%	4,285...	95,833...	-11.8%	108,700...

FIGURE 12.13B *YTD Genre Percentage*

Markets	TP	TP vs LP %Chg	LP	TP TY	TP TY vs TP LY %Chg	TP LY	YTD TY	YTD TY vs YTD LY %Chg	YTD LY
New York, NY									
Albums w/TEA w/SEA-On-Demand Audio	260,441	-1.8%	265,210	260,441			12,901...		
Albums w/TEA	131,438	-3.2%	135,846	131,438	-17.6%	159,607	3,606...	-17.5%	4,371,446
Overall Albums	98,295	-5.7%	10*,257	98,295	-15.5%	116,325	2,732...	-15.8%	3,244,875
Physical Album Sales	50,678	-5.9%	52,881	50,678	-15.1%	59,678	1,485...	-18.3%	1,816,758
Digital Album Sales	47,617	-5.5%	58,376	47,617	-15.9%	56,647	1,247...	-12.7%	1,428,117
Digital Tracks	331,430	4.9%	315,889	331,430	-23.4%	432,822	8,747...	-22.4%	11,265,709
On Demand Audio Streaming	161,25...	-0.3%	161,70...	161,25...			13,140...		
Airplay Audience	7,192...	14.7%	6,2*3...	7,192...	-37.7%	11,551...	259,11...	-12.3%	295,619...
Los Angeles, CA									
Albums w/TEA w/SEA-On-Demand Audio	244,970	-3.0%	252,651	244,970			15,031...		
Albums w/TEA	113,371	-8.4%	123,796	113,371	-27.6%	156,570	3,693...	-17.3%	4,465,093
Overall Albums	85,245	-12.0%	96,879	85,245	-30.3%	122,379	2,977...	-16.2%	3,552,785
Physical Album Sales	44,056	-19.4%	5*,637	44,056	-43.2%	77,621	1,926...	-19.0%	2,377,455
Digital Album Sales	41,189	-2.5%	42,242	41,189	-8.0%	44,758	1,051...	-10.5%	1,175,330
Digital Tracks	281,256	4.5%	269,175	281,256	-17.7%	341,913	7,159...	-21.5%	9,123,081
On Demand Audio Streaming	164,49...	2.1%	161,06...	164,49...			16,284...		
Airplay Audience	6,658...	14.3%	5,824...	6,658...	-37.9%	10,724...	233,22...	-13.5%	269,757...
Chicago, IL									
Albums w/TEA w/SEA-On-Demand Audio	132,649	-1.6%	13*,837	132,649			7,841...		
Albums w/TEA	63,028	-7.4%	68,043	63,028	-21.0%	79,764	1,771...	-18.8%	2,182,224
Overall Albums	49,501	-10.2%	55,136	49,501	-20.5%	62,281	1,427...	-17.5%	1,731,528
Physical Album Sales	29,833	-10.9%	3*,498	29,833	-21.9%	38,206	915,691	-20.5%	1,152,291
Digital Album Sales	19,668	-9.1%	2*,638	19,668	-18.3%	24,075	512,058	-11.6%	579,237
Digital Tracks	135,269	4.8%	129,073	135,269	-22.6%	174,826	3,442...	-23.6%	4,506,961
On Demand Audio Streaming	87,026...	4.2%	83,492...	87,026...			8,695...		
Airplay Audience	3,939...	12.2%	3,5 2...	3,939...	-27.6%	5,438...	135,52...	-6.4%	144,805,...
Philadelphia, PA									
Albums w/TEA w/SEA-On-Demand Audio	114,968	-4.7%	120,617	114,968			4,740...		
Albums w/TEA	54,359	-10.3%	66,582	54,359	-19.1%	67,201	1,480...	-20.2%	1,855,597
Overall Albums	42,088	-13.7%	48,787	42,088	-17.8%	51,199	1,152...	-19.6%	1,432,679
Physical Album Sales	23,589	-19.9%	29,440	23,589	-22.0%	30,231	677,192	-23.6%	886,629
Digital Album Sales	18,499	-4.4%	19,347	18,499	-11.8%	20,968	475,370	-12.9%	546,049
Digital Tracks	122,716	4.0%	117,958	122,716	-23.3%	160,020	3,280...	-22.4%	4,229,182
On Demand Audio Streaming	75,760...	0.9%	75,043...	75,760...			4,607...		
Airplay Audience	2,757...	5.3%	2,6 7...	2,757...	-35.6%	4,285...	95,833...	-11.8%	108,700...

FIGURE 12.14 *DMA Percentage of Business*

SF-Okland-San Jose									
Albums w/TEA w/SEA On-Demand Audio	109,776	-3.7%	114,014	109,776			6,527,...		
Albums w/TEA	57,002	-8.7%	62,423	57,002	-19.0%	70,362	1,819,...	-6.1%	1,937,700
Overall Albums	43,716	-12.5%	49,940	43,716	-18.9%	53,930	1,481,...	-2.1%	1,513,979
Physical Album Sales	20,120	-20.2%	25,214	20,120	-29.0%	28,359	892,065	4.3%	855,130
Digital Album Sales	23,596	-4.6%	24,726	23,596	-7.7%	25,571	589,574	-10.5%	658,849
Digital Tracks	132,857	6.4%	124,829	132,857	-19.1%	164,327	3,375,...	-20.3%	4,237,215
On Demand Audio Streaming	65,967,...	2.3%	64,488,...	65,967,...			6,572,...		
Airplay Audience	2,204,...	8.8%	2,026,...	2,204,...	-41.7%	3,785,...	79,801,...	-15.8%	94,770,492
Boston, MA									
Albums w/TEA w/SEA On-Demand Audio	110,557	0.7%	109,770	110,557			4,973,...		
Albums w/TEA	61,274	1.7%	60,263	61,274	-7.1%	65,987	1,866,...	3.2%	1,808,841
Overall Albums	46,208	-0.1%	46,274	46,208	-4.9%	48,581	1,504,...	11.5%	1,349,189
Physical Album Sales	27,481	4.2%	26,381	27,481	-3.2%	28,391	1,033,...	27.6%	809,903
Digital Album Sales	18,726	-5.9%	19,893	18,726	-7.2%	20,190	471,301	-12.6%	539,285
Digital Tracks	150,664	7.7%	139,894	150,664	-13.4%	174,055	3,620,...	-21.2%	4,596,524
On Demand Audio Streaming	61,603,...	-0.5%	61,883,...	61,603,...			4,275,...		
Airplay Audience	1,863,...	10.6%	1,685,...	1,863,...	-36.7%	2,943,...	65,671,...	-6.7%	70,373,338
Dallas-Ft. Worth, TX									
Albums w/TEA w/SEA On-Demand Audio	106,101	-1.7%	107,917	106,101			6,760,...		
Albums w/TEA	43,762	-6.9%	47,008	43,762	-20.1%	54,799	1,269,...	-14.7%	1,487,377
Overall Albums	32,550	-9.7%	36,058	32,550	-17.9%	39,671	953,446	-13.5%	1,102,724
Physical Album Sales	16,835	-18.8%	20,728	16,835	-18.1%	20,570	550,406	-11.7%	623,212
Digital Album Sales	15,715	2.5%	15,330	15,715	-17.7%	19,100	403,039	-15.9%	479,511
Digital Tracks	112,120	2.4%	109,497	112,120	-25.9%	151,277	3,158,...	-17.9%	3,846,532
On Demand Audio Streaming	77,924,...	2.3%	76,136,...	77,924,...			8,028,...		
Airplay Audience	3,830,...	14.4%	3,347,...	3,830,...	-21.9%	4,908,...	118,36...	-7.3%	127,752,...
Detroit, MI									
Albums w/TEA w/SEA On-Demand Audio	62,432	-3.1%	64,441	62,432			2,662,...		
Albums w/TEA	29,282	-9.1%	32,216	29,282	-37.9%	47,188	870,912	-27.2%	1,197,067
Overall Albums	22,354	-12.2%	25,471	22,354	-41.8%	38,408	693,286	-28.2%	965,236
Physical Album Sales	12,335	-17.1%	14,878	12,335	-55.4%	27,639	438,924	-35.7%	682,359
Digital Album Sales	10,019	-5.4%	10,593	10,019	-7.0%	10,768	254,361	-10.1%	282,877
Digital Tracks	69,276	2.7%	67,454	69,276	-21.1%	87,803	1,776,...	-23.4%	2,318,311
On Demand Audio Streaming	41,437,...	2.9%	40,281,...	41,437,...			2,546,...		
Airplay Audience	1,743,...	15.3%	1,511,...	1,743,...	-39.6%	2,889,...	62,818,...	-13.6%	72,688,104

Washington, DC									
Albums w/TEA w/SEA On-Demand Audio	108,480	-1.9%	110,575	108,480			5,176,...		
Albums w/TEA	50,851	-7.4%	54,932	50,851	-15.9%	60,466	1,321,...	-18.9%	1,629,106
Overall Albums	37,964	-11.0%	42,667	37,964	-13.4%	43,838	994,148	-17.3%	1,201,809
Physical Album Sales	17,616	-18.6%	21,651	17,616	-15.7%	20,906	490,765	-20.6%	618,014
Digital Album Sales	20,348	-3.2%	21,016	20,348	-11.3%	22,932	503,383	-13.8%	583,794
Digital Tracks	128,869	5.1%	122,648	128,869	-22.5%	166,271	3,276,...	-23.3%	4,272,967
On Demand Audio Streaming	72,036,...	3.6%	69,553,...	72,036,...			5,431,...		
Airplay Audience	1,420,...	3.8%	1,369,...	1,420,...	-43.1%	2,497,...	55,023,...	-14.1%	64,023,718
Houston, TX									
Albums w/TEA w/SEA On-Demand Audio	83,673	-0.9%	84,449	83,673			4,704,...		
Albums w/TEA	33,845	-8.0%	36,785	33,845	-19.6%	42,071	929,050	-19.6%	1,155,556
Overall Albums	24,769	-11.5%	27,980	24,769	-17.5%	30,011	689,169	-17.7%	837,363
Physical Album Sales	12,312	-21.7%	15,730	12,312	-17.3%	14,887	364,714	-18.9%	449,498
Digital Album Sales	12,457	1.7%	12,250	12,457	-17.6%	15,124	324,454	-16.3%	387,865
Digital Tracks	90,769	3.1%	88,051	90,769	-24.7%	120,594	2,398,...	-24.6%	3,181,927
On Demand Audio Streaming	62,284,...	4.5%	59,581,...	62,284,...			5,537,...		
Airplay Audience	4,001,...	12.7%	3,551,...	4,001,...	-27.7%	5,537,...	125,36...	-9.0%	137,707,...

FIGURE 12.14 *(Continued)*

Equivalents

		TP vs LP			TP TY vs TP LY			YTD TY vs YTD LY	
	TP	% Chg	LP	TP TY	% Chg	TP LY	YTD TY	% Chg	YTD LY
Albums w/TEA w/SEA On-Demand	15,359...	0.3%	15,317...	15,359...	1.7%	15,108...	382,85...	0.4%	381,324...
Albums w/TEA w/SEA On-Demand Audio	14,584...	0.1%	14,573...	14,584...	12.7%	12,936...	361,20...	9.4%	330,290...
Albums w/TEA	2,131,...	-6.6%	2,281,...	2,131,...	-17.8%	2,594,...	58,328...	-19.3%	72,263,374

Sales

		TP vs LP			TP TY vs TP LY			YTD TY vs YTD LY	
	TP	% Chg	LP	TP TY	% Chg	TP LY	YTD TY	% Chg	YTD LY
Overall Albums	1,645,...	-9.4%	1,815,...	1,645,...	-16.4%	1,967,...	45,546...	-18.1%	55,605,318
Physical Albums	963,750	-13.4%	1,113,...	963,750	-18.5%	1,182,...	27,885...	-20.3%	35,003,872
Digital Albums	681,716	-2.9%	702,420	681,716	-13.1%	784,835	17,660...	-14.3%	20,601,447
Store Strata									
Chain	50,240	-8.6%	54,990	50,240	-47.0%	94,704	1,199,...	-56.0%	2,726,139
Independent	137,388	20.6%	113,950	137,388	-43.6%	243,614	5,693,...	-22.8%	7,374,389
Mass Merchant	247,873	-12.1%	281,970	247,873	-27.8%	343,355	6,767,...	-28.3%	9,442,782
Digital	681,641	-2.9%	702,353	681,641	-13.1%	784,724	17,659...	-14.2%	20,585,683
Non Traditional	522,173	-20.5%	656,886	522,173	7.8%	484,541	14,080...	-3.6%	14,612,900
Internet/Mail Order/Venue	515,745	-20.9%	651,846	515,745	8.8%	473,779	13,960...	-2.2%	14,281,482
Non Traditional Retail Outlets	6,428	27.5%	5,040	6,428	-40.3%	10,762	120,289	-63.7%	331,418
Christian Retail	6,151	8.5%	5,669	6,151	-62.7%	16,506	145,495	-83.1%	863,425
Product Configurations									
CD	614,184	-12.5%	701,613	614,184	-32.8%	913,535	18,507...	-30.2%	26,529,929
LP	345,229	-15.2%	407,329	345,229	32.7%	260,142	9,223,...	11.2%	8,297,379
Digital Album	681,716	-2.9%	702,420	681,716	-13.1%	784,835	17,660...	-14.3%	20,601,447
Others	4,337	-2.9%	4,465	4,337	-51.4%	8,932	154,225	-12.7%	176,563
Digital Tracks	4,863,...	4.3%	4,661,...	4,863,...	-22.4%	6,267,...	127,82...	-23.3%	166,580,...
Singles	36,344	4.7%	34,717	36,344	-29.6%	51,616	1,302,...	-5.6%	1,380,711
Videos	11,933	21.2%	9,845	11,933	-44.4%	21,460	343,892	-57.3%	805,440

FIGURE 12.15 *Equivalents-Sales-Store-Strata*

Weekly data available through: February 4, 2021 | Daily data available through: February 8, 2021 🗓 Data Av... 📷 Ad... 🖨 ...

In My Feelings ⊚⊙
Song by Drake released on July 10, 2018
CASH MONEY (R-CM)

SALES Total	STREAMING Total On-Demand	AIRPLAY Airplay Audience
106 (TP)	**2.7M** (TP)	**0** (TP)
⊕ -8.6%	⊕ +4.8%	-
732K (ATD)	1.7B (ATD)	1.7B (ATD)

Top Albums
Album Sales - This Period (TP)

1	Scorpion	106

Top ISRC
Track Sales - This Period (TP)

1	In My Feelings USCM51800206	68
2	In My Feelings USCM51800207	38
3	In My Feelings FR59R1861811	0

PERFORMANCE   MARKETS   CHARTS HISTORY   RETAILER/PROVIDER

Week Ending: 08/16/2018 🗓   Weekly   Daily

GRAPH EQUIVALENTS SALES STREAMING AIRPLAY ISRCS

Graph: 8 Week Trend

Graph Series   Normalized ∨   📊 ▦

| 06/28/20... | 07/05/20... | 07/12/20... | 07/19/20... | 07/26/20... | 08/02/20... | 08/09/20... | 08/16/20... |

● On-Demand Total   ● Digital Track Sales   ● Airplay Audience   ● Songs w/SES On-Demand

FIGURE 12.16A   *Drake's Artist Single Homepage*

Streaming On-Demand

		TP	% Chg		LP	YTD	ATD
✓ ∨	Total	66,733,456	-24.8%	⊕	88,794,472	580,478,201	580,478,201
∨	Provider	TP ⌄	% Chg ⌄	⌄	LP ⌄	YTD ⌄	ATD ⌄
	AOL Radio	-	-		-	-	-
	MediaNet	-	-		-	-	-
	Napster	78,193	-16.4%	⊗	93,495	1,059,897	1,059,897
	Slacker	3,100	-14.4%	⊕	3,622	37,750	37,750
	Others	66,652,163	-24.9%	⊕	88,697,355	579,380,554	579,380,554
∨	Audio	25,410,515	-16.5%	⊕	30,439,791	301,250,417	301,250,417
∨	Provider	TP ⌄	% Chg ⌄	⌄	LP ⌄	YTD ⌄	ATD ⌄
	AOL Radio	-	-		-	-	-
	MediaNet	-	-		-	-	-
	Napster	78,193	-16.4%	⊗	93,495	1,059,897	1,059,897
	Slacker	3,100	-14.4%	⊕	3,622	37,750	37,750
	Others	25,329,222	-16.5%	⊕	30,342,674	300,152,770	300,152,770
	Video	41,322,941	-29.2%	⊕	58,354,681	279,227,784	279,227,784

FIGURE 12.16B   *Artist Single On-Demand Streaming*

Streaming Programmed

	TP	% Chg		LP	YTD	ATD
⌄ Total	3,977,850	2.5%	⊕	3,881,268	21,632,388	21,632,388
⌄ Provider ⌃	TP	% Chg		LP	YTD	ATD
AOL Radio	-	-		-	-	-
Napster	1,155	11.1%	⊕	1,040	5,748	5,748
Slacker	52,048	122.7%	⊕	23,370	102,712	102,712
Others	3,924,647	1.8%	⊕	3,856,858	21,523,928	21,523,928
⌄ Audio	3,977,850	2.5%	⊕	3,881,268	21,632,388	21,632,388
⌄ Provider ⌃	TP	% Chg		LP	YTD	ATD
AOL Radio	-	-		-	-	-
Napster	1,155	11.1%	⊕	1,040	5,748	5,748
Slacker	52,048	122.7%	⊕	23,370	102,712	102,712
Others	3,924,647	1.8%	⊕	3,856,858	21,523,928	21,523,928
Video	-	-		-	-	-

FIGURE 12.16C *Artist Single Programmed Streaming*

Airplay

	TP	% Chg		LP	YTD	ATD
⌄ Spins	24,973	5.6%	⊕	23,649	98,196	98,196
⌄ Format ⌃	TP	% Chg		LP	YTD	ATD
AC Full Panel	90	20.0%	⊕	75	334	334
Adult Contemporary	0	-		0	4	4
Adult Hits	-	-		-	-	-
Adult R&B	127	21.0%	⊕	105	466	466
Adult Top 40	90	20.0%	⊕	75	330	330
All Video Channel	57	999%	⊕	4	61	61
Alternative	-	-		-	-	-
Christian AC	-	-		-	-	-
Christian Full Panel	-	-		-	-	-
Classic Hits	-	-		-	-	-
Classic Rock	-	-		-	-	-
College Radio	0	-		0	5	5
Country	0	-		0	33	33
Gospel	-	-		-	-	-
Jazz	0	-		1	12	12
Latin	334	-14.6%	⊕	391	1,168	1,168
Latin Pop	234	-2.9%	⊕	241	699	699
Mainstream R&B/Hip-...	5,069	-3.9%	⊕	5,276	27,833	27,833
Mainstream Rock	-	-		-	-	-
Mainstream Top 40	10,561	8.7%	⊕	9,718	37,068	37,068
R&B Full Panel	5,196	-3.4%	⊕	5,381	28,299	28,299
Regional Mexican	2	-		0	4	4
Rhythmic	5,272	0.9%	⊕	5,227	26,438	26,438
Smooth Jazz	3	50.0%	⊕	2	13	13
Top Forty Full Panel	15,838	6.0%	⊕	14,946	63,513	63,513
Triple A	0	-		0	1	1
Tropical	334	-14.6%	⊕	391	1,168	1,168

FIGURE 12.16D *Artist Single Airplay*

Airplay

	TP	% Chg		LP	YTD	ATD
Spins	24,973	5.6%		23,649	98,196	98,196
Format						
Audience	112,184,400	4.6%		107,252,300	470,154,200	470,154,200
Format	TP	% Chg		LP	YTD	ATD
AC Full Panel	252,600	0.2%		252,200	1,001,300	1,001,300
Adult Contemporary	0	-		0	9,400	9,400
Adult Hits	-	-		-	-	-
Adult R&B	769,600	-8.6%		342,000	3,501,300	3,501,300
Adult Top 40	252,600	0.2%		252,200	991,900	991,900
All Video Channel	5,700	999%		400	6,100	6,100
Alternative	-	-		-	-	-
Christian AC	-	-		-	-	-
Christian Full Panel	-	-		-	-	-
Classic Hits	-	-		-	-	-
Classic Rock	-	-		-	-	-
College Radio	0	-		0	4,400	4,400
Country	0	-		0	193,300	193,300
Gospel	-	-		-	-	-
Jazz	0	-		2,800	21,700	21,700
Latin	3,474,300	-6.4%		3,710,600	13,873,700	13,873,700
Latin Pop	2,524,500	-5.4%		2,667,800	9,075,700	9,075,700
Mainstream R&B/Hip-...	29,817,500	-5.4%		31,508,700	180,647,600	180,647,600
Mainstream Rock	-	-		-	-	-
Mainstream Top 40	48,412,400	5.9%		45,721,600	168,198,100	168,198,100
R&B Full Panel	30,587,100	-5.5%		32,350,700	184,148,900	184,148,900
Regional Mexican	1,700	-		0	3,600	3,600
Rhythmic	22,338,100	-2.4%		22,883,300	120,613,700	120,613,700
Smooth Jazz	300	50.0%		200	1,300	1,300
Top Forty Full Panel	70,751,000	3.1%		68,606,000	288,813,500	288,813,500
Triple A	0	-		0	300	300
Tropical	3,474,300	-6.4%		3,710,600	13,873,700	13,873,700

FIGURE 12.16E *Artist Single Airplay*

ISRCs (8)

	TP	% Chg		LP	YTD	2014 TD
Song Total	325,419	-23.5%		425,129	3,409,926	3,409,926
In My Feelings USCM51800...	256,867	-23.2%		334,419	2,708,526	2,708,526
In My Feelings USCM51800...	67,803	-24.5%		89,775	699,510	699,510
In My Feelings FR59R18618...	597	88.3%		317	1,120	1,120
In My Feelings USCMV1800...	152	-75.4%		618	771	771
In My Feelings JPQ3319200...	-	-		-	-	-
In My Feelings USCM51800...	-	-		-	-	-
In My Feelings UKLF219416...	-	-		-	-	-
In My Feelings UK3AZ1820...	-	-		-	-	-

FIGURE 12.16F *Artist Single ISRCs*

Weekly data available through: February 4, 2021 | Daily data available through: February 8, 2021 ▦ Data Av... ⬚ Ad... ⌨ ...

Scorpion ⊚ ⊛
Album by Drake released on June 29, 2018
CASH MONEY (R-CM)

SALES Equivalent: Albums w/TEA	STREAMING Total On-Demand	AIRPLAY Airplay Audience
196 (TP)	**18.3M** (TP)	**0** (TP)
⊕ -23.4%	⊕ +0.9%	-
716K (ATD)	9.9B (ATD)	8.7B (ATD)

Top Album Barcodes
Album Sales - This Period (TP)

1	0602567863182	39
2	0602567874942	39
3	0602567892410-Scorpion	23
4	0602567863212	3
5	0602567879152-Scorpion	1

Top Songs
Songs Sales - This Period (TP)

1	God's Plan	207
2	Nonstop	131
3	Nice For What	110
4	In My Feelings	106
5	Ratchet Happy Birthday	92

Week Ending: 08/16/2018 ▦   Weekly | Daily

GRAPH EQUIVALENTS SALES STREAMING AIRPLAY BARCODES SONGS

Graph: 8 Week Trend   Graph Series Normalized ∨ ▣ ▣

| 06/28/20... | 07/05/20... | 07/12/20... | 07/19/20... | 07/26/20... | 08/02/20... | 08/09/20... | 08/16/20... |

● On-Demand Total   ● Albums w/TEA w/SEA On...   ● Airplay Audience

All ISRC | Album ISRCs Only

Equivalents

	TP	% Chg		LP	YTD	ATD
✓ › Albums w/TEA w/SEA On-D...	119,345	-14.0%	⊕	138,752	3,090,980	3,090,980
› Albums w/TEA w/SEA On-D...	102,104	-12.8%	⊕	117,132	2,796,229	2,796,229
› Albums w/TEA	13,519	-25.2%	⊕	18,066	492,473	492,473

Sales

	TP	% Chg		LP	YTD	ATD
∨ Albums	7,054	-24.3%	⊕	9,324	261,593	261,593
› Retailer						
∨ Store Strata						
∨ Physical	3,879	-23.1%	⊕	5,041	34,114	34,114
Chain	410	-22.8%	⊕	531	3,842	3,842
Independent	541	-24.0%	⊕	712	5,777	5,777
Mass Merchant	2,341	-25.2%	⊕	3,128	18,358	18,358
› Non-Traditional	587	-12.4%	⊕	670	6,137	6,137
Christian Retail	-			-	-	-
Digital	3,175	-25.9%	⊕	4,283	227,479	227,479
∨ Product Configurations						
LP	0			0	0	0
CD	3,879	-23.1%	⊕	5,041	34,114	34,114
Digital	3,175	-25.9%	⊕	4,283	227,479	227,479
Others	-	-		-	-	-
Digital Songs	64,646	-26.1%	⊕	87,423	2,308,802	2,308,802

FIGURE 12.17 *Drake's Artist Album Homepage*

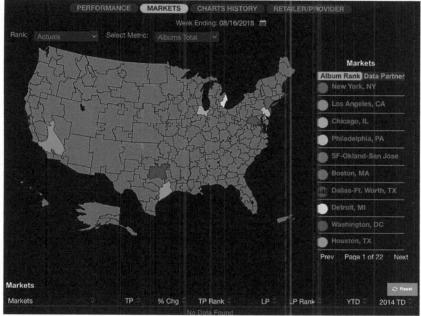

Top Album Barcodes
Album Sales - This Period (TP)

1	0602567863182	39
2	0602567874942	39
3	0602567892410-Scorpion	23
4	0602567863212	3
5	0602567879152-Scorpion	1

Top Songs
Songs Sales - This Period (TP)

1	God's Plan	207
2	Nonstop	131
3	Nice For What	110
4	In My Feelings	106
5	Ratchet Happy Birthday	92

FIGURE 12.18   *Artist Album Market Map*

Figure 12.18 graphically identifies top sales markets for Drake's *Scorpion* album release.

Clearly, syndicated research summaries are capable of revealing overall sales of music, a specific genre of music, a specific artist, a specific market, and many more aspects of the business as it pertains to sales of music.

GLOSSARY

Burnout or burn—The tendency of a song to become less popular after repeated playings.

Spin—The airing of a recording on a radio station one time. "Spins" refers to multiple airings of a recording.

Trades—This is a reference to the major music business trade magazines.

BIBLIOGRAPHY

"Amazon Best Sellers: Best Albums." *Amazon.Com*, www.amazon.com/Best-Sellers-MP3-Downloads/zgbs/dmusic/digital-music-album. Accessed 23 Mar. 2021.

"BDSradio." *BDS Radio*, www.bdsradio.com. Accessed 23 Mar. 2021.

"Billboard Charts Legend." *Billboard*, www.billboard.com/p/billboard-charts-legend.

"MRC Data – Sign In." *Music Connect*, www.musicconnect.mrc-data.com. Accessed 23 Mar. 2021.

"MRC Entertainment: Global Entertainment Company with Businesses Spanning Film, Television, Media and Data." MRC Entertainment | Global Entertainment Company with Businesses Spanning Film, Television, Media and Data, https://www.mrcentertainment.com/pmrc.

"Pollstar | Livestream." *Pollstar.Com*, www.pollstar.com/livestream. Accessed 23 Mar. 2021.

Rhodes, R., Account Executive with MediaBase, interview March 2021 "Spotify Charts." *Www.Spotify.Com*, spotifycharts.com/viral. Accessed 23 Mar. 2021.

"Top 40 (12+) | RateTheMusic | Free Ratings on Songs, Bands, and Artist | AllAccess.Com." *All Access*, www.allaccess.com/rate-the-music. Accessed 23 Mar. 2021.

"TRACtion." *CDX*, cdxcd.com/cdx-traction.

CHAPTER 13

Publicity

INTRODUCTION

Publicity is arguably the most important part of any marketing plan. It lays the foundation on which every other part of the plan is built. Labels often handle publicity for the artist's recording career, as well as for news and press releases about the label itself, through a publicity department. Sometimes an artist will hire a personal publicist to handle other areas of their life and career. By creating awareness of the artist, publicity makes all other aspects of the marketing plan more effective and promotion and sales efforts easier. By definition, publicity is earned media, promotion whose placement is not directly paid for (advertising) or owned (websites), and therefore the most accessible part of any marketing effort regardless of whether you are an independent artist or a major label act at the pinnacle of your career. Since both the publicity department and the publicist do the same job, we will refer to them both as simply the publicist.

PUBLICITY DEFINED

In traditional marketing, **publicity** is part of the public relations function that includes media relations, creating **press kits** and press releases, and lobbying. The AMA Committee on Definitions defines publicity as "the non-paid-for communication of information about the company or product, generally in some media form" (marketing-dictionary.org). Publicity is part of the broader category of public relations. More recently, practitioners have

DOI: 10.4324/9781003153511-13

CONTENTS

FIGURE 13.1   *In Line to See a Favorite Artist*

begun to distinguish promotion by its ownership (Table 13.1). Owned media are the promotion channels the artist or label controls, like their website or Twitter accounts. Paid media is advertising or other promotions where the label pays to have their message strategically placed. Earned media, also known as publicity, is when a story is shared as content (as opposed to advertising). At its extreme, earned media includes word of mouth, word of mouse, or viral marketing. The news is shared directly between individuals via email, text, or social media.

Publicity is distinguished from other forms of promotion by its low cost, but that low cost comes with a sacrifice of control. Because you are not paying for space in a magazine or time on a television network, there is no guarantee your message will get out and, even if it does, there is no assurance that your message will be communicated the way you intended.

The purpose of label publicity is to place nonpaid promotional messages into the media on behalf of the artist's recorded music project. That can range from a short paragraph in *Rolling Stone* to a mention in a music blog to an appearance on *Saturday Night Live*. Mentions and appearances in the new and traditional media contribute to the success of the marketing of the artist and their music. Earned media on behalf of a recording artist has a certain credibility that paid advertising does not. While an advertisement can

Table 13.1	Types of Media				
Media Type	**Definition**	**Examples**	**The role**	**Benefits**	**Challenges**
Owned media	Channel a brand controls	• Website • Mobile site • Blog • Twitter account	Build longer-term relationships with existing potential customers and earn media	• Control • Cost efficiency • Longevity • Versatility • Niche audiences	• No guarantees • Company communication not trusted • Takes time to scale
Paid media	Brand pays to leverage a channel	• Display ads • Paid search • Sponsorships	Shift from foundation to a catalyst that feeds owned and creates earned media	• In demand • Immediacy • Scale • Control	• Clutter • Declining response rates • Poor credibility
Earned media	When customers become the channel	• WOM • Buzz • "Viral"	Listen and respond — earned media is often the result of well-executed and well-coordinated owned and paid media	• Most credible • Key role in most sales • Transparent and lives on	• No control • Can be negative • Scale • Hard to measure

54869   Source: Forrester Research, Inc.

be bought, a feature article or review gets published only because a journalist thought the artist or their music was interesting enough to write the article and a publisher thought it was interesting enough to make the space available to present the story or run the review.

An online article, or one in a newspaper or magazine, suggests to the reader that there is something more to the label's artist than just selling commercial music. Published articles and TV magazine-style stories (for example, "E! News") give credibility to the artist in a way that paid advertising cannot.

There are key differences between publicity from the label and the advertising placed by the label. Label publicists generally create and promote messages to the media that are informative in nature and do not have a "hard sell" to them. Consumers are resistant to paid media (advertising) messages but more receptive to the subtle persuasion of the publicity effort that takes the form of an interesting story or review. Theoretically, the more impressions consumers receive about a recording, the more likely they are to seek additional information about the recording, and to purchase it. Advertising planners and sophisticated publicists use the term "reach and frequency" as they compile a strategy and its related budget. This means they plan an affordable campaign that can "reach" sufficient numbers of their target market with the "frequency" necessary for them to remember the message and act by purchasing. Publicity becomes the foundation, or at least a nice complement, to the advertising strategy without the direct costs of paid advertising.

HISTORY

The earliest music promoters were in the publicity business at the beginning of the 20th century, primarily helping to sell sheet music that was heard on

 *Antique Radio*

(Source: Photo Courtesy of Gordon Rolston)

recording playback devices or at public performances. Those who worked in the publicity profession in the early 1900s relied primarily on newspapers and magazines to promote the sale of music. In 1922, the federal government authorized the licensing of several hundred commercial radio stations, and those in the recorded music business found their companies struggling as a result. People stopped buying as much music because radio was now providing it for "free," and newspapers and magazines were no longer the only way the public got its news. Radio became the entertainer and the informer. But publicists found themselves with a new medium and a new way to promote and quickly adapted to it, as they did in 1948 with the advent of television as a news and entertainment form.

History has a way of repeating itself, but in 1993, the new technology was not a terrestrial broadcast medium but the Internet. Today the work of the publicist involves servicing not only traditional media outlets but online music and entertainment blogs and websites as well. Some of these online outlets are Internet extensions of magazines, newspapers, and video channels, which did not exist 15 years ago (and may not exist 10 years from now), while others are brand new with no traditional media presence. The label publicist also works with print media for feature articles, and they work with television program talent bookers to arrange live performances.

LABEL VS. INDIE PUBLICISTS

Large record labels usually have one or more people on their staff responsible for publicizing their activities and those of their artists, but with the contraction of the industry, much of the work has been outsourced to **independent publicists** on an "as needed" basis. This allows the label or the artist to hire the best person for the job rather than using an in-house publicist. If the in-house publicist lacks the contacts or expertise needed for a particular job, a label will have to hire an independent publicist anyway. Depending on the artist's status, an indie publicist can easily cost several thousand dollars a month and expect a six-month contract. Good publicity is clearly not free in that case.

TOOLS OF THE PUBLICIST

The traditional tools of the label publicity department or publicist have been a Rolodex full of contacts, a press kit, the press release, an artist bio, a couple of 8-by-10 glossy photos and the "three martini lunch." The Rolodex has been replaced by the computer and the smartphone with a database of contacts.

Kevin Akeroyd, CEO of Cision, the parent company of Bacon's Mediabase, said this in an interview on the Cube: "Early in my career, there were 20 influencers that mattered, they were all newspaper reporters or TV folks. There was only 20 of them. I had a Rolodex. So, I could take each one of them out for a three-martini lunch, they'd write something good about me. Now, you have thousands of influencers across 52 channels, and they change in real time, and they're global in nature. It's another example of where, well, if you don't automate that with tech and by the way, if you send out digital content, they talk back to you in real time. You have to actually not only do influencer identification, outreach and curation, you've got to do real time engagement."

A publicist is only as good as their contact list. Many come from media backgrounds and are sought after because their relationships with key people can get a story placed in a particular magazine or get an artist an appearance on a particular show. If you had the contacts (and the time), you could do the publicist's job yourself! So, before you hire a publicist, ask about their contacts and recent successes.

The publicist creates and distributes communications on a regular basis, so the maintenance of a quality, up-to-date contact list is critical to the success of that communication effort. Publicists may maintain their own contact lists or rely entirely on subscription database providers, but usually it is a combination of the two. The value of maintaining a quality database is that the information enables the publicist to accurately target the appropriate media outlet, writer, or producer.

An example of a subscription, or "pay" service, for media database management is Bacon's MediaSource. Bacon, now a part of Cision, updates its online database daily with full contact information on media contacts and news outlets and subjects on which they report in over 200 countries. Services like Bacon's can literally keep a publicity department on target. Most labels and their independent publicist partners maintain keyword, searchable lists within their databases to ensure they are reaching the appropriate target audience with each new message. These lists are used to distribute press releases, press kits, promotional copies of the CD for reviews, and complimentary press passes to live performances.

Cision offers an array of services, including in-depth information on media writers regarding their personal preferences and peeves as journalists. Journalists frequently change jobs, so having a service like Cision can make database maintenance easier for the publicist. Publicist also use paid wire services such as PR Wire and Business Wire for news release distribution for major stories.

 *At the Show*

The most effective way to reach media outlets is through email, because "it's inexpensive, efficient, and a great way to get information out very quickly" (Stark, 2004). A few media outlets still prefer regular mail or expedited delivery services, but the immediacy of the information is lost. An effective publicist learns the preferred form of communication for each media contact. Cision and companies like Meltwater or Propel provide some of that information, but it is always best to check with the journalist.

Internet distribution of press information from a label requires the latest software that will be friendly to spam filters at companies that are serviced with news releases. The most reliable way to ensure news releases and other mass-distributed information are received by a media contact is to ask them to add you to their email contacts. Some companies use services like Emma (www.myemma.com) or MailChimp (www.mailchimp.com) to track whether a news release was received and whether the receiver opened the email containing the news release. It becomes an effective way to be sure news releases are accurately targeted to interested journalists and to be sure they were able to get through spam filters.

THE PRESS KIT AND EPK

The term "press kit" comes from the package of materials that traditionally were prepared, usually in a folder of some sort, to give to the press or the media as a way of introduction of a new artist. As stated previously, the press kit would contain a brief **biography** of the artist, a picture or two, and a press release to go along with a copy of the artist's latest CD. If possible, the publicist would include album or concert reviews.

Physical press kits are used rarely these days. In some cases, "the deck," a PowerPoint presentation, has replaced the EPK. Some labels and publicists will attach the EPK to the end of the deck. The EPK offered several advantages, not the least of which was ease of distribution. The PowerPoint deck is easier still, since nearly every business computer will play PowerPoint slides with no additional software, and music and video can be embedded in the slides. Once created, the deck can be sent to hundreds of outlets with a few clicks of a mouse.

Another major advantage of the deck is the amount of content that it can contain and the ease of customization for each occasion. Because printing costs are eliminated, the variety of high-resolution photos that can be made available so that a media outlet doesn't have to use the same old publicity photo that every other publication has used is limitless. Videos of the artist performing live or in the studio can be included, making it possible for the music supervisor of a television show (or a local promoter) to actually see that artist perform before deciding whether to have them on the show.

The Internet has replaced the DVD and CD as the preferred method of delivering the EPK content. Many labels and artist managers post their artists' EPKs on a password-protected website accessible to the media and music supervisors for film and television. The reason for limiting access is to be sure that the high-resolution pictures and copies of music videos are not misused. Services that provide EPK templates and hosting include Sonicbids.com, bandzoogle.com, wix.com, and powerpresskits.com. Bandcamp Pro and Reverbnation are also good sites for EPK-like information without the password protection.

PHOTOS AND VIDEOS

Publicity photos and videos are not without cost. A full-blown photo shoot with a big-name photographer (think Annie Liebowitz or Randee St. Nicholas) can cost thousands of dollars, like $10,000 or $12,000! If you are going to spend that kind of money, you want to not only get some good head shots for the press kit but photos that can be used in the CD booklet and on the artist's and label's websites.

Why is a photo shoot so expensive? The largest expense is usually the photographer herself, but like so many things, you often get what you pay for. And using a big-name photographer creates its own publicity angle. In addition to the photographer, you will need to hire a makeup artist, a hair stylist, and maybe a wardrobe stylist or costume designer. If your artist is a band, then your expenses can easily double, as each member may require their own hair, makeup, and costumer.

Speaking of clothes, will the artist bring their own, or will they be rented or purchased? If they are a big enough act, you may be able to score some free outfits, as the exposure is mutually beneficial for the designer and the artist. If you want to shoot photos on somebody's private property, you may have to pay for that privilege. The more locations, the more expenses, including transportation. Sometimes location is used to help portray the artist's identity, but studio shots are easier to control (Knab, 2001). Most photographers will do the shoot using digital equipment that will save both time and money; however, some photographers still prefer to use film because of its unique qualities, kind of like how some producers prefer analog over digital recording.

Behind the scenes, there is catering and probably a film crew shooting video of the photo shoot for behind-the-scenes footage that can be used for additional publicity. Since this army of crew and artists may be on location for a day or two and having to travel to a restaurant is disruptive and time consuming, food will need to be provided at the location of the photo shoot.

All these decisions – food, clothing, locations, hair styles – should be determined *before* the day of the shoot by the marketing team, not left up to the artist and the photographer. Remember, these photos will determine or reinforce the artist's image or brand, maybe for their entire career and therefore need to be consistent with that brand.

The music video is an important part of the marketing department's effort to promote the album as singles are released. Production of the video is usually overseen by the label's **creative department** and may be promoted by same people who work the single to radio or by an outside agency like Nashville's Aristomedia. The artist's music video is also a valuable tool used by publicists in securing live and taped appearances in television programs. (Other uses of record label videos are discussed in Chapter 14.)

PRESS RELEASE

A **press release** is a brief, written communication sent to the news media for the purpose of sharing something that is newsworthy. It frames an event or a story into several paragraphs with the hope that media outlets will find it interesting enough to use as a basis for a story they will create. The key concept in the definition of the press release is that the item be newsworthy. Hard-core fans may want to know every detail of the artist's day, but the media and the average fan will not. The publicist and the marketing team must find the balance between too much information too often and too little too infrequently. Some events obviously call for an announcement to the press – the signing of a new record deal, the launch of a tour, the nomination or winning of an award or contest, a marriage, the birth of baby. Best to save the more mundane, day-to-day stuff for the artist's blog or their Twitter followers.

These days, it is not enough to send the press release just to the traditional news media because most of the tastemakers, the influencers in entertainment, are outside the mainstream media. Bloggers are a key target for press releases and a key source of information for traditional media. They are also likely to be more receptive to your story than traditional media, especially if your artist is not a star already. A good publicist will target the right media outlets for each story based on the appropriateness and likelihood that they and, more importantly, their readers or viewers will find the story interesting.

Press releases may be harder to write than you think. There are certain formats to be followed, and they need to be concise and still tell the full story.

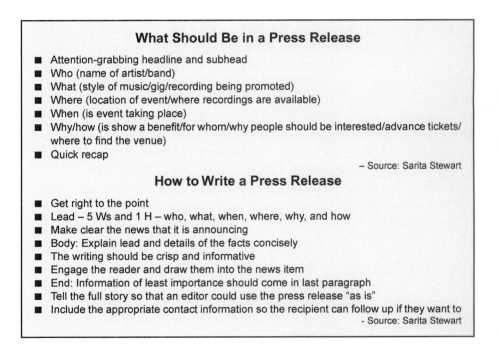

What Should Be in a Press Release

- Attention-grabbing headline and subhead
- Who (name of artist/band)
- What (style of music/gig/recording being promoted)
- Where (location of event/where recordings are available)
- When (is event taking place)
- Why/how (is show a benefit/for whom/why people should be interested/advance tickets/ where to find the venue)
- Quick recap

– Source: Sarita Stewart

How to Write a Press Release

- Get right to the point
- Lead – 5 Ws and 1 H – who, what, when, where, why, and how
- Make clear the news that it is announcing
- Body: Explain lead and details of the facts concisely
- The writing should be crisp and informative
- Engage the reader and draw them into the news item
- End: Information of least importance should come in last paragraph
- Tell the full story so that an editor could use the press release "as is"
- Include the appropriate contact information so the recipient can follow up if they want to

- Source: Sarita Stewart

The press release should be written with the important information at the beginning. Today's busy journalists don't have time to dig through a press release to determine what it is about. They want to quickly scan the document to determine whether this is something that will appeal to their target audience. A well-written headline will contribute greatly to this end.

It is important to include links to relevant photos and videos when the press release is going to Internet-based media so that they can quickly and easily post your story.

THE ANATOMY OF A PRESS RELEASE

The press release needs to have a **slug** line (headline) that is short, attention-grabbing, and precise. The purpose or topic should be presented in the slug line. It is suggested that a sub-heading be placed under the slug line that supports the point of the press release. The release should be dated and include contact information, phone number, email, and links to the artist's social media. The body of text should be double spaced.

The lead paragraph should answer the five Ws and the H (who, what, where, when, why, and how). Begin with the most important information; no unnecessary information should be included in the lead paragraph (Knab, 2003a). In the body, information should be written in the inverse pyramid form: in descending order of importance.

For the electronic news release sent via email, it should include embedded links where appropriate. Some links that should be considered to be included in the text of the news release should take the reader to the artist's website, the label's website, or to any other site that contributes to the journalist's understanding of the importance of the story. The added benefits of embedding links in electronic news releases are that they often become featured in blogs and music websites, which can improve page rankings by search engines on subjects relating to the artist, as well as helping fans find their ways to sites maintained for the artist's benefit.

THE BIOGRAPHY

Writing a **bio**, like writing a press release, takes practice and a special talent. While anyone can put down the facts in some ordered and organized way, a good biographer makes the story compelling and easy to read. A professionally written biography can cost $500 to $1,000, depending on the length. A bio for a press kit needs to be short, usually two pages, maybe three for an established artist. Before writing the bio, research is done on the artist's background, accomplishments, goals, and interests to find interesting and unique features that will set the artist apart from others. (Knab, 2003b) Keep in mind the target readership of the bio. The bio should be succinct and interesting to read (Hyatt, 2004) and create an introduction that clearly defines the artist and the genre or style of music. The hardest decision for the biographer may be deciding what to leave out. If you can afford it, you should hire a professional to write the artist's bio, and if you can't afford to hire a pro, give yourself time to write and rewrite the bio several times until you get the story just the way it needs to be.

An interesting new trend in biographies aimed at more visually oriented consumers is the bio and infographic. Created mostly by third-party sites, infographics are a creative, colorful way to convey otherwise humdrum facts. Biographical infographics have been created for celebrities as diverse as director David Fincher (www.hark.com), Conan O'Brien (sdrscreative.com), and Michael Jackson (www.biography.com).

Press Clippings

Nothing succeeds like success! Every press kit, whether physical or electronic, should contain examples of positive press the artist or their music has received, called **press clippings**. Newspaper articles should be reduced to 8–1/2-by-11-inch paper and the print kept large enough to be readable. One need not include every article but enough recent articles to let the recipient of the press kit know that others are excited about the artist and their music. Finally, it is important to include links to the press clippings on the artist's website.

PUBLICITY AND BRANDING

As noted previously, a major downside of earned media is neither the label nor the artist can control it, or at least not all of it. It takes only one bad decision by the artist, one momentary lapse in judgment, to undo years of work by the publicist and the marketing team building a strong brand. When an artist is signed to a label or embarks on a career with an independent marketing team, the major parties, artist, manager, and label marketing, need to get together and talk about who the artist is, what they want their public reputation to be, and how they will work together to achieve that goal. This is a time for a frank and honest discussion. Pushing the artist to be something they are not will be extremely difficult to sustain in the long run. At their core, the artist has to be who they are. Marketing and management can put some polish and shine on them with media training, wardrobe, and makeup, but in the end, the person on the inside will find his or her way to the surface. Don't try to make an Amy Winehouse into a Taylor Swift or vice versa. It just won't work.

Once the team has a grasp on the image for the artist, everyone must set about the business of building and constantly reinforcing that brand.

Of course, the most important contributor to this effort is the artist themselves. Every outfit they wear, every song they sing, every video they make,
and every word that comes out of their mouth should reinforce the brand.
Some labels hire hair stylists and clothing and costume consultants, some
will pay for dental work, and some are rumored to pay for cosmetic surgery
in order to polish the artist's image to prepare them for their expanded public
career (Levy, 2004). A media consultant may be hired to train the artist to
handle themselves in public interviews and other non-musical performance
occasions.

> Recording contracts which include comprehensive and multiple rights over career man-
> agement of their newest artists create a number of things that impact the image of the
> artist around which the publicist must work. They include:
>
> - The name chosen by the artist
> - Physical appearance of the artist
> - Their recording style and sound
> - Choices of material and songwriting style
> - Their style of dress
> - The physical appearance of others who share the stage
> - The kind of interviews done on radio and TV
> - Appearance and behavior when not on stage
>
> (Frascogna and Hetherington, 2004)

Working with the artist, the consultant prepares the artist for interviews
by taking the unfamiliar and making it familiar to them, teaching them
what to expect, and giving them the basic tools to conduct themselves well
in a media interview.

> *In the long-gone, golden, olden days . . . before TMZ, texting, Twitter,
> cell phone video, and YouTube, image consultants were better able
> to protect both their clients and the egos of journalists, who privately
> agreed to clear-cut parameters in exchange for celebrity access.*
>
> (Carol Ames 2011–2012)

Today's "journalists" may lack the professional training and ethics of the
traditional media. More importantly, with cameras built into almost every
cell phone, every public act, and some private ones, are likely to end up as a
thread on Reddit or a video on YouTube or somewhere else on the Internet.

When an artist is ready to remake their image, the transition, whether intended to be temporary or permanent, must be handled with extreme care and forethought. The history of music is littered with stories of attempts to rebrand artists. Some failed (e.g., Garth Brooks' alter ego Chris Gaines), and some succeeded (Katy Hudson the Christian artist became Katy Perry, the pop star; Darius Rucker morphed from pop to country, and Rod Stewart shifted from rock to singing standards with a big band).

THE PUBLICITY PLAN

The publicity plan is designed to coordinate all aspects of getting nonpaid press coverage and is timed to maximize artist exposure and record sales. The plan is usually put into play weeks before the release of an album. In the case of music magazines, the plan begins months in advance due to the long lead time necessary to meet their deadlines for publication.

Once the genre of music and the target audience for the artist have been determined, publicity planning begins by coordinating with the label's marketing plan and linking the plan's timetable with the marketing calendar. Remember, publicity is just one part of the overall marketing plan for the artist, all of which is coordinated by the marketing department or a marketing director. The media marketplace is then researched, and media vehicles targeted that have audiences, readers, or viewers that align with the artist's target audience. One of the major advantages of an Internet world is that it brings together people from all over the world. There is an outlet for almost every interest on the Internet, and the publicist must seek out those outlets that align with the artist's target market. Next, materials are developed, and the pitching to journalists and **talent bookers** begins. Lead time is the

Table 13.2 The Publicity Plan

The Publicity Plan

- Identify Target Market
- Set Publicity Goals
- Identify Target Media
- Create Materials
- Set up Timetable with Deadlines
- Pitch to Media
- Provide Materials to Media
- Evaluate

amount of time in advance of the publication that a journalist or editor needs to prepare materials for inclusion in their publication. A schedule is created to ensure that materials are created far enough in advance that they can be provided in a timely manner to make publication deadlines. Long-lead publications, mostly print magazines, are particularly problematic for the publicist, as they need to have materials prepared months in advance of the release date, and sometimes those materials are not yet available. If an artist suddenly breaks in the marketplace, it is too late to secure a last-minute cover photo on most monthly publications. Fortunately, today's print media almost always have an online version that can respond quickly to breaking news.

Before the materials are sent out, pitch letters are sent to targeted media requesting publication or other media exposure. The pitch letter is a care-fully thought-out and crafted document specifically designed to grab the interest of a busy, often distracted journalist, TV producer, blogger, or online website editor. It is never emailed to a bulk list but is specifically tailored to each media outlet being contacted (D'Vari, 2003). The pitch letter should begin with a few words presenting the publicist's request and then quickly communicate why the media vehicle being contacted should be interested in the artist or press material. In other words, the publicist will point out why the media's audience will be interested in this particular artist. Prep sheets are also developed and sent to radio programmers and their consultants so that DJs can discuss the artist as they prepare to play the music on the air. Retail and radio are given the first "heads up" about 16 weeks prior to street date for the album. This may be little more than telling a buyer or program director that the artist has an album scheduled to come out on a particular date, but it serves to create initial awareness and, hopefully, a buzz about the forthcoming release. Serious planning with radio and retail begins about ten weeks prior to street. New release materials, including one-sheets, which summarize the information about the new release, are mailed to retailers and media six to eight weeks prior to street in order to make deadlines. Inter-views may be done during this time as well so that articles will be ready for magazines that will hit newsstands the same time as the album hits store shelves.

Review copies are mailed or made available electronically to magazines two months in advance. Major newspapers, having shorter **lead times**, get theirs about three weeks in advance of street date. The artist will begin self-promotion (calling stations, broadcast media interviews, etc.) the week before street date. As the publicity plan unfolds, its success can be evaluated through clipping services and search engines to see how many "hits" the effort has resulted in.

Table 13.3	Example Publicity Timeline for a Major Label	

Jamie Jones's album *Living in Moment*
Release Date: Friday, August 11

Week of	6/5	Announce album via SM, website and press release; Premier first single (exclusive with appropriate media outlet) mid-week
Week of	6/12	Announce tour; begin magazine pitches for release week articles and reviews and for second single exclusive premier
Week of	6/19	Follow up on magazine pitches; first round of magazine articles appear
Week of	6/26	Second single premiers (exclusive), special magazine coverage ("Top Fall Releases")
Week of	7/3	Video premier on MTV Unplugged; follow up on magazine pitches, confirm reviews, and schedule phone interviews
Week of	7/10	Magazine features come out; film guest appearances
Week of	7/17	Final single premiers w/interview; phone interview
Week of	7/24	Album reviews appear on major genre websites
Week of	7/31	Local media day; radio and TV performances, in-person interviews, podcast tapings
Week of	8/7	Jamie sharing 1–2 posts/day on SM. Plug release show. Local papers run interviews and plug release show
Friday	8/11	Release date

Adapted from Stewart et al. (2020)

Table 13.4	Preliminary Publicity Timeline for Indie Artist Prototype

Preliminary Publicity Timeline for Indie artist Prototype

Day	Date	Event
Fri	12/6	Advance promo copies at *The Rocket, UW Daily, Pandemonium, Seattle Times,* and the *Seattle Weekly*
Mon	12/9	Arrange interviews on KCMU Seattle and KUGS Bellingham
Fri	12/13	Deadline for completing database of print/broadcast mailing list
Fri	1/10	Single sent to college radio
Fri	1/17	***Calling It a Day* CD release day**
Mon	1/20	Album mention in *UW Daily/Seattle Times*
Tues	1/21	Album mention in *Seattle Weekly*
Thur	1/24	Tour begins in Seattle
Thu	1/24	Album mention in CMJ/Hits
Tue	1/28	Album mention in *The Rocket* and *Pandemonium*
Tue	1/27	Album mention in Spokane and Tri-City Daily papers
Mon	2/2	Album mention in *Cake* and *Fizz*
Fri	2/6	Album mention in *Flipside* and *Village Noize*
Wed	2/15	Album mention in *Spin*
Wed	2/15	Album mention in *Virtually Alternative*
Fri	2/17	Album mention in *Next*
Fri	2/24	Album mention in *Magnet*
Mon	2/27	Interview on KUGS Bellingham
Mon	2/27	Interview on KCMU Seattle
Tue	2/28	Feature story in *The Rocket*

(Source: Christopher Knab, www.4frontmusic.com)

BUDGETS FOR MONEY AND TIME

Time, energy, and talent can be more important than budget.

(charity founder Scott Harrison)

A budget for the publicity campaign is developed based on the objectives of the project, the expectations of the label for the part publicity will play in stimulating interest in the artist's music, and the degree to which the label is managing the artist's career. If this is the first album for the artist, the development of new support materials may be necessary, such as current photos and a bio. If the new artist is working under a multiple rights contract, tour press support will be necessary. If it is an established artist, budgets could be considerably higher, in part because of the expectations of the artist to receive priority attention from the director of publicity.

Publicity costs include the expense of developing and reproducing materials such as press kits, photos, bios, video, and so forth, communication costs (postage and telephone bills, maintenance of contact lists), and staffing costs. The minimum cost for an indie label would run about $8,000, with $3,000 of that for developing press kits and $2,800 for postage. Adding an outside consultant to the project would add another $1,500 or more per month. For major label projects, an outside publicist can be hired for six months to provide full support to a single and album and tour publicity for $25,000, which includes out-of-pocket costs such as postage, press kits, website maintenance, and anything else the label requires to support the publicity effort.

An equally important part of the plan is to budget adequate time to support the album based on when it will be released during the annual business cycle of the label. If the in-house staffing is adequate, given the timing of the project, the plan can be executed without additional help. If, however, the publicity department is overloaded, the director may consider hiring an independent company to handle publicity for the project. This seemingly removes the burden from the director, but it adds oversight duties, since the director must be sure the outside company is working the plan according to expectations. The ultimate success (and failure) is still the responsibility of the director of publicity.

OUTLETS FOR PUBLICITY

The Internet – One of the negatives about the Internet is how it has fragmented the music market and made it so much harder to have a multi-platinum seller. One of the positives about the Internet is how it has brought the fragmented markets together in one place. No matter how obscure your interests, you can probably find someone else, a blog, an e-magazine, or a website that shares your interest. You will want to scour the Internet for the sites appropriate for your artist and target them for press releases, interviews, and reviews.

Blogs – Bloggers play a critical role in entertainment publicity because they often have greater credibility with young consumers who are actively searching for new music. A recent study found that traditional media is using blogs as sources of information for their own stories more and more often (Messner and Distaso, 2008). Hypebot.com recommends identifying bloggers and writers specifically, rather than sites or blogs, because they have their own musical preferences, and they will do the actual writing of the story (5 Tips for Identifying & connecting with bloggers). Some publicists build relationships with the bloggers and other influencers before they ever pitch them. They study them and what and who they are writing about and target them accordingly. Feedspot.com provides targeted lists of blogger contacts for one dollar or less each. Pitchfork, Consequences of Sound, Tiny Mix Tapes, and Stereo Gum are some of the most influential music blogs. Companies like Agility PR Solutions and Cision's PR Suite provide services that will distribute your message to influencers and monitor the results.

FIGURE 13.4   *Social Media*

Web-zines – Online magazines offer a good way to introduce a label's artist to the target market. MarketingTerms.com defines *e-zines* as electronic magazines, whether posted via a website or sent as an e-mail newsletter. Some are electronic versions of existing print magazines complete with magazine-style formatting, whereas others exist exclusively online or digitally. The web-posted versions usually contain a stylized mixture of content, including photos, articles, ads, links, and headlines, formatted much like a print equivalent. Smaller versions may be emailed to subscribers as a pdf file. Most online magazines are advertiser supported, but a few charge a subscription fee.

Many established music magazines are genre specific or have particular subject areas dedicated to genres. They may feature music news, concert and album reviews, interviews, blogs, photos, tour information, and release dates. As a result, their readers are predisposed to be receptive to new and unfamiliar artists and their music, provided that the artist is within the genre that the webzine represents. A study of the readers of the Americana music magazine *No Depression* found that 90% of their readers learned about new music from an article published either in a print or in an online version of a magazine. It is probably no coincidence that *No Depression* started out as a print magazine in 1995 and added an online version in 1997 on its way to becoming exclusively online in 2008.

What to Send

E-zines are mostly interested in feature articles and press releases pertaining to some newsworthy item (such as an album release or a tour schedule announcement). Label publicists write the article with an assumption that it will appear unedited in the online publication, in the inverted pyramid style. Articles like this also include an attached publicity photo or two along with the article for submission. As with any other publication, label publicists should take care not to send a news release to an e-zine if there is nothing that is considered newsworthy. Like all journalists, those at e-zines will look at the news release to determine its relevance and timeliness for their readership, and an irrelevant news release strains the relationship the publicist has with their contacts.

Where to Send It

The Ezine Directory has a listing of many of the better-known music e-zines, along with descriptions and ratings of each (www.ezine-dir.com/Music). The goal, like any target of publicity, is to find those with the correct target market and submit articles, music, and photographs to the editor, encouraging him or her to include a link to the artist's website. Some e-zines have submission forms available on their website, whereas others are not as specific about their submission policy. When an article does appear online, the publicist will link the artist's and sometimes the label's websites to it.

Resources for influencers and distribution of press releases:

Influencers come and go, as do the websites that track them, so your best strategy is to search the term on a regular basis for the most up-to-date information. Here is a list of sources at the time of the publishing of this text.

The Ezine Directory: www.ezine-dir.com
PRWeb: www.prweb.com (Cision)
PRLog.com
PRbuzz.com
Ereleases.com
Influence.co
Neoreach.com

AMAZON.COM AND WIKIPEDIA AS RESOURCES FOR BASIC INFORMATION

Labels supply and offer product through Amazon.com not because of the sales but because people use it as a resource when looking for information.

Wikipedia.com has over 6 million articles in 287 languages and receives over 18 billion page views per month. "While anyone can contribute to a given article, they must first pass muster from a team of volunteer editors with a particular passion about the subject before the text appears live." Photos and clips of audio and video can be added to the page (not full songs or videos). This is not the place for breaking news because the editing process takes too long, but it is a good place to post the artist's bio and historically oriented information, including the artist's discography and other facts about the artist and their career (Bruno, 2008).

Broadcast – Getting publicity on broadcast outlets will be more of a challenge for a new or unsigned artist because of the limited time, so being persistent (without being annoying) and targeting the appropriate outlet with the right information is important. Many factors go into these booking decisions, but mostly the publicist needs to do their homework to make sure that the artist and their story are a good match for the show.

TELEVISION APPEARANCES

News Shows

Major entertainment television news shows, including syndicated news shows on major network affiliates and cable channels, are most often interested in major acts. Their viewers want to know the latest information about their favorite recording artists. Stars with the highest **Q factor**, that is, those celebrities that are easily recognizable, are most often sought for their entertainment news stories because they draw a lot of interest and big audiences. With major artists filling prime interview opportunities, it becomes a genuine challenge for the record label that is trying to publicize a new artist. In order to compete with the superstars who can easily get airtime, a new act must have an interesting connection with consumers that goes beyond the music. There are more artists looking for publicity than slots on talk shows, forcing label publicists to be as creative as they can to get the attention of producers and music bookers for their new acts.

Television interviews with new artists require a back story that sets them apart from every other "new artist with a great voice" who is seeking the media spotlight. Shows look for that added dimension to a new artist that makes them interesting to the viewers, and they often look for a nontraditional setting in which to present the story. Though at times it is overdone, connecting an artist with their charity work becomes an interesting angle for television.

The challenge to the label publicity department is to find those key personal differences that make their recording artists interesting beyond their music. Label publicists are sometimes criticized for citing regional radio airplay, chart position, or YouTube views as the only positives that make their newest artists stand out. Those in the media say they look for that something special, different, and newsworthy that gives an angle for them to talk or write about. In marketing terms, the media is looking for strong brands that are uniquely positioned against the competition. In that light, it puts the responsibility on the label publicist to find several different angles to offer to different media outlets to generate the interest needed to get a story placed. Writers for major media want their own angle on an artist when possible because it demonstrates to media management that an independent, standout story has been developed, making them different from their competition. Sometimes, though, the story angle about an artist is different enough that it stands on its own and most media will see the value it has for their audiences. Entertainment writers and producers are often self-described storytellers, and delivering that unique story to them is a continuing challenge to the successful label publicist.

Talk-Entertainment Shows

Label publicists may facilitate an artist's appearance on popular talk and entertainment shows, often with the result of introducing an artist to an audience that is not actively seeking new music. This would include shows like *Jimmy Kimmel Live*, *The Late Show*, and daytime shows like the *Ellen DeGeneres Show* and the *Kelly Clarkson Show*.

According to Tompkins (2010), *Saturday Night Live* has the biggest impact on sales after an appearance. This is probably because there are no competing shows in that time slot and the artist gets to play two songs. Shows like *The Late Show* (on NBC) and *The Tonight Show* (on CBS) typically limit the artist to one song, and it is at the end of the show, when many viewers may have already tuned out or fallen asleep. The payoff to the label for an artist's appearances on television shows comes in the form of ticket sales and streams.

Bookings on programs like these are handled by the publicist based upon their relationships with talent bookers on these shows. However, the success of placing the label's new artist on one of these shows is also based on the ability of the publicist to build a compelling story for the artist that will interest the booker. Often the publicist will promise another major artist for a later appearance in exchange for accepting the new artist now.

Major labels have the benefit of their high-profile roster of artists and the financial resources to promote live performances to major shows. Independent labels, with their much smaller promotion budgets, must, by necessity, approach a pitch for a live performance keeping those limited dollars in mind. Cole Wilson was the music booker for *The Late Late Show with Craig Ferguson* and offers these points to the indie publicist seeking a performance on the show.

- The artist needs an online presence where the booker can see performances and read the comments left by fans.

- Talent bookers for late night shows in New York and Los Angeles frequently spend time visiting live entertainment venues in those cities, providing the bookers an opportunity to see a prospective artist.

- If an artist is "different from the norm," it gives the booker an opportunity to present something fresh to the audience. She says an artist who sits on a stool and sings for three minutes lacks visual appeal.

- The publicist should remember that it isn't just the talent/music booker who must be convinced the artist should be invited to the show. Often it is a committee who will want to view the artist from as many perspectives as the publicist can present.

- The artist should remember that an appearance on television does not mean that they can make "outrageous demands" from the show. (Donahur, 2008)

Melissa Lonner, who served eight years as the *Today* show senior producer and entertainment booker, gives the following advice to label publicists pitching an artist for an appearance on the NBC morning show.

- Keep your pitches short over the phone and/or email
- Don't pitch on voicemail
- Send a CD of music with selective press clippings
- Send an email to follow up and recommend a track
- Don't send the deluxe press clippings collection
- Don't say why the artist should be on *Today*
- Provide the music, stats, and facts – not the hype
- Don't say that the artist is the next "----"
- Follow up on pitches via email or phone
- Be kind, calm, and honest
- Don't stalk, threaten, or demand

(Paoletta, 2007)

Award Shows

The value of having an artist perform on an award show is obvious – it provides tremendous artist exposure and sells recordings and concert tickets. These slots are coveted by all the record labels, and lobbying efforts may pay off in a big way. While most awards shows are showing modest declines in viewership, artists who are nominated or who perform on music award shows can see major spikes in the sales and streaming of their music. For example, the day after the 2021 Grammy Awards, Meghan Thee Stallion, awarded the Grammy as the Best New Artist, saw a 178% increase in album sales and an overnight increase in sales of 513% for the "Savage" remix. H.E.R.'s "I Can't Breathe," awarded Song of the Year, saw an overnight 6,771% increase in sales, but the biggest jump was the album (7,800%) and song (13,892%) sales of Black country artist Mickey Guyton. Guyton was nominated for, but did not win, Best Country Solo Performance for her song "Black Like Me."

CHARITIES AND PUBLIC SERVICE

People expect successful artists to give back to society. After all, the world, the fans, has given them so much. But don't wait until your artist wins her first Grammy to do some charity work. Having an artist associated with a respected charity is always good publicity. If the cause is personal, even better. We once worked on a marketing plan for a band whose members had lost their mother to breast cancer. They played every cancer charity gig they were offered. It was personal, not just something they were doing for publicity. As the artist's star rises, the association with a specific charity becomes even more of a win-win, giving both the artist and charity greater exposure. Some artists have become synonymous with their charities: Elton John and the Elton John AIDS Foundation, Jars of Clay and Blood Water Mission, and Bono and Band Aid, for example.

Another way for artists to get involved in charities is through organizations like Global Citizen. This organization fights poverty by getting people to volunteer in order to receive free concert tickets. "The goal is to reward volunteer work with live music. Fans can take part in various social actions, ranging from signing partitions to calling their representatives to earn points they can use to win free concert tickets" (Waddell, 2013). Artists donating tickets to the cause included the relatively unknown Billie Eilish, Miley Cyrus, Bruno Mars, and Usher.

BAD PUBLICITY

Is all publicity good publicity? What comes to mind when you hear these names: Ty Herndon, Janet Jackson, Steven Page, Ozzy Osbourne, Kanye West, Amy Grant, Paul Simon, Ian Watkins? All have received "negative" publicity at some point in their musical careers, and most, but not all, have survived in the business. So, the question is "Why?" Why do some artists bounce back or even thrive despite arrests or drug addiction and others have their careers ruined? Brand image or, more precisely, consistent brand image. When Kid Rock was arrested (Billboard.com, 2005) for punching a strip club DJ, did he lose any fans? Probably not. It hardly made the news. Why? Because the behavior was not inconsistent with his image, his brand. But imagine what would happen if Christian artist Steven Curtis Chapman had been arrested for punching out a strip club DJ! He didn't, but if he had, that would have made the news and probably ended his career because it is not consistent with his image, his brand. For a real-life example one need only look at R. Kelly's career trajectory. At the 1997 Grammy Awards, Kelly won three Grammys for the song "I Believe I Can Fly" and has been nominated more than 20 times since. In 2002, Kelly was indicted and later acquitted on charges of child pornography. In January 2019, a Lifetime channel docuseries, watched by nearly 2 million viewers, detailed allegations of sexual abuse by Kelly and generated over 743,000 social media interactions after the first episode (Hohman, 2019). Public outrage pressured his label to drop him. The next month, he was indicted on ten counts of aggravated criminal sexual abuse, and by the end of the year, he was facing additional charges, including charges for sex trafficking. An interview with Gayle King of CBS News "devolved into a full-blown temper tantrum" according to *USA Today* (McDermott, 2019), and the R&B star's career has never recovered.

What to Do When Negative Publicity Occurs

Every situation is unique and will call for professional judgment. The following is meant to be a general guideline. There is no "one size fits all" solution to negative publicity.

Chances are the first person the artist will call when something bad happens is their personal manager. Managers are usually easier to face than an angry spouse and may have more experience with bailing people out of jail. Then the manager will call the publicist. Before anyone makes a public statement, get the facts about what happened. Talk to your artist and anyone else that was in their entourage when the "event" occurred. Make sure you are getting the truth and not just the artist's version of what happened. You don't want to be surprised later when contradicted by other witnesses. Every cellphone has a camera in it. Look for videos of the event to show up on Instagram, YouTube, TMZ, and the evening news.

Once you have the facts, you can decide how you and the artist should respond. You have two basic choices: Ignore it or respond to it. You can safely ignore the situation, that is, not have an official response, if the event simply reinforces the artist's reputation. If your Blues artist was seen smoking an illegal substance but not arrested, and the story makes it to the tabloids, fine. The behavior is not unexpected or contrary to his image. On the other hand, if your artist is involved in an accident or hurts somebody, you will need to respond publicly. Be direct and as open and honest as possible. If the artist has a drug problem or other reoccurring behavioral problem, then management will probably want to direct them to a rehab program. Fans are usually sympathetic when the problem rises to the level of disease or addiction, and sometimes managers use this to their advantage, even if the artist is really just a badly behaved spoiled brat! The rehab program will take the artist out of the public view for weeks or even months, allowing the furor to subside and for them to get their act together. And with any luck, they may even come out a better person and be inspired to write a hit song about it.

CONCLUSION

Publicity, or earned media, is one of the most powerful tools that a record label marketer has at their disposal. Unfortunately, much of it is also out of their control. For publicity to be successful, the label needs the cooperation and assistance of the traditional and internet media. The most successful publicity will go viral and spread directly from fan to fan. This will be discussed further in the Internet and social media chapter.

GLOSSARY

API—The acronym for application programming interface, which is a software intermediary that allows two applications to talk to each other. Each time you use an app like Facebook, send a text message, or check the weather on your phone, you're using an **API** (Mulesoft.com).

Bio—Short for **biography**. A brief description of an artist's life and/or music history that appears in a press kit or other publicity material.

Creative department—This is a department or division at a record label that handles design, graphics, and imaging for a recorded music project. Also called creative services.

Clippings—Stories cut from newspapers or magazines.

Discography—A bibliography of music recordings.

Electronic press kit—An electronic version of a standard artist press kit that includes digital images, documents, audio and video files, and PDF versions of all documents and news clippings. Some may contain video clips that can be used on-air and magazine-quality images for reprint purposes.

Independent publicist—This is someone or a company that performs the work of a label publicist on a contract or retainer basis.

Lead time—Elapsed time between acquisition of a manuscript by an editor and its publication.

Press kit—Collection of printed materials detailing various aspects of an organization, presented to members of the media to provide comprehensive information or background about the artist.

Press release—A formal printed announcement by a company about its activities that is written in the form of a news article and given to the media to generate or encourage publicity.

Q Factor—A term used to indicate the overall public appeal of an artist in the media. A high Q factor means an artist is able to draw large television audiences.

Slug—A short phrase or word that identifies an article as it goes through the production process, usually placed at the top corner of submitted copy.

Talent bookers—These are people who work for producers of television shows whose job it is to seek appropriate artists to perform on the program.

Tech stack—the combination of technologies a company uses to build and run an application. A **tech stack** typically consists of programming languages, frameworks, a database, front-end tools, back-end tools, and applications connected via APIs (https://heap.io/).

BIBLIOGRAPHY

Akeroyd, K. (2019) "Kevin Akeroyd, Cision," *Cube Conversation*, March 14, 2019.

Allen, P. (2007) *Artist Management for the Music Business*, Boston, MA Focal Press.

Ames, C. (2011–2012) "Popular Culture's Image of the PR Image Consultant: The Celebrity in Crisis," *Image of Journalist in Popular Culture Journal*, pp. 90–106.

"Creative Marketing: Making Media Waves: Creating a Scheduled Publicity Plan. Music Business Insights," (Issue I:3) www.musicianassist.com/archive/newsletter/MBSOLUT/files/mbiz-3.htm.

DeSantis, R. (2021) "Global Citizen Launches Recovery Plan for the World: A Path Forward,' Says Priyanka Chopra Jonas," *People.com*, February 23, 2021.

D'Vari, M. (2003) www.publishingcentral.com/articles/20030301-17-6b33.html. How to Create a Pitch Letter.

Dommeruth (1984) *Promotion: Analysis, Creativity, and Strategy*, Belmont, CA: Wadsworth, Inc.

Donahur, A. (2008) "The Indies Issue: How to Get on a Late Night Show," *Billboard Magazine*, June 28, 2008, Vol. 120 Issue 20, p. 27.

"Five Tips for Identifying & Connecting with Bloggers" (2012) http://hypebot.com/hypebot/2012/11/5-tips-for-identifying-connecting-with-music-bloggers.html.

Frascogna, X., and Hetherington, L. (2004) *This Business of Artist Management*, New York: Billboard Books.

"Grammy Awards 2013: Sales Soar for Mumford & Sons, Fun. and Goyte in Wake of Awards Show Glory," www.nydailynews.com/entertainment/music-arts/mumford-fun-gotye-enjoy-post-grammy-spike-article-1.1268860 (retrieved May 1, 2014).

"Grammys Spike Sales," www.grammy.com/blogs/grammys-spike-sales (retrieved May 1, 2014).

Havighurst, C. (2007) "All Things Considered: Nashville Band Leaves Label and Thrives," *National Public Radio*, February 7, 2007.

Hohman, M. (2019) "Twitter Explodes Over Shocking R. Kelly Doc: 'About Time' Many Say as Others Defend Singer," January 4, 2019. https://people.com/music/r-kelly-documentary-twitter-reactions-about-time/ (retrieved March 28, 2021).

Hurst, J. (1995) "Rising Star's Drug-Related Scandal Gets Mixed Reviews," http://articles.chicagotribune.com/1995-07-02/entertainment/9507020153_1_airplay-country-music-music-industry (retrieved May 1, 2014).

Hyatt, A. (2004) http://arielpublicity.com. How to be Your Own Publicist.

"Kid Rock Arrested on Assault Charge" (2005) www.billboard.com/articles/news/64079/kid-rock-arrested-on-assault-charge, February 16, 2005 (retrieved May 1, 2014).

Knab, C. (2001) www.musicbizacademy.com/knab/articles/. Promo Kit Photos.

Knab, C. (2003a) www.musicbizacademy,com/knab/articles/pressrelease.htm. How to Write a Music-Related Press Release, November, 2003.

Knab, C. (2003b) www.musicbizacademy.com/knab/articles.

Lathrop, T., and Pettigrew, J, Jr. (2003) *This Business of Music Marketing and Promotion*, New York: BPI Publications.

Levy, S. (2004) CMA's music business 101, unpublished.

Messner, M., and Distaso, M. W. (2008) "The Source Cycle," *Journalism Studies*, Vol. 9, Issue 3.

Paoletta, M. (2007) "As Seen on TV," *Billboard Magazine*, April 21, 2007, Vol. 119, Issue 16, p. 27.

"Publicity," Marketing Dictionary (2017) marketing-dictionary.org.

Stark, P. (2004) personal interview.

Stewart, K., et al. (2020) "Music Publicity: A Practical Guide. Maeve McDermott R. Kelly's 'CBS This Morning' Interview Was Disgusting – But We Couldn't Look Away," *USA Today*, March 6, 2019.

Tompkins, T. (2010) "The Impact of Late Night Television Musical Performances on the Sale of Recorded Music," *MEIEA Journal*, Vol. 10, Issue 1.

Waddell, R. (2013) "Global Cause," *Billboard*, May 11, 2013.

Walker, J. (2009) personal interview.

Yale, D. R. (1993) *The Publicity Handbook*, Chicago: NTC Business Books.

APPENDIX: STRUCTURE FOR A PRESS RELEASE

FOR IMMEDIATE RELEASE or FOR RELEASE DATEFOR MORE INFORMATION: Name, phone, email, City/State, Date

HEADLINE

- Considered the most critical part of the release.
- You have 20 seconds to grab the reader's attention.
- Often the only part of the press release that the media reads.
- Use short, clear, and hard-hitting one-line summaries to identify what you are promoting. Don't be afraid to be dramatic.
- Sub-headings are also used to attract attention and provide information.

FIRST PARAGRAPH

- Purpose of the first paragraph is to alert the media and to inform them of what you are promoting – who, what, when, where, why, how.
- Make sure that the first paragraph has no more than three or four sentences.
- Needs to set forth all of the main points covered in the release.

BODY OF PRESS RELEASE

- This is an opportunity to provide the details of the story.
- Press release should be written in the third person.
- Don't weigh the piece down with extraneous details.

END PARAGRAPH

- Should summarize the story
- #### or – 30-About Us Contact Information

    Source: Sarita Stewart.

CHAPTER 14

Digital Marketing

Digital marketing uses electronic devices like computers and phones to engage with consumers. Digital offers many advantages, but two of the biggest are being able to finely target the audience and measure the success of promotional campaigns.

There are several major components of digital marketing efforts regardless of what product or brand being promoted – paid advertising, website design, search engine marketing and search engine optimization, email and newsletters, social media, content marketing, online PR, blogs, mobile/texting, web analytics, and optimization. We have already reviewed paid media and publicity in chapters dedicated to those topics. Here, we will also look at what other digital marketing efforts are being embraced regularly by recorded music marketers.

Let's first review paid, owned, and earned media and discuss how these types are presented in the digital landscape. **Paid media** refers to any publicity or awareness that is gained through a paid ad placement such as ads on social media platforms, radio and television, search ads, pre-rolls on YouTube, banners, outdoor advertising, and print ads. For digital marketing, top paid efforts generally center around ads on social media and search ads (including contextual ads that show up on various websites based on keywords that are targeted by the marketing team). **Owned media** refers to the promotion channels on which a brand controls the messaging such as an artist's website, email, blog, mobile apps, or social accounts, like a Facebook Page, Instagram, or Twitter account. **Earned media** refers to positive publicity generated through user-generated content such as word-of-mouth

CONTENTS

DOI: 10.4324/9781003153511-14

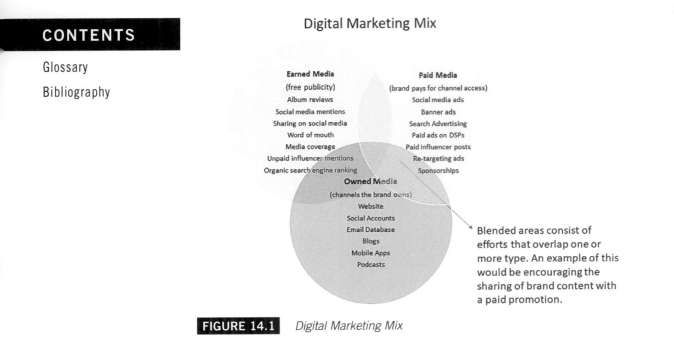

FIGURE 14.1 *Digital Marketing Mix*

promotion, music or live show reviews, shared blog posts, and comments and shares on socials.

SOCIAL MEDIA: CONNECTING WITH FANS

So much interest today is being placed on social media efforts to market products. Artists and music companies are using social media to create closer ties to their fans, generate buzz, and drive traffic to their websites for more information or to a streaming site to listen to a track.

Music fans have always been prone to share their favorite acts and tracks with their friends, so social media platforms provide an easy way to facilitate that sharing and enable artists and labels to build direct-to-consumer fanbases. Social media sites like YouTube, TikTok, Facebook, Twitter, Instagram, and Snapchat are platforms used widely in the music industry and serve as hosts where many conversations about the artist and music are taking place.

Social media is an important component for an artist to build and maintain a two-way relationship directly with their fans. And many artists engage regularly with their most loyal supporters on socials to build a deep and long-lasting connection. A central theme throughout this textbook has been the importance of understanding the target audience, and that is particularly true in developing a successful social media strategy. Many marketers feel they must be hyperactive on every popular platform, and while that might be ideal for artists with large teams, it is not always possible for small marketing groups or individual digital marketers trying to balance the workload by themselves. Instead, a better strategy might be to focus largely on the two or three platforms that are most meaningful and spend more time cultivating those audiences.

Generally, goals for a company or brand on social networks include one or more of the following: distributing content to the masses, informing a community or audience, building the brand, and monitoring the community's reaction to news around the artist, thus being able to more quickly manage any negative sentiments that might damage the brand. For the marketing of artists, social media is extremely important in helping increase awareness and building an artist's fan community.

Social Media vs. Social Media Marketing

Social media refers to all online communication channels – like Facebook, TikTok, Instagram, Twitter, and Snapchat – through which users create online communities where they can share thoughts, ideas, and content they like. **Social media marketing** refers to the process of gaining attention or increasing traffic to a website through social media sites. It is important to always think about how social activity may be used to drive traffic back to the artist's website because that is the central hub that is fully owned and controlled. It is often suggested to think of the artist's website as "home base."

Super Fans

Although social media allows an artist to connect to every fan, the artist team must realize that not all fans are equal. Some are very passive, while others would be considered "super fans" who are the most engaged followers on social media and the most active contributors to the artist's financial success. The super fans are motivated to pre-save new music on Spotify, purchase merchandise, pay for an exclusive opportunity with the artist, and buy tickets to a live show. The "80/20 rule" is a term used across industries to refer to the fact that 80% of all sales by a company can be attributed to only 20% of that company's consumers. The same is true in music – the biggest fans are often responsible for the largest portion of an artist's financial revenue. In addition to contributing to the bottom line,

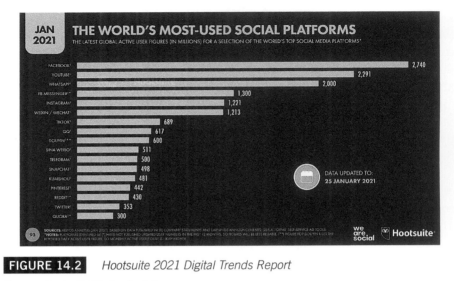

FIGURE 14.2   *Hootsuite 2021 Digital Trends Report*

(Source: Hootsuite/We Are Social)

though, the super fans also serve as an artist's brand ambassadors – sharing information and evangelizing about the artist's music. Passionate fans are important to both the artist and record label, as they serve as both financial supporters and ambassadors who happily introduce the artist and music to others.

Social Media Platforms

Because there are so many established social platforms and new ones that emerge every day, it is important to figure out which platforms most closely align with the artist's target audience, because that is where marketers will want to spend most of their time and resources. Although most artists certainly have a presence on each of the major platforms, the time and energy they spend in developing content and cultivating the audience may be utilized on the top few networks they feel are key to driving growth and achieving business goals. For most artists and brands, Facebook, Instagram, and YouTube are core platforms simply due to their high number of monthly average users (MAUs); however, TikTok has quickly become a critical platform for music promotion, particularly for those artists and genres targeting teens and tweens.

In fact, the biggest social media story in recent years must be China's ByteDance-owned TikTok. This short-video app is now favored among

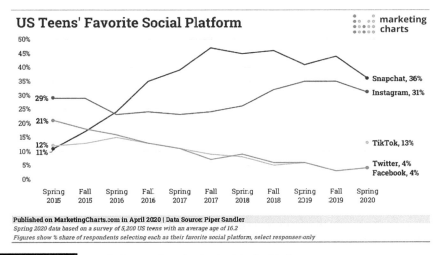

US Teens' Favorite Social Platform

marketing charts

Snapchat, 36%

Instagram, 31%

TikTok, 13%

Twitter, 4%
Facebook, 4%

Published on MarketingCharts.com in April 2020 | Data Source: Piper Sandler
Spring 2020 data based on a survey of 5,200 US teens with an average age of 16.2
Figures show % share of respondents selecting each as their favorite social platform, select responses only

FIGURE 14.3   *U.S. Teens' Favorite Social Media Platforms*

(Source: Piper Sandler)

teens second only to Snapchat, according to an annual study by Piper Sandler. Although Instagram has been a continued platform of choice for Millennials and teens with its Stories feature that has lured younger audiences away from Snapchat, Snap is certainly feeling the pressure of Tik-Tok's rapid rise in popularity. To thwart the growing threat of TikTok to Instagram, Facebook released a copycat version of TikTok called Reels that lets users make short videos such as lip syncing or dancing. Reels has now been rolled into the Facebook app as well with a dedicated section in the Feed.

What is important is to keep focus on the primary target audience and which platforms most appeal to them. Social is not a "one-size-fits-all" option. It is also critical to stay on top of shifts in consumer usage to know not only which social networks the audience is using today but which ones will they be using a few short years from now.

Social Media Goals

When companies utilize social media as a marketing tool, they are attempting to gain attention or traffic to their sites by creating and publishing content their readers will want to share with their friends, which results in electronic word of mouth. Again, social media marketing is considered a type of earned media, although it is important to note here that although the artist controls the messaging, the platform is essentially "rented" from the social media company. Social media platforms may face regulatory threats such as TikTok did in 2020; social media users may change their preferences, as happened quickly with MySpace and Vine; or a social platform may decide to shut down an account that it feels has violated its terms of service. Thus, any audience built through social media may be lost in the unlikely event that the platform decides to delete the artist's account or if regulatory issues force changes or removal of the platform altogether. Savvy marketers cross-promote to their fully owned platforms such as the artist's website and email list to ensure they always have access to the audiences they have built on socials.

A key step in determining what type of content would be best for the marketing team to share on socials is to define the social media goals. These goals will be different for each artist and could even vary by genre or the level of the artist in question (emerging artist, mid-tier, or superstar). And, of course, it is always helpful to have any goals set be specific and measurable. "Increase traffic to my website by 10% within three months" is much better than "Increase traffic to my website."

Social Media Strategy

The following are key steps involved in developing an effective social media strategy:

1. **Know and understand the audience**. Use the artist's and other research sources to fully identify the audience by age, gender, geography, likes, and interests. Analyze website and social media insights, plus use secondary sources such as genre-based studies and consumer databases like MRI-Simmons.

2. **Analyze the network options and choose wisely**. Spend some time analyzing which networks the fans utilize and where they are most active. Also understand the demographics and wide uses of each platform. While specific data points change each year, for example, TikTok and Instagram are preferred platforms of younger audiences, Twitter users tend to skew male, and Facebook offers broad reach across multiple age groups.

3. **Identify your social media goals and make them measurable.** Is the primary goal to build awareness, increase engagement, or encourage conversions? Establish and monitor **key performance indicators** to track success. The overall goal of social media marketing efforts will largely determine the metrics that are tracked across all accounts.

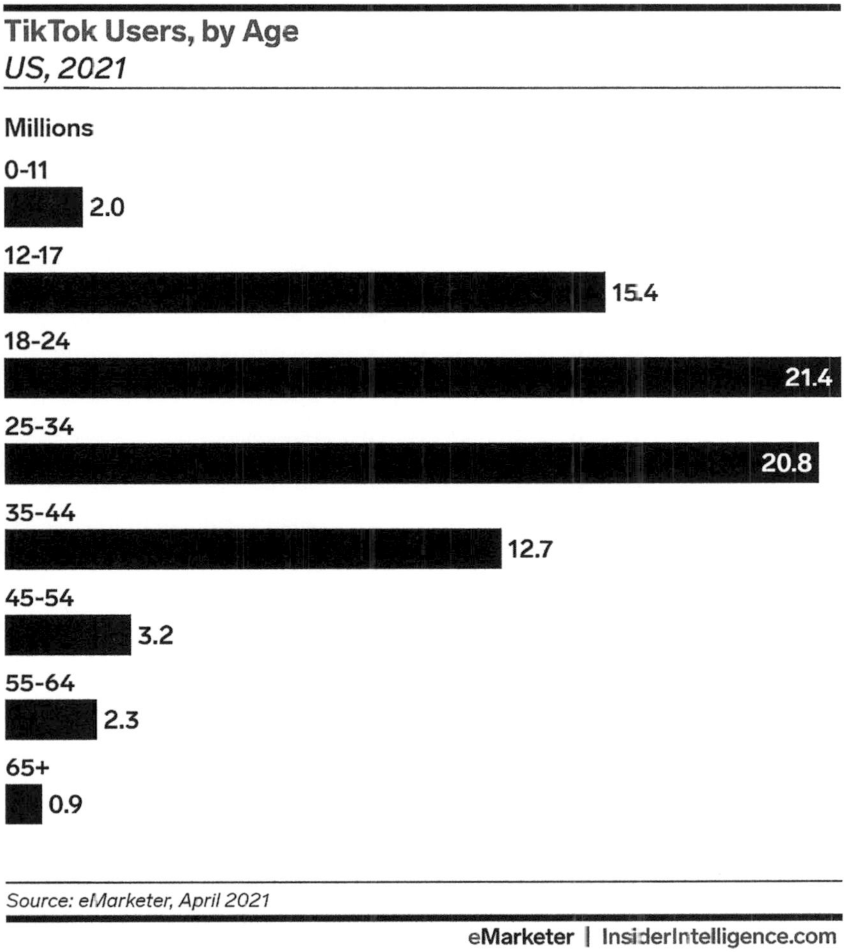

TikTok Users, by Age
US, 2021

Millions

0-11

2.0

12-17

15.4

18-24

21.4

25-34

20.8

35-44

12.7

45-54

3.2

55-64

2.3

65+

0.9

Source: eMarketer, April 2021

eMarketer | InsiderIntelligence.com

FIGURE 14.4   *TikTok Users by Age*

(Source: Courtesy of eMarketer)

Marketers interested primarily in building brand awareness, for example, will want to track follower count, impressions, and overall reach; someone using the platforms to build a strong community would be better served to track metrics such as comments, shares, and other forms of active engagement such as reactions and Story replies, whereas someone concerned with driving sales and monetization might want to follow metrics such as the number of times a website button or call-to-action button was clicked.

Key performance indicators are quantifiable measurements that help a company assess how successfully they are meeting their goals. KPIs should be SMART: specific, measurable, achievable, relevant, and time-based. For example, a company could track the number of impressions generated on one social media platform and analyze how the number grew (or didn't) over a certain period of time. Another KPI might measure the click-through rate or engagement rate over time.

4. **Conduct a social media audit**. Review every social account for the artist to determine what is working and what is not. Research artists in similar genres or at similar career levels to identify the platforms on which they are focused and how they are using them. How does the artist's brand compare to others? How is the audience using the platform?

5. **Social media listening.** Monitor channels to determine what has fans excited, identify influencers, monitor industry conversations, and follow leading publications for industry news, entertainment outlets, and prominent insiders.

6. **Post engaging content.** Seek to inform and entertain vs. sell. Use the analytics tools built into the platforms as well as others to help measure results and apply key learnings to future efforts. Although many marketers desire for their content to go **viral**, planned virality is rarely realized and can't be forced. Marketers can, however, focus on developing compelling content that truly resonates with their audience. This approach gives any content the best chance at virality.

Viral content is an image or video that spreads rapidly through social communities by being shared. Some content becomes viral within just a matter of a few hours, while others may go on for days or weeks, continuing to build during that time. One example of viral content was the #10-year challenge that was shared in 2020 with artists and other celebrities showing photos of themselves 10 years younger alongside a current image. The challenge was meant to highlight how much they have changed. The challenge resonated with the general population, and soon people in all demographics were sharing photos of themselves and noting the changes in their own lives. Another popular example in 2020 was that of Tik-Tok user Nathan Apodaca skateboarding to Fleetwood Mac's "Dreams" while drinking Gatorade. The video quickly spawned copycats (including versions from Stevie Nicks and Mick Fleetwood) and introduced a new generation of music fans to Fleetwood Mac's music. The exposure sent the 1977 hit catapulting back to the top of the music charts.

Social Media Best Practices

The right approach to social media marketing for any artist will vary based on the artist's style of communication and what information they like sharing, but there are a few rules, or "best practices," that are consistent across all.

- Content is king on socials and consistency is the queen. When deciding what type of content to post, it is helpful sometimes to think as a fan. If you were a fan of an artist, what type of content and information would you want to see? Breaking news, behind-the-scenes content, posts about the artist's personal life, and throw-back Thursday–type content are all appealing options. Approximately 80% of published content should be compelling content to build the artist's brand and develop a deeper connection with the fan base instead of sales-oriented messages. Think content marketing rather than self-promotion. Seek to inform and entertain rather than sell. Creating great content on a consistent basis takes resources that sometimes may not be available, though. Curating content (or linking to relevant content that other people have created that the team thinks would be of interest to the artist's fan base as well) is one way to lessen the load.
- Vary the types of content posted. Mix it up with a combination of polls, photos, videos, and questions to keep it interesting for the fans. When trying to manage multiple platforms for several artists, it can get very confusing. Creating a **content calendar** is enormously helpful to keep all posts organized, and it enables marketers to see any holes or gaps in content that need to be filled and to ensure a good mix of content types.
- Being consistent with the artist's brand as well as tone and voice is critical. This rule also applies to any representatives that might be posting on socials for the artist. A more reserved and professional tone is fine if that's the company's "brand," but it is important to remember that social is social and even more conservative brands should be less formal on social platforms than any other communication channel. This is not traditional advertising, so the same stuffy and sales-y language used in other efforts will not work here. Keep in mind that everything that an artist does and says communicates their brand to others. Brand elements including colors and key art components should be cohesive and consistent across all social media platforms.

- One of the most important rules of social media is to be authentic and keep branding top of mind in everything that is said and done. Every comment, post, and share an artist makes reflects their brand. Fans develop deep connections with artists who have strong personalities and are not afraid to say what they think. Successful artists seek to deepen their connection to fans by engaging in a two-way conversation.
- Posting the same content and messages across all social platforms might be easy for the social media manager and the creative team, but it is really boring for the fans that follow the artist on all of them. It is best to vary the content to keep fans engaged across the spectrum (for example, a photo might be shared on Instagram and a video on Facebook). Optimize content for each audience, and keep in mind that usernames tagged on an Instagram post might not be the same on Facebook. Hashtags might be different on each as well, and vertical images work better on some platforms than others. Keep in mind the commonly accepted practices for post frequency and engagement. Posting more than one of two times a day on Facebook might seem excessive, but tweeting the same content multiple times a day on Twitter would not be problematic.
- Focus on engagement. Respond to fans directly and quickly to build a strong relationship and trust. Ask questions regularly to stimulate conversation. Show emotion by using emoji or emoticons to increase engagement. According to HubSpot, emojis elicited 25.4% more engagement in tweets, as well as 57% more likes and 33% more comments on Facebook posts. Respond to comments on a regular basis, but it is not necessary or expected to try to respond to every post on a page with thousands of followers.
- Remember that although social platforms may be used to spread good news rapidly, the reverse is also true. Having a plan for a social media crisis is important so the team may react quickly and appropriately when they need to minimize any damage to the artist's brand.
- Understand the importance of mobile. According to Lyte Marketing, mobile accounts for approximately 80% of total time spent on socials, so short, easy-to-read messages will be more appealing than longer-form content. Make every word count and ensure every link posted is to a website that is optimized for mobile viewing.
- Analyze results to determine the best time to post and incorporate those key learnings into future efforts. Use **A/B (split testing)** to identify the calls to action, copy, and media types preferred by the fans.

Platform-Specific Best Practices
Facebook

While younger demographics have shifted to more trendy platforms such as TikTok and Snapchat, Facebook remains the most widely used network behind YouTube in terms of global reach and total active users. In fact, approximately 2.7 billion use it monthly (Facebook, 2020). In the United States, Pew Research reports that approximately two-thirds of all adults are users, and it is the third most visited website after Google and YouTube.

One important note to make is that all artists should create a Facebook Page as opposed to building an audience through their personal profile, as there are many benefits that come from doing so. Personal profiles are limited to 5,000 friends, so creating a page for fans is necessary to prevent having to constantly unfriend followers to add new ones. Having a Page also allows brands to have access to two very important tools:

1. Ads Manager (the comprehensive suite of advertising tools to be utilized for both Facebook and Instagram).

Importantly, the swift decline in organic reach (the percentage of a brand's posts that are seen by its followers) is now reported to be only 5.17% of a Page's total number of likes (Hootsuite/We Are Social, 2020). If an artist has amassed 100,000 fans to like their Page, only around 5,000 of them are likely to see any given post made by the artist without some type of paid spend to support it. Further, the average percentage of fans that engage with any piece of content is a paltry 0.08% for accounts with more than 100,000 followers. Even with a minimal paid campaign, the average post reach increases to 28.19%. To put it simply, Facebook has become a pay-for-play platform, as advertising generates the bulk of its profit.

2. Facebook Insights (the analytics that let artists understand more about their fan base and the type of content that resonates with them). The following are a few of the most important information and metrics available:

   Audience Demographics – Insights provides excellent demographic information to help accounts better understand their audience such as gender, age range, and locations, as well as the days and times that the most fans of the Page are online.

   Top Posts Area – shows top-performing content to help marketers quickly understand the type of content that is resonating with the community. Key learnings should be applied to future efforts. For example, if most of the top posts include video, that could be a good indication that is the delivery method preferred.

Top Post Types – illustrates the success of various post types (video, links, image, Live, etc.) based on average reach and engagement across all posts.

Engagement – seen as perhaps the most important Facebook metric, engagement is defined as any time a fan acts on a post. Not only is engagement a sign that the team is publishing great content that fans enjoy, but it is also a trigger to the algorithm that the post contains "good" content, which could prompt it to share the post more widely. Most people think of engagement as likes, reactions, shares, and comments, but it also includes liking an image, viewing a video, clicking on a link in a post, or even liking a comment.

Reach – the number of people who saw either organic or paid content. Organic-only reach is also available. As stated previously, even if the artist has a large audience, it is likely only a fraction of them may see the posts without some sort of paid support. It is important to study the elements of a post that was underperforming as well as those that overperform on reach to analyze the differences in media, copy, and other components used that could provide insight on what to incorporate into future posts to improve results.

Page Likes and Follows – these are considered "vanity metrics," as they do not reflect the success of the artist's Page. Despite this fact, Page likes are still important to track as one piece of the total audience of the artist's brand. Also, it is important to be on the lookout for large numbers of unlikes and unfollows on the artist's Page, as that could indicate offensive content or events that shouldn't be repeated.

Competitive Information – Facebook allows accounts to track up to five "Pages to Watch" for which it will quickly provide insights on things such as the top posts from those accounts. These metrics help accounts keep tabs on what is working for similar Pages and how their efforts compare.

Video Metrics – some of the video metrics available include retention that shows where in a video fans may lose interest as viewership might drop off. Video engagement is also available to show how many viewers click to play the video vs. simply seeing

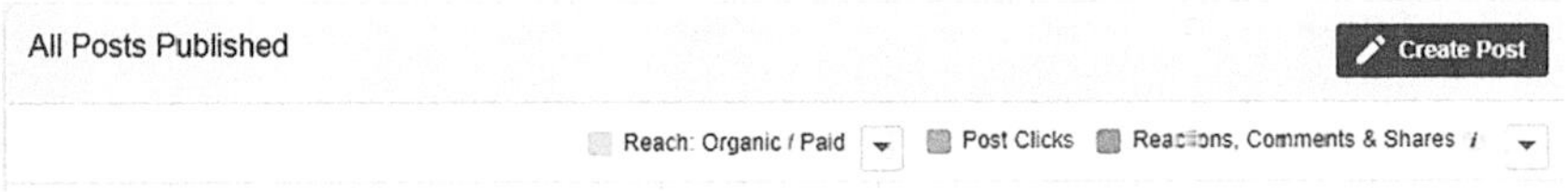

All Posts Published							✏ Create Post

☐ Reach: Organic / Paid ▼ ☐ Post Clicks ☐ Reactions, Comments & Shares / ▼

FIGURE 14.5A *Facebook All Posts Header*

FIGURE 14.5B *Facebook All Posts Data (Artist Disguised)*

it play automatically in the feed. Although many people like to promote the number of times a video was viewed, one should keep in mind that Facebook currently defines a video view as three seconds or more of viewing, and many of those might have occurred automatically.

Ad Metrics – more about ad metrics is covered in the paid media chapter, but analytics to measure Facebook ads include the click-through rate (CTR), cost per click (CPC), cost per action (CPA), and cost per one thousand people reached (CPM).

Key Platform Best Practices Include

Use video. Video posts get higher engagement than text and image posts. Videos should be uploaded directly to Facebook (referred to as "native video") rather than sharing a link to a YouTube video. While the team would certainly want to add videos to the artist's YouTube channel as well, there are many advantages to native video, including a bigger video preview window and the ability for the video to auto-play in a feed (this feature is not available when sharing a YouTube link), and the algorithm clearly prioritizes native video. In fact, a study by Quintly found that native videos received 477% more shares than those linked to YouTube. In addition to native video uploads, Facebook Live is a very popular tool that is favored in the algorithm and is an excellent content type for Q&As and to share behind-the-scenes content. Facebook Live Events also has a Watch Party feature that can be used for fans to experience special events together, such as the release of a new video.

Use Facebook Stories. Like Instagram Stories, these are photo or short video posts that appear at the top of a user's news feed and disappear after 24 hours. Stories also provide a way to essentially bypass the algorithm.

Use Reels on Facebook. In late 2021, Facebook blended Reels with its Facebook app. The short TikTok-like clips appear in a dedicated News Feed section, as well as alongside Stories and Rooms. Editing tools include a music library, augmented reality (AR) effects, speed adjustments, and multi-clip stitching to create a single reel from several clips.

Pin important information. Utilize the Pinned Post feature to highlight important information such as new music release dates and tour announcements.

Boost exposure with paid posts. As noted in the paid media chapter, Facebook offers a plethora of paid advertising options through Ads Manager, and it also enables posting to both Facebook and Instagram at one time.

Ask fans to mark the Page as a favorite. Facebook gives fans the ability to identify up to 30 friends or Pages as a "favorite," which essentially prioritizes the content from those accounts in the user's news feed.

How Social Media Algorithms Work

Every major social media network uses an **algorithm** to determine what content a user sees in their feed and in which order they see it. To do this, the networks rely heavily on machine learning models that are built to predict which posts a user will be post interested in seeing. Therefore, a social media user might feel they are seeing posts from the same ten people, even though they have hundreds of followers. In other words, it presents first the content it deems will be most valuable to that individual rather than posting everything in chronological order. If a user rarely sees a post from a page they liked years ago or a friend they knew in high school, it isn't because they aren't posting; it's because the algorithm has a limited amount of space to show the posts it thinks the user wants to see. If they are not actively engaging with certain friends or brands, the algorithm assumes they are not interested in the content and stops showing it to them. If a user goes to their page and likes or comments on a post, it is likely they will magically start seeing content from that person or brand again, as the algorithm believes the user is once again interested in seeing it.

The same process happens on artist pages. At first, the network might show a new post to a small percentage of the artist's followers (the ones who interact most often). If the content is well received as evidenced by many likes, shares, and engagement, the algorithm might push the content to a wider percentage of the artist's followers, and so on.

This process of evaluating thousands of data points for posts happens each time a user refreshes their feed. The goal of most social media platforms in using the algorithm is to keep users interested in their feed for as long as possible during the day so they can show more ads and generate more money for the company.

The Facebook Algorithm

Over the years, Facebook's data points have been added, removed, or adjusted, but the following are some of the most consistent factors for the algorithm:

Relationship to the user. Posts by friends, family, and groups take priority over posts by brand pages, particularly those that are highly promotional in nature. Even within these clusters, the algorithm favors "close friends" with whom a user interacts often. For brand pages to have a chance at pushing its organic content, they must now show high engagement (likes, reactions, comments, shares, etc.) to get the algorithm to recognize and reward their effort with improved reach.

Past user behaviors are evaluated to determine the likelihood of engagement. If the user, for example, liked a brand Page five years ago but has never interacted with one of their posts, it is likely that any post from that Page will be demoted in the feed. Facebook has also implemented tools to quickly identify posts using click bait and those are removed or downranked. The remaining posts are then evaluated and scored for each user based on what the algorithm knows about their preferences. For example, if the algorithm records that a user tends to react with a love emoji to 75% of all music video posts she sees, but only "likes" 10% of more business-oriented content, it will assign a higher ranking to all music videos. Importantly, Facebook did implement the "See First" feature to allow users the option of identifying the brand page posts it wants to see so those are prioritized above others.

Content Type. Which type of content the user interacts with the most is important, as is which content type the platform is pushing at any moment in time. When the platform introduced Facebook Live, it prioritized that type of media not only because it was earning more watch time than recorded video but also because it wanted to increase awareness of the new feature. In 2019, the network prioritized "high-quality, original video" that holds people's attention for longer than one minute.

Popularity. What kind of reaction is the content garnering once posted? If the algorithm identifies a post that is immediately generating a lot of interest (reactions, comments, or shares), it sees the content as being "good" and is more likely to then share it with more people. Sometimes, even the reactions themselves are evaluated with emojis, such as the heart or angry face being assigned more weight than Page likes.

Recency. Newer posts are favored over older posts.

Again, each network uses its own set of factors (and they generally keep the details secret) to determine how it prioritizes content. Although the previous themes seem to run true across most of the social media platforms that utilize algorithms, marketers must be vigilant in keeping track of changes that are announced and in monitoring their own reach for each post type to ensure their content strategy is up to date.

Instagram

For reaching Millennials, there is perhaps no better platform than Instagram. The Meta-owned social media network's largest demographic is women 18–24, but overall, it boasts more than 1 billion monthly active users, and 500 million of those are using Instagram Stories (Instagram, 2021). Although Instagram's origins focused heavily on quality images, video posts now receive more engagement than photos. In addition, Instagram is also cited by 90% of U.S. marketers as being the more important platform for influencer marketing, with top post types being general posts (78%) and Stories (73%). And data from Forrester showed this platform offers the highest engagement rates between fans/customers and brands of any major social network. Instagram provides enormous opportunity to connect with a brand's audience.

To use Instagram for professional purposes, switch the personal profile to a business one. With a business account, music marketers will be able to create and track advertising campaigns, use call-to-action buttons, and be verified by the platform. A business account is also required to access Instagram Insights, the platform's native analytics tool that shows how content is resonating with followers. Following are a few of the important metrics available to track within the app (desktop access for Insights is not available):

Audience Demographics. Accounts with a minimum of 100 followers may find information here about their followers including gender, age, and top locations, as well as top hours and days for followers to be most active.

Engagement Metrics. Insights provides information on interactions on overall activity such as website clicks, email button taps, and call button taps at the account level, as well as detailed engagement metrics for each post.

Reach. For accounts whose main goal includes increasing brand awareness, reach is an important metric to track as it shows the number of unique followers who saw a post or Story. In addition, the percentage of people who were exposed to the post who were not already following the account is listed. This type of information may be used to inform future content decisions. A post with abnormally high reach might provide insight that could be applied to future posts. For example, were certain hashtags used in the post? Was there anything different about the type of image or video that was used? Was a new CTA used that could have prompted more interest?

Instagram Story Metrics. Because Stories, the Snapchat-like feature, are such an important component to success on the platform, additional attention should be given to the metrics that are specific to that post type, including exits (the number of people who abandoned the Story before finishing), replies, and the people insights available to better understand the audience makeup of the people who are watching. It is critical to note that insights for Stories are only available for up to 14 days after they are published, so it's important to capture those quickly and save them in a different file for later analysis or to identify trends over a period of time.

Top Posts Area. Shows top-performing content to help marketers quickly understand the posts and Stories that resonated the most with the community over a defined time period.

These native insights available with an Instagram Business Account provide a basic overview and give brands a jump start to quickly understand what is working for them on the platform and what actions they might take to improve their performance. Many brands, however, choose to utilize additional measurement tools that provide more robust information on their Instagram efforts. Social Baker, for example, provides information on top filters used and robust competitive data, while Keyhole reports optimal post length and the top hashtags for engagement, and Iconosquare offers social media monitoring trends and more robust insight on Stories.

The Instagram Algorithm

Although very similar to the Facebook algorithm in that high-quality, engaging content and posts from family and friends are prioritized, there are some differences that should be noted when developing a content strategy here. Instagram has identified six main factors:

1. Interest: the algorithm looks at previous behavior to identify the accounts and types of posts a user interacts with the most, and those are given priority in the ranking. As with all social media networks, engagement is key, so it is best to consistently post engaging content to give more followers opportunities to connect.

2. Relationship: a software engineer at the network has noted that the platform determines how close one user is to an account by looking at what content the user likes, whom they direct message (DM) the most, people the user searches for, and people they know in real life.

3. Timelines/recency: current posts are favored over older ones.

4. Frequency: a person who opens the app frequently might see content differently than someone who opens it once a week.

5. Follower counts: A user who followers many accounts might be shown a single post from several of them, whereas someone who follows only a handful of accounts would likely see more posts from a single brand.

6. Usage: If the user spends short bursts or time with the app or sits down for long periods to explore the content more thoroughly, that might influence the content rank.

Platform best practices include:

Quality is valued over quantity. Although Instagram Stories may have a rawer feel to them, posts in the feed often include stylized images or those that offer a cool and different perspective. They tell the brand's story visually. Create eye-catching content with filters and effects that represent the brand creatively.

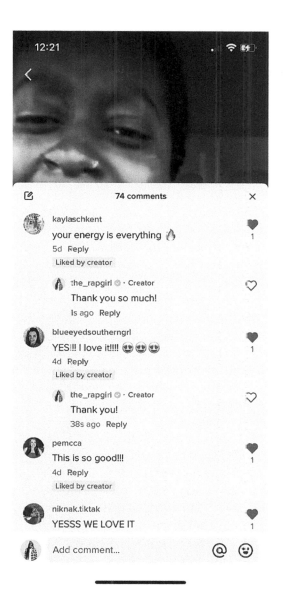

FIGURE 14.6   *Daisha McBride Social Media Interaction*

Use relevant hashtags. Hashtags are very popular on the platform and
could boost discoverability. Users often search for content using hashtags, and they also have the option of following hashtags, so include
those that are most relevant in the bio and posts

Optimize the bio. Although the bio is not searchable, so keywords aren't necessarily of paramount importance, it is critical that words be chosen carefully to capture the essence of the artist's brand in the 150-character space allowed. Also, make sure to apply for an Instagram Verified badge.

Use popular tools to increase engagement. Boomerangs (three-second looping videos that play forward and backward), polls, emoji sliders, and question stickers drive interactions. Also, consider including a CTA such as "tag a friend" or "share your thoughts" in post captions.

Use Instagram Stories. Stories are a powerful engagement tool. Stories allow short video messages or photo loops that appear at the top of the follower's screen in the app. Stories disappear after 24 hours and may contain text, links, or GIFs. The Stories are ranked for each individual so that those from accounts where the fan interacts the most often (comments, likes, story views, reactions, or Direct Messages) and those that are most timely appear first. Story Highlights (that do not expire) may be added to a profile page. Within Instagram Stories is the option to use link stickers taking users to external websites when they tap on them. Creators using link stickers may choose different styles, re-size the sticker, and place it anywhere in the story. This feature is currently only available to verified users or those who have at least 10,000 followers.

Utilize Instagram Video. In 2021, Instagram TV (IGTV) and feed videos were combined into one format: Instagram Video. All videos shared to the Instagram Feed can be up to 60 minutes long, with standard editing features available for longer videos, including trimming capability, filters, and location tags. Users may create 60-second previews for longer-form content. For an artist, compelling videos might include a "track by track" feature or the making of an album.

Instagram Reels is a short-form video option like TikTok that allows videos of up to 15 seconds to be used. Select Instagram creators who have their content set to public may have their Reels recommended on Facebook based on the user's interests. Reels offers many editing tools including a music library, augmented reality effects, speed adjustments, and hands-free recording with a countdown feature.

Explore Instagram Shopping. Prior to the launch of Instagram Shopping in 2020, brands who wanted to sell products directly through Instagram had to either utilize the swipe-up feature previously available to link followers to a website or place a link in the bio section of the account profile. With Instagram Shopping, brands may sell any number of physical products within the app and utilize shopping tags that work similarly to tagging people in a post. Artists may feature several items in their store, including albums, merchandise, and even tickets to events.

It should also be noted that Instagram has become a social media platform of choice for influencer marketing. With this content, influencers with either large numbers of followers or niche followings that are appealing to certain brands will either use or interact with a product in exchange for a fee. The main benefit to the brand, of course, is being exposed to a difference audience. Although this is a relatively new technique in promoting new music releases, some artists have seen success with influencer campaigns. More on this is discussed in the paid media chapter.

YouTube

Google-owned YouTube is a behemoth with more than 2 billion users each month. In fact, it is the second most visited website behind its parent company, and it is used by a full 33% of the world's 13+ population. YouTube has an enormous global audience, with only 15% of its total traffic coming from the United States. Younger demographics love this network, with 81% of U.S. 18–25-year-olds reporting usage (Pew Research, 2021).

To utilize YouTube for business purposes, switching to a brand channel is necessary. Not only does a brand channel allow for multiple users to access it simultaneously, but it also allows a single user the ability to manage multiple channels. Also, a brand channel provides access to the analytics tab to track results as well as to learn more information about the channel's audience.

The platform offers artists and their teams a fantastic hub called YouTube Studio that includes many in-depth courses to teach creators of all kinds how to optimize their channel and improve their results on the platform. From courses on monetization to analytics to how to build and grow an audience, the platform has put its best tips and tricks in one location for easy access. It truly is a must for any music brand serious about leveraging YouTube.

Platform best practices include:

Researching the competition. Use Google's Ad Planner tool or simply type the brand's targeted keywords into YouTube search to see what other channels are competing for the space. Also, marketers may use YouTube Trending Videos for insight on content that is popular at any given time, as well as other elements common in top-performing videos such as length, text overlays, and thumbnail designs.

Optimize both the channel and the videos uploaded to it. It is reported that approximately 70% of what people watch on YouTube is determined by the algorithm; in fact, a Pew Research study supports that as well in finding that 81% of YouTube users say they regularly watch videos the algorithm recommends to them. To that end, optimizing the channel elements as well as the videos themselves is an important step in ensuring the brand's videos are served to the right audience.

- Complete the account profile. Ensure the channel description provides a compelling and complete overview on the artist and the type of content that will be featured on the channel. Weave relevant keywords throughout the copy. Social network links should be added to the About section as well.
- The channel icon should be a crisp logo or other element; the channel art should include a call-to-action and social network links. Branding of the channel should be consistent with other social media accounts.
- Create a channel trailer for new visitors and returning ones to welcome them and give them a glimpse into what they may expect to see.
- Organize videos into playlists. While it is true that most videos viewed are served to users via the algorithm, there are some users who find their way to the channel itself. Making sure playlists are easy to decipher will help visitors find the content they seek.
- Optimize the day and time videos are posted. Review channel analytics to determine the optimal time to reach as many viewers as possible, then utilize YouTube Studio to automate the scheduling process.
- Strong titles are a primary signal to the algorithm and should use relevant keywords. While the platform allows for up to 70 characters to be used, it is recommended to limit those to 60 to prevent text from being cut off in results pages.
- YouTube will only show the first two or three lines of a video description before presenting the "show more" button to continue reading. As such, marketers should take care to ensure the first 100 characters of the description include targeted keywords.
- Create a thumbnail that stands out and is compelling. YouTube reports that "90% of the best performing videos on YouTube have custom thumbnails" that often feature a strong call to action or highlight a key element or benefit in the video.
- Add captions and translations. Including a video transcript not only helps devices of impaired users find the content, but it also improves

overall search engine optimization, as it provides more keywords for the engine to crawl. While YouTube does have an automatic translation tool, it's important to note that the accuracy depends upon several factors (diction of the person speaking, for example) so editing of those is critical.

- Add cards and end screens to cross-promote other content and drive traffic to an external site. Cards are the rectangular notifications that appear briefly as a video is being used. These are sometimes presented as quick polls to increase engagement or clicks to related content. Up to five cards are allowed per video. End screens appear within the last 5 to 20 seconds of a video, and up to four may be utilized per video. End screens will often direct users to continue engaging with the brand by encouraging them to watch another video, subscribe to the channel, or visit an external website.

Utilize YouTube Live to provide real-time comments and access to live events.

Maximize results with YouTube Analytics. As mentioned earlier, YouTube provides robust analytics to brand channels. Watch time (the amount of time a user watches a video before stopping) is one of the ranking factors for the platform's algorithm. Other important analytics include retention rate (the average percentage of a video an audience watches per view), traffic sources (this shows if users are finding the content via YouTube search, Twitter, external websites, or through paid ads, for example), demographics (age, gender, and geography), and engagement (what viewers are sharing, clicking, commenting, and promoting).

The YouTube Algorithm

Historically, YouTube's algorithm rewarded videos with high clicks and watch time. Currently, however, the approach is much more personalized and utilizes machine learning. For example, just as results are personalized for two individuals who query Google using the same search term, so would be the results for two users who enter identical search terms into YouTube. The algorithm not only tries to match the video metadata (the title, description, and keywords), but it also looks for which videos have the highest engagement (likes, comments, watch time), recency/timeliness of the video, and how quickly the potential videos it might serve have grown in popularity. As with other platforms, these elements are all indicators to the algorithm that help it prioritize what it feels is "good" content. It then goes further to review the individual user's patterns of behavior to pinpoint more finely what it feels the user is seeking to find. For example, if the user generally chooses videos that are less than three minutes in length, the algorithm might prioritize the videos that match that description over more long-form videos.

TikTok

In just a few years, and despite regulatory intervention that threatened to shut down the service, this social network has exploded onto the social media scene and has grown to include 1 billion monthly active users. Although 60% of total users reside in China, it has quickly amassed a legion of teens and tweens who use the platform to create and share videos. Although known for viral dance challenges and reaction videos, there are many opportunities with the platform to reach niche audiences with these 15-second clips. Although many genres of music may be found on the platform, hip hop/rap and pop dominate the landscape.

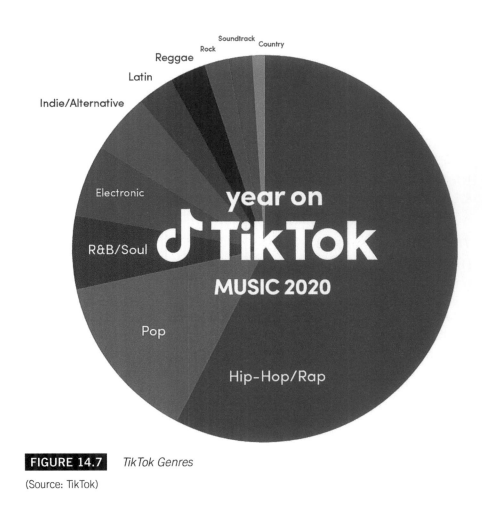

FIGURE 14.7   *TikTok Genres*

(Source: TikTok)

The following are just a few best practices to consider:

Consistency. To gain momentum on the platform, artists must regularly post compelling content.

Respond quickly to trends. Keep updated on hot topics by frequenting the Discover tab and For You Page. The hashtags that are trending are provided by location, so various users' Discover content will appear differently based on where they are geographically located. Type relevant keywords like "jazz music" into the search bar to find the top sounds, users, videos, and hashtags around that topic.

Be authentic, but have fun with it. The content that works best on TikTok is generally light-hearted.

Text overlays are utilized often to help tell a brand's story and communicate processes simply.

Use analytics to inform strategy. Review analytics to better understand what is resonating, then apply those learnings to future videos.

Transition videos are very popular. Take some time to learn how to master this content type.

Add a link to the website in the bio section, and make sure links to the artist's Instagram and YouTube account are added as well.

Utilize TikTok's Creator Marketplace that helps connect brands with platform creators.

Keep updated on changes to the algorithm. It changes quickly, but as of the time of this writing, follows and views are weighted heavily.

The TikTok Algorithm

TikTok has not released many details about how its algorithm works, but many brands and users with large followers on the platform have noted the following insights:

- Exposure appears to be based on the performance of each individual video as opposed to how many total followers a user has. For example, one video that quickly gains traction can become a viral success, even though the user might have only a handful of followers.
- Follows, rewatches, shares, comments, likes, and whether viewers watch the entire video are strong indicators to the algorithm.
- The algorithm seems to favor videos using trending hashtags and sounds.

Although advertising on the platform is new as of the writing of this book, there are several advertising options that are available, including brand takeovers and branded hashtag challenges. See the paid media chapter for more information.

Snapchat

With Instagram's adoption of filters and some of the other features that made Snapchat so popular with younger demographics, the platform saw a bit of a migration of its users; however, in recent years, its popularity seems to have rebounded, with 218 million daily active users being reported in early 2020, which was a year-over-year increase of 17%. According to the Digital 2020 report, 82% of people on Snapchat are 34 or younger, and 61% of total users are women. Interestingly, 38% of Snapchat users are not on Instagram at all.

For some brands that prefer to project a very polished appearance in everything they publish, this platform can be a challenge. Snapchat Stories are compilations of images and short videos that may be viewed for only up to 24 hours; then they disappear unless they are saved. Rather than highly produced content, Snapchatters prefer authenticity and material that is not picture perfect. To use the platform for business purposes, a business account is needed. With a business account, brands may advertise through the Ads Manager as well as target by age and location.

Platform best practices include:

Create a Snapcode. These badges work much like a QR code in that they allow a user to scan it with their phone to easily locate an account.

Have fun with augmented reality (AR) lenses. These lenses are largely what made Snapchat so popular and allowed it to differentiate itself from its competitors when it launched. AR allows for the superimposing of digital effects or animations on top of a real-life image.

Use geofilters for special events. While geofilters are not free, they can be relatively inexpensive and are used frequently by festivals to provide users with a branded filter to add fun and excitement.

The Snapchat Algorithm

Snapchat has followed Instagram's lead with the algorithm in that it is personalized and prioritizes stories and messages from those with whom the user interacts the most. Messages and Stories from those the algorithm deems close friends appear to the left of the camera, while Snaps from brands and celebrities appear to the right of the camera icon in the Discover section, with publishers and creators the user actively subscribes to positioned at the top and any others the algorithm feels the user might be interested in positioned at the bottom. As with all other social platforms, providing compelling and timely content that keeps subscribers engaged is key in making sure a brand is positioned favorably.

Twitter

The go-to social network for breaking news and all things political is Twitter. The platform boasts an average of 145 million daily active users, with 20.4% of those residing in the United States. Twitter's power users dominate, with 80% of all tweets coming from 10% of U.S. users.

Platform best practices include:

Optimize the account profile, header image, and photo. Making sure the handle used for multiple social media platforms is consistent is ideal. Otherwise, make them as similar as possible. The profile picture should be crisp and clear, particularly if it includes text of any kind. Header images may be used to reinforce a brand message or highlight a current campaign. The biography section is capped at 160 characters, so use the limited space wisely by showcasing the brand's personality or perhaps a slogan. Generally, the URL should link back to the main website.

Keep messages short. With a character limit per tweet of only 280, it is important to be concise and clear with messaging. Remember to also check tweets for errors before posting. While other networks provide an edit option, Twitter does not. Consider sharing links using a link-shortening service such as bit.ly, which also provides a great way to track results.

Be consistent and post content at optimal times. Because tweets move so quickly through a feed, it's important to have a consistent presence on the platform and to schedule tweets for the times that are optimal for viewing by followers. Also, multiple studies that have tested the optimal number of tweets per day have consistently reported anywhere from three to five tweets as being the ideal quantity to maximize exposure and avoid reduced engagement due to too much content.

Share great content and use an authentic voice. Share interesting stories and articles about the industry, the artist's genre, or even the artist's personal interests. Also, the voice used on all tweets should be consistent and reinforce the brand, regardless of who is posting on behalf of the artist.

Use Lists. The Lists feature in Twitter gives accounts the ability to tune in to conversations and people that are relevant to the business. Twitter feeds can be very cluttered and noisy, so it is helpful to be able to focus their attention on what a select group of the audience is saying. For example, music marketers may want to segment their followers by creating lists for bloggers, top genre artists, influential genre leaders, and entertainment outlets. Just remember that lists are public, so use great care in what name is assigned to each. Fun tip: follow Twitter lists that others create!

Master hashtag use. Hashtags are often searched on Twitter as users seek the topics of importance to them. Marketers may use the Explore page to see trending topics and hashtags or create their own branded hashtags, then encourage followers to use them as well. Generally, one to two hashtags per tweet is sufficient.

Use Twitter Polls. Twitter polls are an easy and fun way to gain fan interactions as well as to provide quick feedback. While the findings would never replace a comprehensive research study, they can provide a quick pulse for consumer preferences that could then be confirmed with an additional study.

Utilize Twitter Advanced Search. This tool allows for a more in-depth query to find users or search for relevant hashtags.

Pay to promote. As discussed in the paid media chapter, Twitter offers options to promote an account to gain new followers or to promote a tweet to increase engagement and awareness.

The Twitter Algorithm

As with most other platforms, Twitter does not release details on its algorithm, but the following are the main factors that influence it:

Recency. This includes not only how fresh the content is that is being tweeted but also how quickly an account responds to replies and mentions.

Engagement. How much attention is the content receiving in terms of likes, comments, and retweets?

Rich media. Tweets with images, video, and GIFs are generally favored.

Activity. How long it has been since the user was last on the platform may affect the order in which tweets appear.

Live Streaming

Live streaming was already gaining traction prior to the coronavirus pandemic, but the quarantine periods implemented to try to mitigate the spread of COVID-19 gave an unprecedented push to artists to utilize these platforms to stay connected to their fan communities. From John Legend taking requests in his bathrobe to the Verzuz Battles that pitted leading DJs or singers against each other on Instagram Live, music fans everywhere appreciated the creativity and effort artists took to engage with them and keep the music playing.

While there are many livestreaming platforms, Facebook Live, Instagram Live, YouTube Live, Twitch, Vimeo, and Mandolin have emerged as

FIGURE 14.8   *Jackie Venson Live Streaming*

FIGURE 14.9A   *John Smith* The Fray *Album Art*

John Smith Live at Yellow Arch Studio

Sunday, March 28th

Sunday, April 11th

Time: 3pm EST/8pm BST

Album Release: Sunday, March 28th

> *Join me as I perform The Fray live for the first time. I've only really played these songs in recording studios, straight into a microphone. One of the joys of performing live and interacting with real people, is the excitement of rediscovering the songs as they breathe and change shape. I've been desperate to play them in a live context, so I'm thrilled at the chance to play the album in its entirety and stream the show around the world.*

> *Behind the Music: Sunday, April 11th*

> *A unique livestream in which John delves into the songs and talks about the writing process. He'll discuss how he puts the pieces of an album together; from choosing guitars, approaching lyricism and writing riffs, to how he orchestrates his recordings in the studio. Please note each stream will offer a 48 hour replay.*

FIGURE 14.9B *John Smith* The Fray *Live Stream Invite*

the leaders in the music industry space. For many artists who experimented with livestreaming during the lockdown, the search for added features, more robust analytics, and a better user experience with high-quality audio and video has led them to Amazon-owned Twitch and full-service streaming platform Mandolin. On Twitch, fans subscribe to channels hosted by the "creators" (artists) and can access all the channel's content. Mandolin, meanwhile, allows artists and venues the opportunity to host one-time virtual concerts the fan pays to attend.

British folk artist John Smith took to livestreaming several events to promote the release of his sixth studio album, *The Fray*. Smith utilized livestreaming company Mandolin to host the event, providing fans three options: the full-album performance from $15, a behind the music segment starting at $15, and a combo deal from $25.

Livestream best practices include:

- Have a consistent schedule of livestreams.
- Perform a "test run" to ensure there are no technical difficulties or adjustments needed prior to going live.
- Pay particular attention to lighting in the room to ensure it is accurate.
- Experiment with the length and topics to discover what fans want to see.
- Acknowledge some fans by name who connect.

FIGURE 14.10   *John Smith* The Fray *Live Stream Options*

■ Promote the event to the fanbase well in advance. Send the event details to the artist's email list, and post extensively on all social accounts. Provide insight into what the artist will be playing/doing during the livestream, as well as any special guests who are planned to make an appearance during the event.

SOCIAL MEDIA MANAGEMENT AND LISTENING TOOLS

There are many tools available now that provide an easy way for digital marketers to streamline their overall social media activity. As with other digital products, social media management platforms make scheduling, distributing, monitoring, and analyzing social media activity a snap. These platforms offer different features, and each has its drawbacks. The optimal solution will depend upon the specific needs of the marketer, how they prefer the information be presented, whether they are working in teams, and if they are focused on multiple social media platforms or just one.

Most social media management tools serve the following purposes:

■ **Scheduling.** Schedule posts in advance, see gaps in the posting schedule, and push content simultaneously to multiple platforms.

- **Engagement.** See posts to the timeline, comments and replies on the posts, and messages to easily engage with the fans.
- **Tracking.** Monitor strategic keywords the team is targeting and keep tabs on what competitors and industry leaders are posting.
- **Reporting.** Access how the content is performing.

Examples of items to track include:

- The artist's name (and nicknames or common misspellings, if applicable)
- News pertaining to the genre(s)
- Album names
- Track titles
- Names of similar artists (the "competition")

Popular Social Media Management Tools

Hootsuite

Hootsuite is one of the most popular options for social media management for music and non-music brands alike. It offers a free version that allows a single profile/user the ability to track and post on up to three social media accounts, as well as to schedule up to 30 messages. Additional functionality is available via higher-tier subscriptions. The tool connects with multiple networks, including Facebook, Instagram, LinkedIn, Twitter, YouTube, and Pinterest. Unfortunately, it does not currently integrate with Snapchat or TikTok.

Hootsuite uses streams to manage content rather than inboxes, allowing marketers to customize their own dashboards and even organize streams into groups with tabs. For example, a user's dashboard might include a stream on Twitter scheduled posts, one for hashtags being tracked, a stream for mentions, and separate streams for each Twitter list. If a team approach is being utilized, the program allows for the assigning of tasks, and it provides some metrics on Facebook, Instagram, and Twitter, as well as the option to export data. Hootsuite enables marketers to exclude certain associated terms with a keyword they enter in the search. For example, a music marketing team promoting the band Anthrax would want to track any mentions of the term "Anthrax" except for those that also mention the word "infection" using the Negative Keyword function. A separate app, Hootsuite Insights, offers social listening capabilities.

Buffer

Buffer is a cloud-based platform that offers an easy-to-use interface with many of the same features as Hootsuite. One of the main differences between the two is that Buffer does not integrate with YouTube.

Socialoomp

Another option with a free version is Socialoomp. This platform offers unlimited posts by a single user, but only to one profile. For those focused on one social media channel, this is a great option.

SocialPilot

Unlike Hootsuite, this platform allows the use of multiple link shorteners such as Bit.ly. It also has a free content curation and influencer tool.

AgoraPulse

This paid option has all the standard features offered by its competition, with enhanced team collaboration capabilities.

CONTENT MARKETING

What Is Content Marketing?

In an age where attention from consumers is brief and the messages targeting them are so many, companies and their marketers are looking for ways to rise above the competition and to make a deep and lasting connection with their audiences. These companies understand that the best way to draw customers (or potential customers) to them is to create valuable and compelling content that will serve them in some way (make them laugh, educate them, or inspire them) rather than just aggressively pushing their products and services. This process of creating and distributing high-value information to build the trust and respect of an audience is referred to as **content marketing** and is a popular tactic used by marketers, but the concept has been used for years by leaders in every industry.

For example, in 1895, John Deere launched *The Furrow*, a magazine that provided information to farmers on how to become more profitable. Then, in 1900, Michelin offered drivers information on automobile maintenance and lodging on the road in its Michelin Guide. And in 1904, Jell-O salesmen went door to door, distributing their cookbook for free, driving sales of its products.

What started off primarily as printed pieces has today turned into a plethora of digital content marketing opportunities to create goodwill by helping customers succeed. Most content marketing pieces today are distributed via social media, email, podcasts, blogs, video, and audio. Customer relationship management platform Hubspot, for example, has become an authority in the social media space due to the large number of quality posts it publishes on topics that would be of interest to its current and potential customer base, from SEO tips to social media statistics and best practices. The idea behind content marketing is that providing customers with non-promotional content they find interesting or informative will increase interest in the product or service naturally and consequently increase loyalty. Additional sales will naturally follow over time (without asking for them directly) because of the value the company or brand has provided to the consumer. Marketers use content to educate, overcome resistance, establish credibility and trust, tell a story, build buzz, and inspire consumers.

In fact, such is the popularity of content marketing that in a 2020 study conducted by SEMrush of more than 1,500 international companies and organizations, a full 84% of respondents stated that they have a content marketing strategy. Of those actively engaging in the practice, 75% reported a goal of using it to drive website traffic, 57% reported an interest in improving their brand's reputation, and 47% reported a focus on increasing engagement and loyalty with their customer base.

Whereas many business-to-business and consumer brands will focus their efforts on white papers, how-to posts, infographics, blog posts, case studies, and question-and-answer lists, content marketing plans in the music industry might include any of the following:

- Behind-the-scenes videos of the making of the album
- A track-by-track video or podcast where each episode features the artist or songwriters talking about the stories behind each track
- Links to free music downloads
- Concert livestreams
- Newsletters via email that include previously unpublished information about the project
- A blog or vlog series that walks the fans through each step of a music video from concept creation to final editing
- A playlist of the songs that inspired the artist's latest project (curated by the artist)
- Facebook or Instagram Live Q&As with the artist

Information shared could also be more personal in nature and give fans a glimpse into the artist's background and life with content such as:

- A social media piece with the artist speaking about their childhood or hometown
- A post that expresses the artist's passion for their support of a cause that is close to their heart
- A video that expresses thanks to one or more of the artist's biggest influences (musical or other)

The possibilities to nurture the relationship with the artist's fanbase are numerous, and the rewards are meaningful. Increased social media followers due to the publishing of engaging content results in increased YouTube views and streams on Spotify, more tickets sold to live and virtual events, and improved website traffic and merch sales. It all starts with a focus on giving freely for the long-term benefit. For example, rather than posting "please pre-save my new music" or "tickets on sale now" message, offer value by providing an asset, for example, "I thought you guys might like to hear how the idea for my next single came to me." Ultimately, creating compelling content enables marketers to convert browsers into buyers and convert casual fans into ambassadors.

As consumers, we all know that marketing online is all about serving up great content. The marketers who are winning the brand race are the ones who are great storytellers who have their own special voice and who know how to engage with consumers in a very personal way. As stated previously, content is king. The goal is to consistently produce compelling content that will make fans want to help share the message.

With all content marketing efforts, consider the following to get started:

- Know the audience. What types of content will be most appealing?
- Determine the objective. Is the main goal to attract new fans or further endear the brand to existing customers?
- Speak in a conversational tone.
- Do not be afraid to re-purpose content. An in-depth blog, for example, can be chopped up into smaller segments to distribute on Twitter.

FIGURE 14.11 *Content Marketing Best Practices*

- Offer a variety of content whether the pieces shared be curated or created in-house.
- Ask fans what type of content they would like to see and utilize user-generated content (UGC) such as concert photos fans upload to a photo-sharing site.
- Monitor keyword searches and trending topics to see what people are naturally talking about, then try to join the conversation if it is relevant. Look to see what other artists are doing for ideas.
- Look at industry news, go behind the scenes, blog an industry event, or create compelling "how-to" content.
- Write and post blogs on a regular basis to keep the audience engaged. Also keep in mind that blogs are an important piece of an effective search marketing strategy.
- Let users help populate the content of the Frequently Asked Questions (FAQs) page and consider making it searchable to better serve the audience. Strive to solve problems and answer questions honestly.
- Keep in mind that search engines like Google and Bing use blended search, meaning that they display both standard web pages as well as videos and other materials in their search results. In fact, a video is 50 times more likely to appear on the first page of search results than text-based content, according to Forrester Research, so start recording and be sure the descriptions are complete, with extra attention given to keywords.
- When it comes to photos, be sure to tag people. And instead of using stock images, consider mixing it up with Flickr's Creative Commons license and other free photo-sharing sites that allow use of photos with proper recognition to the original photographer.
- Always keep in mind ownership of materials and if additional permissions are needed to legally use them. Of course, using the artist's music as a bed for video is ideal; however, there are many websites, such as Pond5, Audio Jungle, and Digital Juice, where it is easy to find professional stock video very inexpensively, in addition to royalty-free music.
- Use a content calendar to make sure there is a good mix of content to keep it interesting for fans.
- Above all else, *focus on providing value, not selling.*

Content may be original pieces created by the brand or curated content created by other people, then shared, but the goal is helping consumers find the information they are seeking, educating, or entertaining them rather than attempting to encourage any type of act (buying tickets, for example) that benefits the artist.

CONTENT CALENDARS

A **content calendar** visually displays the content that is planned for any given timeframe and is generally organized by platform. This tool is used by digital marketers, bloggers, publishers, and other groups to organize and plan content that is to be distributed across different media. One might also see the term **social media content calendar**, which has a similar layout but is limited to content distributed via the brand's social media channels.

Content calendars are an excellent tool to help determine the type of content that will be shared, the date and time the team will share it, and where it will be distributed.

Types of Content to Share

Find out the type of content that the audience wants to see that is relevant to the artist's brand. Reviewing social media insights is a good place to start with this process. With Facebook Insights, for example, a marketer can quickly see which post types have received the most views or engagement on the Page.

When to Post Content

Once again, reviewing social media analytics can be very helpful with this issue to determine when the artist's fans are online and most likely to engage with the content. Which days and times are optimal for each platform?

Where to Share

This is another area where research of the target audience is critical. On which social media platforms do most of the artist's fans spend their time? Which platforms are strategically important in terms of growing the audience?

Content calendars take many shapes depending on a brand's needs, but for most, a simple Excel or Google Sheet is preferred. Common items to track on a calendar include:

- Platform
- Date and time that content is to be published
- Copy for the post
- Content type (e.g. Story, video post, etc.)
- Links to assets needed to compile the post (photos, videos, etc.)

- Approvals (if an internal team member or supervisor reviews it prior to posting)
- Character count
- Partners to include (sponsors or other promotional partners referenced in the copy)
- Notes

Creating a content calendar is a must to ensure a consistent flow of relevant and interesting content, but also a good mix of content. Daily, or at least several times a week, provide updates on socials, forward interesting articles or videos the team finds, respond to blog comments, and post user-generated content. Planning reduces errors and allows marketers sufficient time to conceptualize content tie-ins to important big events such as award shows and festivals.

CURATED CONTENT VS. CONTENT CREATION

The challenge that many artist teams have when creating their own content to publish is in the resources that are needed to produce quality work. Hiring a video crew to capture those behind-the-scenes moments in the studio and edit them into a series of videos or producing a high-quality live stream takes skilled professionals, money, or both. Thankfully, many artists have successfully produced low-budget pieces through capturing video on their mobile devices. Another possibility for many artist teams is to use **curated content** (content that has already been created and published that might be valuable to the fans). In addition to Twitter lists, curation tool sites like Scoop.it and Curata help marketers quickly filter, organize, and share content already available online. With these services, finding the most relevant content for a specific audience is streamlined. For example, a tracked keyword of "hip hop music" on Scoop.it might generate a wide range of popular content around the topic such as a top gymnast performing a routine to hip hop, historical notes, and new music releases, as well as other relevant articles, pictures, and videos. Social media management tools such as Hootsuite and Tweet-Deck can also be excellent tools to find curated content.

Design Tools

Thankfully, there are many design tools available (and many of them have free versions) to make creating original content much easier than it has been in the past. GIMP is a free alternative to Photoshop, but if it is easy social media shareables or posters the team is looking to create, Canva, DesignBold, and Snappa are top choices for "drag and drop" photo editing. Stencil, Piktochart, Easel.ly, Visme, and Biteable are often used for infographic designs such as an artist one sheet; Giphy is a top choice for GIFs; video editing is made almost easy using iMovie; and Boomerang and Powtoon are top choices for animation.

THE SEO BENEFITS OF CONTENT MARKETING

Content marketing executed properly is time consuming but worth the effort. Consumers use search engines such as Google, Bing, and Yahoo to find information and inform their purchase decisions. The more relevant and compelling content the team posts online utilizing some of the keywords consumers would use in their search, the more successful the team will be in attracting fans. Simply stated, the search engines recognize and reward good content by pushing the artist's brand higher in the search engine results page (SERP). Content helps new fans find the artist sites as the artist's visibility in searches increases and as it generates a greater number of inbound links from other sites – particularly authoritative sites like mashable.com and Wikipedia.

SEO

Search engine marketing (SEM) and search engine optimization both fall under the umbrella term of search marketing. Both SEM and SEO focus on efforts to improve a website's search ranking, but only SEM involves the act of *paying* to improve a brand's search visibility. The search engines sell that opportunity to the person or company that places the highest bid on their targeted keywords or key phrases. Ads in search are sold on a cost per click or pay per click (PPC) basis. We discuss search advertising options more extensively in the paid media chapter.

SEO focuses on organic (non-paid) efforts to help improve ranking. SEO is the process of making changes to the website structure and content to improve the site's positioning in search engine results. More exposure by improving the brand's ability to appear higher on the SERP translates to more visitors to the artist's websites, which then gives the marketer the opportunity to turn those casual visitors into passionate fans.

SEO involves three main pillars:

1. **Site Structure/Architecture**. Optimize sites by using accurate page titles, implementing page tags, and ensuring clear site navigation.

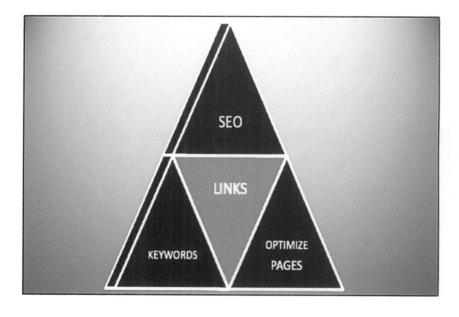

FIGURE 14.12   *Search Engine Optimization*

2. **Content**. Featuring plenty of great content that includes appropriate keywords is a primary focus.

3. **Linking**. How many other reputable or authoritative sites are linking to the artist's? The quantity of links can help, too, but it is the quality of the links that search engines value. So, a link from Wikipedia is assigned more value than a local website. Wikipedia is an important site for SEO purposes because it is considered an authoritative site by the search engines. The more inbound links from authoritative sites, the higher the artist's search rank will be.

While the programmers and other technical specialists at a company are mostly concerned with the site structure and linking components of SEO, the focus of marketers is on the content piece, and that's what we are going to primarily focus on in this book. But first, let's take a step back to understand the role the algorithms play in trying to help the search engines deliver applicable results when a fan enters information into the search bar. When a user types a set of keywords into Google, the algorithm quickly searches with the goal of delivering the most relevant and current information. When a fan searches for something like "pop singer baggy clothes," the search engine will scour the internet to find the content it deems most relevant to that search. The results factor in the words or phrases entered, the order of the words, and recency. The recency element is why so many marketers find blogging helpful to improve search engine rank. Whereas content might sit stagnant on the website for months, a recent blog containing targeted keywords is seem by the search engines as "current" and is often favored over content that is years old. When one is searching for an article on a particular topic, for example, it might be noticed that many of the most current postings tend to be positioned at the top of the search results page. Relevancy and recency are both important. Another very important factor is authority. Search engines assess the website's authority on certain topics by evaluating other websites that link to the artist's. Respected or industry-focused websites such as Billboard.com, Rollingstone.com, and Grammys.com would be viewed by the search engines as **authoritative sites**, so a link from those websites to the artist's, known as backlinks, will weigh more in the algorithm than links from bobsfavoritemusic.com (no offense to Bob intended).

When results are delivered from a user's search query, they typically are presented with paid, or sponsored, content at the top of the results, with organic (non-paid) results below. Depending on the search engine, results could be a blended display featuring just text and web pages, video, an image or a visual carousel, news, or maps.

SEO Best Practices

Research and Know Targeted Keywords

Prior to optimizing a website, the marketer needs to identify the keywords and phrases with which they want to be associated. We have all typed a phrase into a search engine to find a product or service only to realize that we need to get more specific when the initial search results don't come back as planned. For example, someone who types in "rap music" might be served results including notifications of national rap tours, top music videos, or personal news for top rap artists. They may then type in "rap music concerts in Los Angeles" to get closer to their desired outcome.

> "Use Keyword Planner Tools to discover new keywords and see historical search volume."

Using Google's Keyword Planner Tool or similar platforms can help aid a marketer in identifying search volume for keywords or phrases and estimated cost for paid ads, but one of the main benefits is that it may generate additional search terms that are similar that might not have previously been considered. Most would target the obvious keywords such as the artist or band name, as well as the name of the album and tracks, but using

☐ Keyword (by relevance)	Avg. monthly searches	Competition	Top of page bid (low range)	Top of page bid (high range)
Keywords you provided				
☐ christian music	100K – 1M	Low	$1.45	$3.01
Keyword ideas				
☐ worship songs	10K – 100K	Low	$1.47	$3.10
☐ christian songs	100K – 1M	Low	$1.19	$2.41
☐ toby mac	10K – 100K	Low	$0.30	$0.65
☐ jesus songs	1K – 10K	Low	$1.70	$2.52
☐ lecrae	10K – 100K	Low	$1.49	$25.17
☐ chris tomlin	10K – 100K	Low	$0.49	$3.39
☐ praise and worship songs	10K – 100K	Low	$1.55	$3.11
☐ hillsong songs	1K – 10K	Low	$1.47	$3.92
☐ worship music	10K – 100K	Low	$1.14	$3.52

FIGURE 14.13 *Keyword Planner Tool*

a keyword tool can help identify additional terminology to consumer. For example, a search for "Christian music" in Google's Keyword Planner tool yields suggestions such as worship songs, Toby Mac, Chris Tomlin, praise and worship songs, and LeCrae.

Another important point is to understand non-related but similar terms that might be generated with common names or keywords. For example, "Pink" returns results including Victoria's Secret and websites associated with the color pink, in addition to videos and links pertaining to the artist.

Assess the Artist's Website Regularly and Optimize It

Review the artist's website regularly to see where it ranks for the targeted keywords. Review the targeted keywords currently being used on the website to identify opportunities for improvement. Programmers should make sure page titles, tags, and meta data for videos are optimized.

Load the Website With Compelling Content

Content really is key to a successful SEO strategy. Make sure biographies include keywords such as the artist's influences, genres, and music titles. Also make sure all images and videos have alt text to help crawling search engines identify the content of the media.

Secure Inbound Links From Quality Websites

Remember that authoritative sites are weighted more heavily than others. Having reputable publications, bloggers, and industry leaders pointing to the artist's website tells the search engines that the site is trustworthy. Make sure all social media accounts link back to the website as well.

Secure a Google Knowledge Panel

A **knowledge panel** (also referred to as a knowledge graph) is a panel that displays details and media for an artist or band keyword search. It appears on the right-hand side of the first results page and includes images, the artist's website address, DSPs where the music is available, biographical information, movie and television placements, upcoming events and tour dates, songs and albums, and social media links.

Although Google's full criteria for which artists receive a knowledge panel is unknown, there are steps that have been known to help:

1. Register the website address with Google Search Console (available within Google Analytics).

2. Make sure the artist or brand has a verified Google brand account and an official artist channel on YouTube.

3. Build a website for the artist.

4. Create a Wikipedia and Wikidata.

5. Register with Music Brainz, a music encyclopedia that collects and disseminates music metadata.

POPULAR SEO TOOLS

In addition to Google's Keyword Planner and the Google Search Console, there are now many tools available to help even the non-technical professional improve their search engine results, including:

Google Trends allows marketers to find topics that have recently been trending. It also helps marketers visualize when people start searching for information around special events such as award shows. This enables marketers to better time advertising and social media campaigns to capitalize on that increased interest.

Ubersuggest is a search tool that displays which URLs rank for targeted keywords and the competition for those keywords.

SEMrush shows ads competitors are using for targeted keywords.

Moz Keyword Explorer is a paid tool that provides additional competitive information above what is available through other services. This insight allows marketers the opportunity to quickly identify which targeted keywords might be too much of a competitive challenge.

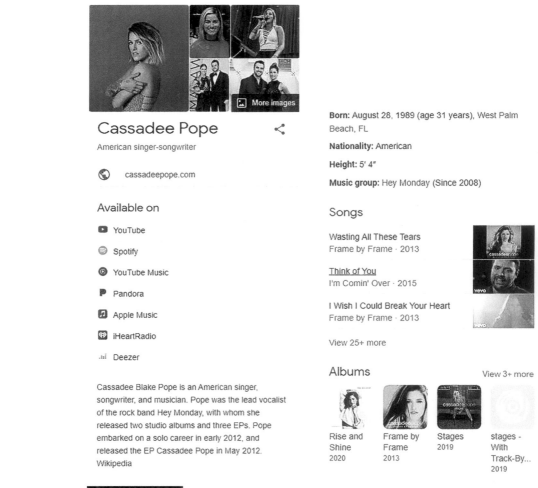

FIGURE 14.14A *Cassadee Pope Knowledge Panel*

WEBSITE DESIGN

Focusing on keywords and site structure to improve search visibility through proven SEO tactics is an important component of building a successful website. For most organizations, the artist's website should be home base, with all other activities – blogs, advertising, search efforts, and social efforts – driving traffic to it. In recent years, there have been clear examples of personalities who have amassed a significant audience on a social platform, only to see it disappear through violations of the platform's policies. This is a stark reminder of the differences between platforms marketers rent like social media and those they own such as websites, email lists, and blogs. Although we will not go into detail here about website design and development, we will review some common best practices:

- **Secure a strong domain name.** Keep it clean and simple and be sure it corresponds to what is being used on the artist's socials.
- **Present clear architecture and smart navigation.** Make sure it is easy for the consumers to get to the page they are trying to find. Make sure visitors have easy access to tour information, media coverage, biographical information, merchandise, the press kit, and, of course, links to hear and buy music, the artist's latest videos, and contact information for the artist's team.
- **Create a captivating homepage.**

  - Make sure to incorporate key marketing components like calls to action such as "Get Spotify for free" or "Pre-order now." A **call to action** is a short statement that directs consumers to take an action. A button on a site that says "Listen to the track" is an example of a call to action. Since building an email database of fans is often a top initiative with artist teams because email offers generally result in the largest response rate (yes, still!), the homepage is often seen as a perfect opportunity to capture that information in exchange for a free track or other incentive.
  - The homepage of a website is key real estate and, as such, everyone in the company wants to have their message displayed front and center. Navigating those political issues requires a clear vision of the website layout and content, as well as fierce negotiation and diplomacy skills. Each artist team approaches the homepage differently, but common elements found here include a quick snapshot of the artist, key highlights of the artist's career or a strong quote from a reputable media outlet, links to the artist's social media accounts, links to hear the music and/or embeds of the latest video. **SmartURLs or Pivot Links** are often used here to direct fans to the artist's music on their preferred music streaming or download site.

- **Keep mobile optimization in mind.** Gone are the days of having to code a website from scratch. Even large companies now are turning to platforms like WordPress because of their ease of setup and nearly unlimited amount of plug-ins. After all, if design cost can be reduced, that provides more money to focus on developing great content and engaging with the audience. Just remember that regardless of which design tools used to build the site, having one that is **mobile-optimized** that alters the layout to best fit whatever screen size is being utilized by the consumer is key. Mobile optimization ensures that when someone visits the site from a mobile device, they can easily read and navigate through the content without having to enlarge the screen multiple times.

- **Be consistent with branding.** It is important that the design of the website be consistent and use complementary design elements and colors with the email newsletters, social media accounts, and other communication channels.
- **Feature compelling content.** We have already discussed the importance of offering valuable and interesting content through all an artist's platforms, and the website is certainly the central delivery channel for that. Well-written biographies of the artist should be presented in at least two forms: a short, succinct bio that is approximately 100 words or fewer, and a more in-depth, long-form biography that tells more of the artist's story, complete with album and song titles, the genre, key influences, and other pertinent information. With each type of biography, and with other content on the site, make sure to keep strategic keywords in mind.

Mobile/Texting

Mobile is a form of digital marketing where a promotional message is displayed only on mobile devices like smartphones and tablets. Texting is great for "just in time" promos, and it is very personal in that most people do read every text message they receive, so open rates are extremely high. Promotional messages could be intended to drive online merch sales, encourage pre-saves of a new release, sell event tickets, or even promote a free event. Texting is also a good way to let fans be the first to hear about important news related to the artist. Marketers may also be able to use it for geotargeting if they keep in mind that many users are taking their phone numbers with them when they move away from the city where the area code is associated.

Best practices for texting include:

- Always secure the proper opt-in.
- The space in which marketers must place a text message is very short. Be concise with copy and use link shorteners for additional tracking.
- Select a short message service (SMS) that meets the team's current and future needs.
- Be selective on what messages are sent via text, as it is easy to annoy fans with too many or with content they don't find useful. Either of these scenarios could prompt them to unsubscribe.

FIGURE 14.15   *Laine Hardy* Tiny Town *Smart Link*

BLOGGING

Blogs are an important piece of digital marketing efforts because they allow an artist to establish expertise and recognition, reach broad audiences, and improve the search ranking of the artist's website. As discussed in the SEO section, inbound links and recency are two important elements of search strategy, and blogging improves both of those. Search engines like Google, Bing, and Yahoo show favor toward blogs because they are timely, the keywords contained within a post are indexed, and posts may be picked up and shared by authoritative sites the search engines value. The more quality blogs that the team posts containing the targeted keywords, the greater the likelihood that the posts will be shared and that search engines will reward the great content by positioning it higher in the search results.

As a long-form media, blogging allows the artist or artist team to go deeper than a standard social post, as well as to add compelling images and video to help tell the story. It is a great place to showcase promotions, promote ticket sales, feature merchandise deals, and more. It also helps connect recorded music marketers to other bloggers, who love to write about musical events and provide reviews of albums. A blog allows the ability to share the artist's thoughts and viewpoints with their fans, but unlike editorial, the marketer is in control of what the fans get to see; as such, blogging is referred to as **owned media**.

The key to blogging is consistency. One must maintain somewhat of a regular schedule and post at least twice per month. Also, make sure branding and tone is consistent with the artist's other platforms. Be an active member of the community by commenting on other blogs. Blogs may cover a wide range of topics, including:

- Favorite concert venues
- Personal stories and interests
- Inspirational figures in the artist's life
- Tour and new music release information
- Behind-the-scenes photos and videos
- Favorite instruments to play

Blogging Platforms

There are many blogging platforms from which to choose, and most of them offer a basic, free version to get started. These include Squarespace, Wix, Tumblr, and Blogger. Wordpress.com is yet another option that powers 34% of the top 10,000 blogs worldwide. It is easy to use, offers thousands of plug-ins for customization opportunities, and has a large community of users from which to draw support. For greater customization of the blog name, pay for a domain name. A free account on Wix, for example, might be AngleaJonesMusic.wix.com, whereas owning a domain would allow the artist to use AngelaJonesMusic.com, and that is the optimal approach for SEO purposes as well.

EMAIL MARKETING

Why Email

We live in a world that is obsessed by social media, so it might come as a surprise that email remains one of the most effective ways to engage with fans and is seen as a key element of any marketing strategy. It is an outlet on which artists can reach all fans in one place, and the connection point is secured because it is fully owned by the artist and label. Email also provides a direct link to fans without having to worry about an algorithm choosing which fans see the messages and which get buried in a fan's social media feed.

Although email may not be as top-of-mind as social media, it still delivers consistent results. In fact, more than half of Internet users (58%) report checking their email as the first thing they do in the morning, while only 14% responded that they check social media first. Email still boasts one of the highest ROIs by marketers globally. According to a study by Campaign Monitor, every dollar spent on email marketing results in $44 in return.

Email newsletters are a way to build a fan base and increase brand awareness of the artist. They also drive traffic back to the website and are an excellent way to sell product – from merchandise to concert tickets – as 60% of consumers report signing up for emails to get promotional messages. Fans may be segmented by demographics or behaviors such as those who attended a concert recently or people who engaged with certain content in previous newsletters.

The CAN-SPAM Act of 2003 was enacted to set a national standard for regulating spam email. Because of all the rules around email marketing, most marketers use an email service like MyEmma or MailChimp to ensure they are in compliance with the Act. Among other things, the Act:

- Prohibits the use of copy and headlines that are deceptive or misleading.

- Requires the sender to include identifying information such as a physical address.

- Provides clear opt-out options and prohibits the continued sending of emails after a consumer has requested that communication be ceased.

BENEFITS OF NEWSLETTERS

- Drive traffic to the artist's website
- Showcase new music and videos
- Build customer loyalty and encourage engagement
- Share promotional offers

Best Practices for Email Newsletters

1. **Be consistent**. Send emails on a regular basis, and research the artist's data points to determine the optimal timing for distribution. Multiple studies have collectively cited Tuesday as the best day of the week to send an email, with Thursday being the next-best day.

2. **The subject line is the most important part of the email.** Use compelling and concise wording with no more than 50 characters (including spaces) to avoid the text being cut off on mobile devices. Also, consider using incentives in the subject line, as this has been shown in some studies to increase the open rates.

3. **Write compelling copy**. Use a casual, relaxed tone. Remember to inject fun and personality into the newsletter. Use the newsletter to get the fans excited! Just as one would on social media, make sure the copy is compelling and focused on entertaining and educating, not necessarily selling. Position the primary message and call-to-action above the fold (the area presented on a screen without scrolling down). Always use **anchor text** to allow subscribers to jump to a different website or article with a single click. Avoid using all capital letters, too many exclamation points, special characters, or certain trigger words to avoid spam filters. Rather than using No Reply (noreply@companyname.com) as the address when sending an email, use a first name instead (Britteny@companyname.com). Not only does "noreply" violate the CAN-SPAM laws, but it also seems impersonal to the subscriber and could hurt the open rate.

4. **Create a strong design.** Use high-resolution images, video, and infographics to create visual interest. Use high-quality images and embedded links to video services such as YouTube or Vimeo. Stick to a cohesive color theme and use no more than two to three typefaces or fonts.

5. **Optimize for mobile**. Most email service providers offer templates that are mobile responsive (it is adjusted automatically to display content in the best way possible for the specific device on which it is being utilized).

6. **Personalize the content**. Use merge tags to automate the process of personalizing an email by incorporating the subscriber's name or other information. Also, segment the newsletters to tailor content based on the subscribers' interests.

7. **Test, test, and test again.** Always send a test email to a small number of people on the team to check for broken links and buttons, to ensure merge tags are working properly, and to ensure proper alignment of copy and other elements. Test on multiple devices as well (laptop, smartphone, desktop). Track the performance of the copy and components included in the newsletter to better understand the kinds of content to which the subscribers are likely to respond and use that insight to inform future email campaigns. Test the effectiveness of various subject lines, color and font options, and calls-to-action by performing **A/B tests**.

8. **Give subscribers control**. Try to use re-engagement newsletters with subject lines such as "We Miss You" to prompt former enthusiasts into getting re-connected. Importantly, make sure it is easy to unsubscribe in every campaign to remain in compliance with CAN-SPAM laws. More than 50% of people who unsubscribe report the main reason as being "too many emails," so give the option to continue to be a subscriber, but at a reduced frequency. Also, use the tools provided by the email service provider to let the fans choose the type of content they want to see (tour news, promotions, new music releases, etc.). Clean the email list regularly and remove subscribers who have not opened an email within a pre-determined timeframe. Leaving uninterested subscribers in the database is likely to have the undesirable effect of lowering the open rate.

9. **Make the collection of fan email addresses a priority**. Collect email addresses via a content form on the artist's website, paper sign-ups at events, or via cross-promotion on the artist's social media accounts, but make sure to adhere to all CAN-SPAM opt-in laws. Be careful to not break privacy laws or use words or images that could flag a spam filter.

AVERAGE OPEN RATE 15–25%

Average click-through rate: 2.5% (vs. the average CTR on Facebook of just .07%).

Source: Campaign Monitor

Top Words and Phrases to Avoid in Subject Lines

Although there are hundreds of words and phrases that sensitive spam filters may target, the following is a list from email service Emma on top expressions to avoid using in an email subject line.

- Do you like/have/want?
- Act now
- Order now
- All natural
- Affordable
- Amazing stuff
- Cash Bonus
- Compare rates
- Credit
- Guarantee
- Home-based
- Increase sales
- Info you requested
- Incredible deal
- Limited time offer
- No investment
- Obligation
- Almost any phrase with the word "free" in it
- Cheap
- Promise
- Thousands

What to Include

- A recap of the latest show, including a video with highlights or a thank-you message to the fans for coming out to support the artist.
- Tour information such as dates/markets, when tickets go on sale, and how to purchase tickets.
- Information about upcoming artist appearances in television shows or music that is integrated into film projects.
- Links to new songs or new music videos.

Calls-to-Action

Every newsletter should have a clear and strong call to action with the goal of either building the artist's fan community or encouraging action that will drive revenue in some way. Examples of CTAs include:

- Invite the fan to attend a concert.
- Encourage them to tune in to a livestream.
- Offer a discount on artist merchandise.
- Ask them to listen to the music on Spotify or pre-save the new album.
- Cross-promote the artist's social platforms by asking subscribers to follow the artist on Instagram or watch a video on YouTube.
- Encourage them to vote or review the artist's music.

Choosing a Platform

There are many key considerations when it comes to selecting the right email marketing tool for the team's needs, but some of the main ones include:

- Cost
  - Does the cost fit within the team's budget?
- Company size and number of subscribers
  - How many individuals on the team will need accounts, and how many fans are signed up for email?
- Customization and automation needed
  - Most email service providers have some level of this support that enables welcome emails to be sent automatically upon sign-up or based on several other actions taken by the subscriber.
- Double opt-in requirements
  - If the fan signs up at an event, does the marketing team have to get them to opt-in again to confirm it?
- Analytics
  - Open rates, click through rates, geographic area of the subscribers, and so on.
- Customer service support
  - Is help available via phone, email, or online chat?

Email providers service many functions:

- Analytics.
- Ensure the team is abiding by legal requirements.
- Templates that are visually appealing.
- Automation (welcome messages, based on previous purchases, geotargeting). For example, many services allow the ability to segment an audience to enable sending not only general messages that apply to all subscribers to the full list but also the ability to send smaller, more targeted emails to those in certain geographic areas or only to fans of a particular music genre.

Top Email Marketing Tools

Email service providers help marketers with the tasks of obtaining, organizing, and managing emails collected. They also offer attention-grabbing templates, automation tools, and analytics. There are many email providers available, but the following are a few popular options in the music industry.

MailChimp is a leading provider. The platform is very easy to use and offers geotargeting capabilities for sending offers to fans in a specific region. The service is free up to 2,000 subscribers, but only paid accounts have access to automation tools and advanced features.

Emma is often chosen by music companies that need an enterprise-level solution.

Fanbridge offers integrations with SoundCloud, Spotify, and Bandsin-Town, among others.

Constant Contact is another popular option used by many businesses of various sizes.

CONCLUSION

With so many fans now consuming music online through streaming platforms, the migration of marketers from using radio, out-of-home, television, and other outlets to now utilizing digital media is a move that makes sense. The rate of digital adoption has been swift across all categories, but it has brought with it some very exciting opportunities for marketers to reach and engage with their fan base.

GLOSSARY

A/B tests—The idea of testing two emails or social media messages that are identical except for one element. This allows the marketer to measure the effectiveness of the variable to determine which achieves better results. The key learnings from the test are then applied to future campaigns.

Algorithm—A set of rules that helps a computer accomplish its tasks. In social media, algorithms are used by platforms to sort posts in a users' feed based on relevancy instead of published time.

Anchor text—The clickable words in a hyperlink that take the reader to a targeted web page. These words typically appear as blue and may be underlined.

Call to action—A short statement that directs users to take action such as "sign up for our newsletter" or "buy tickets now."

Content calendar—Visually displays the content that is planned for any given timeframe and is generally organized by platform (Facebook, Instagram, email, etc.).

Content marketing—The concept of delivering high-quality content that fans want to see rather than pushing sales messages.

Earned media—Positive publicity generated through user-generated content such as word-of-mouth promotion, music or live show reviews, or social media comments.

KPIs or key performance indicators—Quantifiable measurements that help a company assess how successfully they are meeting their goals.

Mobile optimization—Alters the layout and positioning of websites based on the screen size the viewer is using.

Owned media—The promotion channels on which a brand control the messaging such as an artist's website or email list.

Paid media—Any publicity or awareness that is gained through a paid ad placement such as ads on social media platforms.

Pivot link or smartURL—Used to direct fans to the artist's music on their preferred platform.

Social media—Online social communication channels such as Facebook and TikTok.

Social media marketing—The process of gaining attention or increasing traffic to a website through social media sites.

BIBLIOGRAPHY

Auxier, Brooke, and Monica Anderson. "Social Media Use in 2021." *Pew Research Center: Internet, Science & Tech*, Pew Research Center, 31 Jan. 2022, https://www.pewresearch.org/internet/2021/04/07/social-media-use-in-2021/.

"Best Practices for Email Campaigns." *Mailchimp*, mailchimp.com/help/best-practices-mailchimp-email. Accessed 29 Mar. 2021.

Chen, Jenn. "36 Essential Social Media Marketing Statistics to Know for 2021." *Sprout Social*, 3 Feb. 2021, sproutsocial.com/insights/social-media-statistics.

"Content Marketing Statistics You Need to Know for 2021." *Semrush Blog*, www.semrush.com/blog/content-marketing-statistics. Accessed 29 Mar. 2021.

Cooper, Paige. "How Does the YouTube Algorithm Work? A Guide to Getting More Views." *Social Media Marketing & Management Dashboard*, 18 Aug. 2020, blog.hootsuite.com/how-the-youtube-algorithm-works.

Cooper, Paige. "How the Facebook Algorithm Works in 2021 and How to Work With It." *Social Media Marketing & Management Dashboard*, 23 Feb. 2021, blog.hootsuite.com/facebook-algorithm.

"Digital in 2020." *We Are Social USA*, 5 Oct. 2021, https://wearesocial.com/us/blog/2020/01/digital-2020-us/.

Easton, Jonathan. "Almost 4 Billion Hours Watched as Live Streaming Industry Benefits From Lockdown." *Digital TV Europe*, 14 May 2020, www.digitaltveurope.com/2020/05/14/almost-4-billion-hours-watched-as-live-streaming-industry-benefits-from-lockdown.

Ellering, Nathan. "What 14 Studies Say About the Best Time to Send Email." *CoSchedule Blog*, 24 Sept. 2018, coschedule.com/blog/best-time-to-send-email.

"Facebook Reports Second Quarter 2020 Results." *Investor*, investor.fb.com/investor-news/press-release-details/2020/Facebook-Reports-Second-Quarter-2020-Results/default.aspx. Accessed 29 Mar. 2021.

Hott, Allison. "40+ Email Marketing Statistics You Need to Know for 2021." *OptinMonster*, 6 Jan. 2021, optinmonster.com/email-marketing-statistics.

HubSpot. "Instagram Marketing: The Ultimate Guide." *Hubspot*, www.hubspot. com/instagram-marketing. Accessed 29 Mar. 2021.

Hyatt, Ariel. "Email Newsletter Best Practices for Musicians." *Cyber PR Music*, 20 Sept. 2020, www.cyberprmusic.com/email-newsletter-best-practices-musicians.

Kemp, Simon. "Digital 2021: The Latest Insights Into the 'State of Digital.'" *We Are Social*, 27 Jan. 2021, wearesocial.com/blog/2021/01/digital-2021-the-latest-insights-into-the-state-of-digital.

Kusek, Dave. "Why Blogging Is Still One of the Most Powerful Music Promotion Tools." *DIY Musician*, 10 Mar. 2021, diymusician.cdbaby.com/music-promotion/why-blogging-is-still-one-of-the-most-powerful-music-promotion-tools.

Lee, Kevan. "Best Frequency Strategies: How Often to Post on Social Media." *Buffer Library*, 24 July 2020, buffer.com/library/social-media-frequency-guide.

Letang, Shaun. "23 Content Marketing Ideas for Musicians, and What That Even Means." *Music Industry How To*, 29 Dec. 2020, www.musicindustryhowto. com/23-content-marketing-ideas-for-musicians-and-what-that-even-means.

Marmer, Daria. "These Emojis Can Increase Click-Through Rates, According to New Data." *Hubspot*, blog.hubspot.com/marketing/best-emojis?zd_source=mta&zd_campaign=15103&zd_term=vanditagrover. Accessed 29 Mar. 2021.

McCabe, David, et al. "TikTok Deal With Oracle and Walmart Trips Over U.S.-China Feud." *The New York Times*, 21 Sept. 2020, www.nytimes.com/2020/09/21/technology/tiktok-bytedance-deal-walmart-oracle.html.

Mohsin, Maryam. "10 TikTok Statistics You Need to Know in 2021 [March 2021]." *Oberlo*, 22 Mar. 2021, www.oberlo.com/blog.

Nemeth, Cole. "How the Twitter Algorithm Works in 2020." *Sprout Social*, 12 Mar. 2020, sproutsocial.com/insights/twitter-algorithm.

Occhino, Lisa. "The Complete Guide to Live Streaming for Musicians." *Bandzoogle Blog*, 23 Sept. 2020, bandzoogle.com/blog/the-complete-guide-to-live-streaming-for-musicians.

Olafson, Karin. "100+ Social Media Demographics That Matter to Marketers in 2021." *Social Media Marketing & Management Dashboard*, 27 Jan. 2021, blog. hootsuite.com/social-media-demographics.

Omnicore. "Omnicore: Medical & Healthcare Digital Marketing Agency." *Omnicore*, 8 Jan. 2021, www.omnicoreagency.com.

Patel, Neil. "The Only 6 Keyword Research Tools You Need to Use." *Neil Patel*, 26 Mar. 2021, neilpatel.com/blog/necessary-keyword-research-tools.

Peisner, David. "Concerts Aren't Back. Livestreams Are Ubiquitous. Can They Do the Job?" *The New York Times*, 21 July 2020, www.nytimes.com/2020/07/21/arts/music/concerts-livestreams.html?auth=linked-facebook.

Santora, Jacinda. "Email Marketing vs. Social Media: Is There a Clear Winner?" *OptinMonster*, 20 Dec. 2019, optinmonster.com/email-marketing-vs-social-media-performance-2016–2019-statistics.

"Social Media Marketing & Management Dashboard – Hootsuite." *Social Media Marketing & Management Dashboard*, blog.hootsuite.com. Accessed 29 Mar. 2021.

studioID. "StudioID: Content Marketing and Brand Studio." *StudioID*, studioid. com. Accessed 29 Mar. 2021.

TikTok. "Year on TikTok: Music 2020." *Newsroom | TikTok*, 17 Dec. 2020, newsroom.tiktok.com/en-us/year-on-tiktok-music-2020.

Tumidolsky, Becky. "10 Instagram Best Practices You Should Be Following in 2020." *Social Media Marketing & Management Dashboard*, 4 Nov. 2020, blog.hootsuite.com/instagram-best-practices.

Vaughan, Pamela. "17 Email Marketing Best Practices That Actually Drive Results." *Hubspot*, blog.hubspot.com/blog/tabid/6307/bid/23965/9-email-marketing-best-practices-to-generate-more-leads.aspx. Accessed 29 Mar. 2021.

Walker-Ford. "15 Jaw-Dropping Email Marketing Stats You Need to Know in 2020 [Infographic]." *Social Media Today*, 16 Jan. 2020, www.socialmediatoday.com/news/15-jaw-dropping-email-marketing-stats-you-need-to-know-in-2020-infographic/570372.

Walls, Wes. "10 SEO Tips for Musicians." *Bandzoogle Blog*, 19 Jan. 2021, bandzoogle.com/blog/10-seo-tips-for-musicians.

Warren, Jillian. "The Ultimate Guide to TikTok Marketing." *Later Blog*, 13 May 2020, later.com/blog/tiktok-marketing.

Warren, Jillian. "This Is How The Instagram Algorithm Works in 2021." *Later Blog*, 26 Mar. 2021, later.com/blog/how-instagram-algorithm-works.

"What Is the CAN-SPAM Act?" *LII/Legal Information Institute*, www.law.cornell.edu/wex/inbox/what_is_can-spam. Accessed 29 Mar. 2021.

CHAPTER 15

Paid Media

PAID MEDIA BASICS

Given the vast media options and intense competition for consumers' attention, it is difficult to imagine a music world without paid opportunities to put your artist front and center and allow your message to rise above the collective noise. The advertising space is intensely competitive, and that is especially true for digital media. As consumer behavior has dramatically shifted online, advertisers have followed them. In fact, eMarketer reports that total spending on digital ads by marketers is expected to increase to $278 billion by 2024. With so many placement choices and a vast number of competing advertisers, music marketers must optimize their utilization of targeting tools to stand out in the crowded online marketplace.

Paid media is a form of marketing communication that is used to promote a brand's goods or services by trying to influence consumer behaviors or beliefs. It is different from all other forms of marketing communications in that the content and placement of the messaging are created and controlled by the brand or company in exchange for compensation. The brand develops the creative and determines the desired placement, but the messages appear on channels the brand does not own or control. Paid media is used to amplify content and driving exposure across a wider audience than what might be possible through organic efforts alone. In fact, with organic reach on social media platforms often reaching only 1% of the brand's audience, the need to pay for additional exposure becomes a critical one.

While traditional options like television and radio advertising, billboard, sponsorships, and search engine ads are certainly key components in this

DOI: 10.4324/9781003153511-15

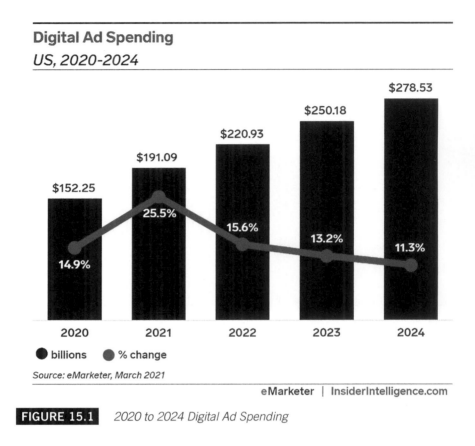

Digital Ad Spending
US, 2020-2024

FIGURE 15.1 *2020 to 2024 Digital Ad Spending*

category, recent years have ushered in additional paid offerings including paid influencers, social media advertising, sponsored content and ads within podcasts, advertising via digital service providers such as Spotify and Pandora, and product placement opportunities in television and film. It is even possible now to place paid messages in personal devices such as Amazon's Echo and Google Home. Ask Alexa for the date, and she might follow up her answer by asking if you want to hear about ideas for a birthday gift. Paid media is meant to persuade our audience to act on products, services, or ideas, whether our campaign goals be to build awareness, increase streams, generate traffic to a website, or drive product sales. Our target audience – or who we are trying to reach with these messages – drives which mediums and vehicles we use.

A **medium** refers to a class of communication carriers such as digital, television, radio, print, and out-of-home. A **vehicle** is the specific carrier within the group, such as Spotify, the local radio station, *Rolling Stone* magazine, or Amazon.com. Brands determine where to place their advertising budget based on the likelihood that the paid effort will generate enough in incremental sales revenue or accomplish another predetermined objective to justify the expense of the campaign or, in other words, the **return on investment**.

Paid media generally targets two basic market segments: (1) **consumers** and (2) **trade**. Campaigns that are targeted to consumers are generally intended to create demand for a product that pulls consumers into music outlets to consumer the product, while trade advertising targets decision makers in the industry who may influence an artist's success. Consumer campaigns utilize mediums such as digital, television, and radio to reach the end-users of a product, whereas a trade campaign might target industry insiders such as journalists, music retailers, bloggers, DSP executives, radio station programmers, and more. These individuals are gatekeepers who may be favorably influenced by an ad in a way that helps an artist. For example, a playlist curator who sees an ad in *Billboard* or on *MusicRow*'s website touting the success of a new release might then be influenced to listen to the track or add it to their own playlist. **Trade advertising** is typically conducted through targeted digital efforts, direct mail, and trade publications (usually magazines or email). Occasionally, it may include a targeted out-of-home effort such as a billboard in an area that is heavily populated by trade professionals. Trade advertising is not common through broadcast outlets like television and radio because they are too general in nature and not targeted enough to effectively reach the industry.

Consumer Media

Media planning involves the decisions made in determining the mediums you are going to use and when the schedule will run across each. It consists of a series of decisions made to answer the following questions:

1. How many potential consumers need to be reached?
2. In what medium should the ads be placed?
3. When and how often should these ads run?
4. What vehicles should be used?
5. Which geographic markets should be targeted?
6. What choices are most cost effective?

It also involves analyzing the costs to determine which mediums are most cost effective.

When considering a media buy, labels look at many factors:

- Target market
- Campaign objectives
- Media placement options
- Budget
- Timing
- Corporate partners and others who might help fund the campaign

Paid media is usually secured by the marketing department at a label placing the buy directly with the media outlet (for example, placement of a sponsored story on Instagram or promoted tweets) or with an **advertising agency** or marketing agency that purchases media for many clients for a fee.

Target Audience

First and foremost, music marketers must target the right audience with their campaign, as that information will help determine the paid media platforms that are most appropriate. Although the targeting options are different across each platform, a general profile of your targeted consumer that includes demographics, geographics, psychographics, and behaviors is essential to help ensure you are connecting your message with the group of consumers that are most likely to engage.

Demographics (age, gender, household income)
Geographics (grouping customers by country, region, state, city, or neighborhood)
Psychographics (activities, interests, opinions)
Behavioral (purchase and usage, loyalty, benefits sought)

It is helpful to note that most broadcast media are sold targeting demographic groups. Popular age brackets for music ad campaigns are 18–34, 18–49, and 25–54.

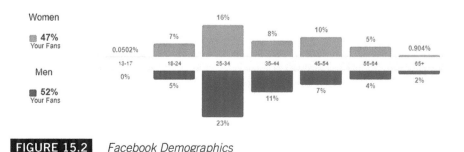

FIGURE 15.2   *Facebook Demographics*

Insights available from social media and DSP accounts can be an important research tool in the development of a target market. Spotify, Facebook, Instagram, Twitter, YouTube, and other outlets provide detailed accounting of the people who follow and interact with an artist's account, so pulling that information and cross-analyzing it between the various platforms used can often give a solid starting place for audience age, gender, geographic location, and even the types of content the broader audience tends to prefer.

Media Strategy

Determining the purpose of an ad (drive sales or build brand awareness, for example) is key, as there are some media that work better than others for each objective. For example, a simple message to alert consumers to a new album release date might be easily accomplished with a billboard in strategic areas, but using video ads online, radio spots, or television might be better options for brand-building efforts.

Another key issue is which medium(s) are most likely to capture the attention of the intended target audience. Understanding the media preferences of the target consumers will be very important in helping to guide placement decisions, but these have only become more complicated with the explosion of options that are now available. Television advertisers, for example, used to have three basic options (NBC, CBS, and ABC) who collectively reached the majority of American households. Now, according to Nielsen, the average American receives 189 cable channels. Also, many consumers now get their news and entertainment from thousands of online outlets. The choices for where to place our promotional messages are almost limitless.

It is not uncommon for today's music marketing campaigns to focus 80%–95% of their non-radio ad dollars on digital placement, although this varies by genre. Within the digital category, search advertising, social media ads on popular platforms like Facebook, Twitter, Instagram, and YouTube, and ads on DSPs such as Spotify and Pandora tend to dominate the music advertising landscape. The creative marketer, however, needs to be always aware of other media options that are available should the need and opportunity arise, so we will give some time to discussing these non-conventional outlets as well. Audience profile information should be received from the media company prior to purchasing to ensure the intended target audience aligns with the individuals reached by the ad platform you are considering.

Media Options

Digital

Digital, or online, media refers to ads that are displayed digitally through digital devices or over the internet. Digital advertising allows marketers to

Table 15.1 Advantages and Disadvantages of Different Media

Media	Advantages	Disadvantages
Digital/Online	<ul><li>Highly targeted<ul><li>Psychographics</li><li>Demographics</li><li>Geographics</li></ul></li><li>Potential for audio and video sampling; graphics and photos</li><li>Can be considered point-of-purchase if product is available online</li><li>Can use cost-per-click or cost-per-action instead of impressions for setting rates</li><li>Detailed analytics; easy-to-measure KPIs</li><li>Rich media options help break through the clutter</li></ul>	<ul><li>Cluttered environment</li><li>Internet is vast and adequate coverage is elusive</li></ul>
Radio	<ul><li>Is already music oriented</li><li>Can sample product</li><li>Short lead, can place quickly</li><li>High frequency (repetition)</li><li>High-quality audio presentation</li><li>Can segment geographically, demographically, and by music taste</li></ul>	<ul><li>Audio only, no visuals</li><li>Short attention span</li><li>Avoidance of ads by listeners</li><li>Consumer may not remember product details</li></ul>
Television	<ul><li>Reaches a wide audience but can also target through use of cable channels and programming options</li><li>Benefit of sight and sound</li><li>Captures viewers' attention</li><li>Can create emotional response</li><li>High information content</li></ul>	<ul><li>Short lifespan (30–60 seconds)</li><li>High cost</li><li>Clutter of too many other ads; consumers may avoid through DVR usage</li><li>Can be expensive</li></ul>
Out-of-Home (OOH): Billboards, Street Furniture, Transit, Alternative	<ul><li>High exposure frequency</li><li>Lower cost</li><li>Can segment geographically</li></ul>	<ul><li>Message may be ignored</li><li>Brevity of message</li><li>Not targeted except geographically</li><li>Environmental blight</li></ul>
Direct Mail	<ul><li>Highly targeted</li><li>Large information content</li><li>Not competing with other advertising</li></ul>	<ul><li>High cost per contact</li><li>Must maintain accurate mailing lists</li><li>Associated with junk mail</li></ul>
Magazines	<ul><li>High-quality ads (compared to newspapers)</li><li>High information content</li><li>Long lifespan</li><li>Can target audience through specialty magazines</li></ul>	<ul><li>Long lead time</li><li>Position in magazine is uncertain unless premium positioning is secured</li><li>No audio for product sampling except with the use of a link or QR code</li></ul>
Newspapers	<ul><li>Good local coverage</li><li>Can place quickly (short lead)</li><li>Can group ads by product class (music in entertainment section)</li><li>Cost effective</li><li>Effective for dissemination of information, such as event details</li></ul>	<ul><li>Poor-quality presentation</li><li>Short lifespan</li><li>Poor attention-getting</li><li>No product sampling</li></ul>

finely target the audience, distribute compelling messaging using both video and audio, and access detailed analytics to quickly identify what is working (and what's not) so adjustments may be made before campaign funds are exhausted. Disadvantages are that the environment is extremely cluttered, and many consumers have trained themselves to ignore most standard ad units. Also, the Internet is vast, so finding adequate coverage to the target market may be challenging.

There are four main categories of digital advertising that are core components in music marketing plans: search engine marketing, display ads, social media ads, and ads placed in and around song streams and broadcasts. Some overlap, however, does occur. For example, display ads may appear on social media networks.

> Search marketing is an umbrella term that refers to both search engine marketing and **search engine optimization.** While SEM uses paid strategies to improve search results, search engine optimization focuses on improving your website to boost your organic (non-paid) search ranking. SEO involves using strategic keywords on a website, link building, and using metadata that is crawlable by search engines such as Google and Bing. SEO is covered more thoroughly in the digital marketing chapter.

Search Engine Marketing

Search engine marketing and search engine optimization both fall under the umbrella term of search marketing. Both SEM and SEO focus on efforts to improve a website's search ranking, but only SEM involves the act of *paying* to improve a brand's search visibility. Artists and labels pay for their ads to appear either above or below search results when a user types in certain keywords that match those the advertiser has selected. When the user types in those keywords, they are shown an ad by the brand. These ads may be on the search engines themselves (Google, Bing) or on other sites that have agreed to allow the ads in exchange for a fee from the search engine. The most common formats for search engine ads are text (words only), image, or video.

How does this work? While working within the **Google Ads** platform, for example, the advertiser enters in a series of *keywords* (search terms or words their customers are likely to use in a search engine when looking for a particular product or type of website). Advertisers place bids to have their ad strategically located in the sponsored links category on search engine results pages or in a featured carousel. Ad positions are determined by several factors including the quality of the ads and landing page, the context of the person's search, and the price the advertiser is bidding. Although first-page results are highly coveted by marketers, being specific with the intended target is key. For example, targeting the words "blues music" may capture the attention of blues fans from all over the world, which would be great for brand-building efforts, but if the goal is to increase ticket sales for a local live show, including descriptors of the location such as "East Texas" or filtering within a specific geographic region would provide better results. Finding the right keywords and combination of keywords may take a bit of trial and error at first, but a keyword search tool can help. Paid and free options such as Google's Keyword Planner, SEMrush, Wordstream, and Moz's keyword tool provide detailed information on searches that are critical to planning a campaign such as the number of monthly searches of a particular keyword or keyword phrase and the average cost of each keyword you are considering. These tools also provide additional "suggested keywords and keyword phrases" you might want to pursue.

Ads that appear in search engine results are sold on a **cost-per-click** basis. This is also referred to as **pay-per-click**, which simply means that an ad might be seen by hundreds of people, but the advertiser is only charged each time a person clicks on the ad unit; however, it is important to note that while search engine marketing generally utilizes PPC advertising, not all PPC ads are SEM. PPC can also be utilized in social media or display ad marketing.

There is often confusion between the products **Google AdSense**, **Google Ads**, and the **Google Display Network**.

- Google Ads is for *advertisers* who wish to promote products or services on Google's search site. This program is used to drive traffic to a site when a fan types a relevant keyword that has been selected for targeting by the marketer.

FIGURE 15.3   *Google Keyword Planner*

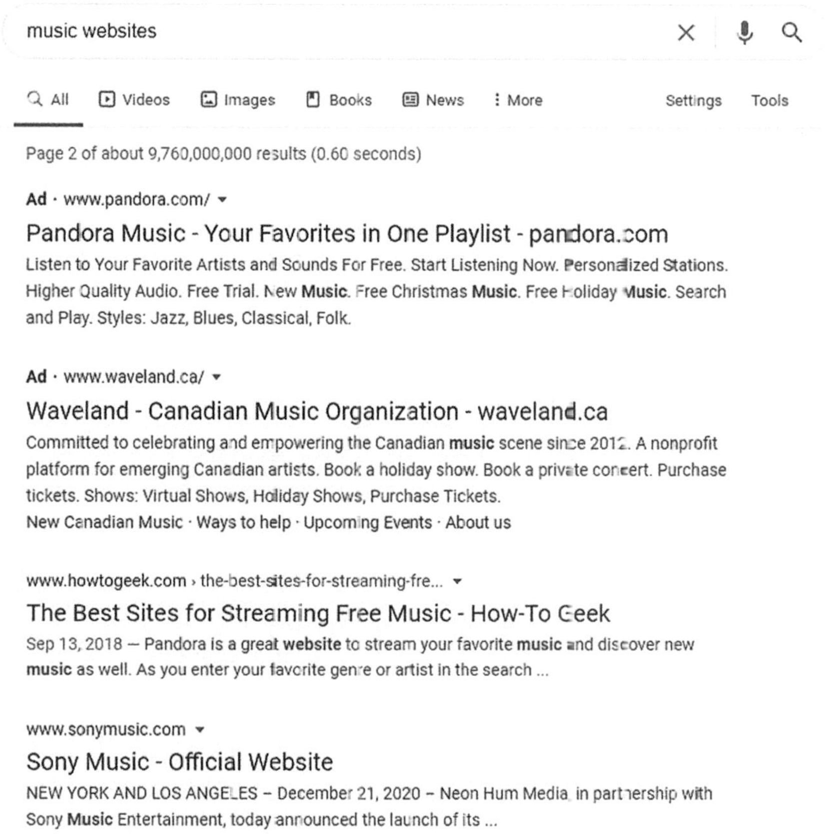

FIGURE 15.4   *Google Search Results for the Keywords "Music Websites." Notice the two ads positioned at the top*

- Google Display Network (GDN) also uses Google Ads as an entry point, but instead of displaying an ad within Google's search results, it places the ad on any number of GDN partner websites, mobile apps, and videos.
- Google AdSense is a platform that allows *publishers* of websites, blogs, or other forums to monetize their websites and apps. Google pays publishers a commission to place text, image, video, or interactive ads on the publisher's property. The amount of money the publisher earns depends on how many clicks or impressions the ads placed on the site generate.

Display Ads

Display ads are an image-based way of promoting goods and services and they appear on other websites and apps. Text, images, and a URL are used to

FIGURE 15.5   *Searching for "Christian Music" on Google. Ads that fit the keywords as well as organic results from the search, such as this one of Christian artist Lauren Daigle, are lifted to the top of the results page*

link consumers to a website where they may learn more about the product being advertised. Display ads are often used to increase brand awareness or when an advertiser needs to make a visual impact.

- Standard/static banner ads – generally use one image and allow one function (for example, the user clicks on the banner and is taken to a website).
- Animated banner ads – a type of **rich media** ad. This ad type uses amination and possibly audio.
- Floating ads – another type of rich media ad that suddenly appears on a screen then disappears after 10–30 seconds. These might be standard ads or more creative offerings such as a bear dancing across the screen.
- Pop-up ads – units that open a new browser window.

- Interactive ads – offer viewers the chance to change the ending of an ad or play a game within the ad unit. These rich media offerings seek to actively engage the audience.
- Expandable ads – automatically enlarge to cover most of the screen.
- Homepage takeovers – ads cover the entire homepage and eliminate any other sponsor on the page.
- Video ads.
  - Although video ads are often served through YouTube and social platforms, they may also be distributed through display ad networks. Video posts have been shown to drive higher engagement rates than using images. On Twitter, for example, Social Media Today reports ten times more engagement for tweets containing video than for those including images or text.
- Interstitial ads – appear between content segments. For example, when a gamer finishes playing a level, an ad is shown while the next level is loading.
- Remarketing/retargeting – these are often display ads on websites or apps. These ads use "cookies" to follow people around the web. The ad units may be shown to people who have visited the artist's website pages or who have downloaded a brand's app within a designated time period. If a consumer looked at tickets for an upcoming concert on a website and then a day or so later had an ad for that tour served to them while they were perusing social media or another site, that's remarketing at play. The advertiser is attempting to remind the consumer of their interest in the concert and to convince them to come back to the site to complete their purchase. Platforms that specialize in retargeting campaigns include AdRoll, Retargeter, Fetchback, and Chango. Retargeting campaigns are also possible using Facebook's Custom Audiences tool.

FIGURE 15.6A *Lilly Hiatt Walking Proof Banner Ad*

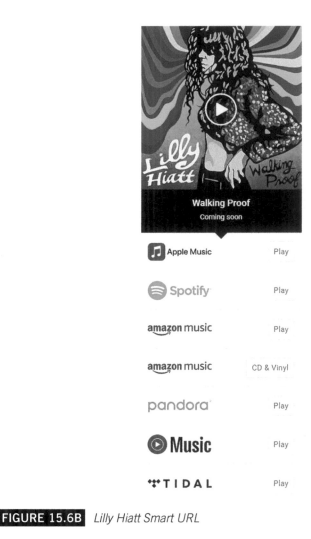

FIGURE 15.6B   *Lilly Hiatt Smart URL*

The Lilly Hiatt "Walking Proof" ad was a click-through banner ad that rotated on No Depression's home page. As an Americana artist, this banner ad placement made perfect sense, notifying the target market that the release is now available. To click through the banner, the reader could instantly become a consumer through the use of a **smart, or pivot, link** which directs fans to a landing page that enables them to choose the streaming platform or retail of their choice.

Display ads are often sold through a service like **Google Display Network**. These ads are placed by going through the Google Ads platform. GDN matches keywords, topics, your language, and location, as well as a visitor's recent browsing history to more than 2 million websites and apps that have agreed to let Google sell advertising for their site. Ads are sold on a cost-per-click, cost-per-thousand impressions, or cost-per-action basis. The process that matches ads to relevant content on sites is more broadly referred to as **contextual targeting**.

- *Contextual advertising* is defined as advertising on a website that is targeted to the specific individual who is visiting the website based on the subject matter of the site and then featuring products that relate to that subject matter. For example, if the user is viewing a site about playing music and the site uses contextual advertising, the user might see ads for music-related companies such as music stores or instrument manufacturers. Google has added AdSense as a way for publishers (website and app owners) to feature relevant advertising on their sites and share in the revenue generated by sponsors. The source of these ads comes from the AdWords program, so that those who sign up for AdWords can specify if they want their ad to appear on these related websites – the *content network*. Google's website states, "A content network page might be a website that discusses a product you sell, or a blog or news article on a topic related to your business."

Additional contextual advertising examples:

Della Mae New Music Bluegrass

 Buy Della Mae CDs, Vinyl Records, Digital Music, T-Shirts, Bandanas, Stickers and More!

FIGURE 15.7   *Della Mae Google Network Ad*

- A blog covering rock music might be served an ad pushing ticket sales for a local rock band.
- An ad for soup might appear alongside an online article for crock pot recipes.
- An article highlighting summer travel tips might feature an ad promoting tickets to a music festival.
- An ad for eyewear might appear on the website of a bookstore.

GDN offers several payment options, and the one you choose will often depend upon your objective. According to social media marketing and analytics website Hubspot, cost-per-thousand (CPM) is often preferred for campaigns that are intended to generate awareness of a project, while CPC is chosen for generating traffic to a website or app, and cost-per-action is utilized when the goal is driving sales conversions.

Social Media Ads

Advertising options on social networks continue to grow. Sites like Instagram, Facebook, Twitter, TikTok, YouTube, and Snapchat provide a cost-efficient way to reach a preferred target market.

Facebook dominates the social media advertising landscape not only because it has the most robust targeting tools within Meta's **Ads Manager** platform, but it also extends those same options to company-owned Instagram. From Facebook carousel ads to sponsored stories on Instagram, there are numerous paid opportunities to amplify social media messaging across each of these platforms. Other popular social media ad options include promoted tweets on Twitter and advertising on TikTok. Google-owned YouTube is also a critical component of any music marketing campaign. The platform offers two options: (1) advertise on YouTube or (2) advertise on other platforms via Google Ads.

Facebook

Even though Facebook is not as popular with the younger audience as it as once was, the network's massive reach and the power of the Ads Manager platform make it one of the main components of many music marketing paid campaigns. Ads are created through Ads Manager, which requires that the marketer have a Facebook Page (a Page is a business account rather than a personal profile).

Although anyone can "boost" or "promote" a post on a Page, marketers must utilize the Ads Manager platform to unlock the extensive options available for targeting. Simply put, "boosting" a post on your Page only offers a small fraction of the targeting choices in Ads Manager.

Facebook can provide laser-focused targeting by utilizing its vast data on user activity. The platform has made it easy for even novices to implement an ad campaign by first asking advertisers to identify an objective (i.e. brand awareness, reach, traffic, engagement, conversations, etc.), then walking the marketer through the options step by step. It also provides the option to create an **A/B test** that allows for the ability to test two campaigns against each other to see which strategies perform best.

Once of the most important considerations in setting up a campaign is selecting a targeted audience. In addition to age, gender, location, and language, it is necessary to add detailed targeting to help the platform find the most likely prospects to be receptive to the ad. Detailed targeting is available in three broad categories:

- Demographics
- Interests
- Behaviors

For example, under demographics, the marketer could select targets according to education, financials, life events, and even parental status. They might also target people who list their job title as "marketing director."

The interest options are vast. The Pages a user follows or engages with in any way (commenting on a post, for example) are stored and used later to help direct advertisers to a likely prospect. Examples include targeting people who like Pages related to concerts, Twitch, or *Billboard* magazine.

Detailed Targeting

Include people who match ❶

Interests > Additional Interests

Americana (music)

🔍 Add demographics, interests or behaviors	Suggestions Browse
American folk music	Interests
Folk music	Interests
Singer-songwriter	Interests
Folk rock	Interests
The Avett Brothers	Interests
Songwriter	Interests
Nathaniel Rateliff	Interests
Sturgill Simpson	Interests

 FIGURE 15.8   *Facebook Targeting*

Marketers may browse the broad interest categories or type in their own. The search "Americana Music" returns an audience estimate of 2,778,210 of people who "have expressed an interest in or like Pages related to Americana Music." When clicking on the "Suggestions" tab, the system drills down further into Americana by offering other searches to consider such as "singer-songwriter," "folk music," "Sturgill Simpson," and "The Avett Brothers." Behavior options include digital activities, type of mobile device used, purchase behavior, travel tendencies, political preferences, and more.

Facebook's **Custom Audiences** feature may be used to by inserting an existing database (e.g. a list of email subscribers) to target those people on the platform, or retarget visitors to a website through the use of a **Facebook pixel.** One of the most powerful and popular ways to use Custom Audiences, though, is to set up a **lookalike audience**. Lookalike audiences are used to find people who share similar traits. The platform analyzes the demographic information and interests of the people who like a Page, and they then deliver the ad to that audience.

Facebook offers different payment options such as CPC and CPM, depending upon which objective is chosen for the ad. Display options are numerous as well. The marketer may either choose to let the platform choose where it places the ad (it uses its data to determine the best prospects who would be most likely to engage with your ad), or they may choose to select placement manually. Choices there include whether the advertiser wishes the ad to appear in both Facebook and Instagram and whether they want the placement to be shown within the news feed, while they are watching videos, within Stories, and more.

Twitter

As mentioned, Facebook has the most extensive targeting options available across socials. Twitter looks much like Ads Manager, only with fewer options. The platform follows a similar approach to Facebook in that they ask advertisers to first state their objective (i.e., awareness, video views, pre-roll video, app installs, driving traffic to a website, acquiring followers, etc.) which will determine the type of ad and placement, of which there are three main types:

1. Promoted Tweets – tweets that show up in the targeted users' timelines, profiles, and at the top of search results.

2. Promoted Accounts – these ads often show up in the "who to follow" section, but they may also appear in users' timelines or in search results. Instead of the ad being a Tweet, however, your brand

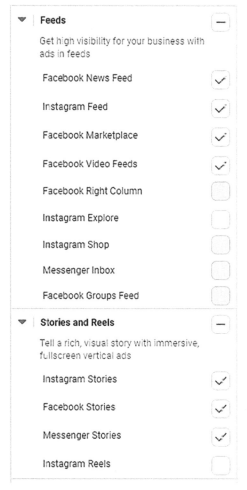

Placements

Feeds

Get high visibility for your business with ads in feeds

Facebook News Feed

Instagram Feed

Facebook Marketplace

Facebook Video Feeds

Facebook Right Column

Instagram Explore

Instagram Shop

Messenger Inbox

Facebook Groups Feed

Stories and Reels

Tell a rich, visual story with immersive, fullscreen vertical ads

Instagram Stories

Facebook Stories

Messenger Stories

Instagram Reels

FIGURE 15.9   *Facebook Manual Placements*

is promoted. These ads are usually detectable because they feature a "follow" button.

3. Promoted Trends – these ads appear within the Trending Topics section of Twitter. Promoted Trend Spotlight is similar, but in addition to static images, Spotlight supports video and GIFs

For Promoted Tweets, Twitter considers the quality of tweets, in addition to the budget, in determining with ad units to show. Generally, favored tweets include those from accounts who have demonstrated engagement with consumers (do they re-tweet, favorite, or reply often to the tweets?), is the tweet related to subjects the user is interested in, and is the tweet timely. The last one is important, as more timely tweets receive priority.

Twitter also has robust targeting options including age, location, gender, language, interests, movies, television shows, events such as the Sundance Film Festival or the Grammy Awards, and recurring topics such as #Music-Monday. Targeting by device and carrier is also available, as is developing custom audiences that enable you to target based on activity with your app, an email list of customers, and activity with a website. An advertiser may choose to focus on keywords used in search queries, recent tweets, or tweets with which users engaged. Lookalike targeting is very similar to that in Facebook, as it searches for people with similar interests to an account's follows. Enter @countrymusic (the Twitter account for the Country Music Association), for example, to target people likely to be interested in that genre.

Instagram

A professional account is required to place Instagram ads. Two options are available to place ads on the platform:

1. Promote an existing post within the app (this is like the "boost post" feature in Facebook).

2. Utilize the Ads Manager.

Instagram ad options include Stories, Photos, Videos, **Carousel ads** (a popular option that features two or more scrollable images), Collections, Explore, IGTV, and Instagram shopping.

Boosting a post within an app is easy and achieved by simply clicking on the "Promote" button under a post and walking through the prompts, or Ads Manager may be utilized for the most customizable options. Marketers simply walk through the steps in choosing an objective, identifying the target audience, selecting the preferred placement and schedule, specifying budget and payment details, and finalizing ad creative. One important note for Instagram ads is to not forget to add the popular call-to-action buttons such as "Book Now" or "Sign Up." Also, keep in mind that space is limited, so keep the text used to a minimum. Only two rows of text are displayed without being cut off.

TikTok

Unsurprisingly, the TikTok Ads Manager looks strikingly like the other platforms. Users are prompted to include a payment option as the system walks

Della Mae Instagram Shopping Carousel Ad

an advertiser through the account creation, but if they click "next" at that prompt, they may bypass that request and pay for the ad after it has been created.

Advertisers start by choosing from among several advertising objectives for the campaign: reach, traffic, app installs, video views, and conversions. Promotion types offered are app and website. The TikTok pixel may also be utilized to track an individual's movement on the advertiser's website after the visitor clicks on the ad.

Placement options are automatic (advertisers let TikTok place the ads where they think they would perform best, or marketers may choose the manual option where they manually select target placement). This includes the in-fee ads on the "For You" page or News Feed apps like TopBuzz. Not all placement options are available in every region, as the app is popular with users globally. The system does allow the option of allowing users to comment on your ads and to download your video ad.

As with other platforms, a customer list such an email file may be uploaded to create a Custom Audience. The Custom Audience may then be used to create a Lookalike Audience, which targets people on TikTok who have similar traits to the Custom Audience uploaded.

Location, gender, age, language, and device preferences are available. Advertisers may also target users by interests as well as behaviors such as users who watched videos to the end, liked, commented, or shared. Under the Bid Control section, advertisers may choose whether they want to cap the amount they will pay per result or let the system utilize the budget based on the optimization goal. Campaign metrics may be viewed and filtered by audience segment. A/B, or split, testing is also available on the platform.

YouTube

YouTube is the only social media platform with a reach as vast as that of Facebook. In fact, Pew Research Center notes that 73% of U.S. adults use the app, surpassing Facebook's 69% and earning the survey's top spot. This massive platform not only boasts extensive reach, but it also offers tremendous targeting capabilities. Because YouTube is owned by Google, the amount of data available to pinpoint a potential target is unmatched.

Main types of YouTube ads:

- **Skippable in-stream** ads allow the viewer to skip the ad after five seconds. These ads may be "pre-roll" (play before the video) or "mid-roll" (play at some point during the video) or even "post-roll." Because the advertiser does not pay if the user skips immediately after the five seconds expire, this can be an effective strategy to use a relatively small budget to reach a broad audience if the messaging is simple enough to be promoted in the first five seconds. For example, award shows have used this strategy to communicate tune-in information. A label could do something similar and use this option to a release date for their artist's next track.
- **Non-skippable in-stream and non-skippable bumpers** ads must be watched by the user before their selected video will play. These units are generally 15 seconds in length and typically placed by national ad

agencies and brand marketing firms. Bumper ads are six seconds in length and also non-skippable. These units work well for short, very succinct messages. For either option, advertisers are charged on a cost-per-thousand basis.

- **Video Discovery ads** – display ads that are designed to reach users who utilize relevant keywords in the search box. For example, if a user were to search for "rock music," an ad pushing tickets for a rock music festival might appear alongside the results.
- **Non-video (display/overlay) ads** – display ads that appear to the right of the selected video and above the suggestions list. These ads generally use an image, text, and call-to-action. **Overlay ads** are the in-video, semi-transparent units that appear to be floating on the lower third of the video.

Just as a marketer would to place a Google search campaign, they also need to access Google Ads to place YouTube advertising. Remember that Google Ads provides access to the full suite of Google products including search ads, so one of the ad type options that supports video must be selected in order to place a video ad on YouTube.

Tip: Like Ads Manager for Facebook and Instagram, YouTube offers a guided system for placing ads. Marketers familiar with setting up campaigns will click on "switch to expert mode" to have access to additional options. For example, when choosing "Product and Brand Consideration" as an objective, expert mode then offers the advertiser the option of setting up an ad sequence that is not highlighted in the guided walkthrough."

In "expert" mode, setting up a campaign is like the steps reviewed earlier for other platforms. Advertisers start by identifying an objective such as generating website traffic, product and brand consideration, or brand awareness and reach. The marketer is then prompted to choose from among several bid strategies, confirm budget, note the language, and add the location.

New audiences may be created through the Audience Manager. A Google Ads tag may be added to a website or app to enable the tracking of users who engage with an ad. Demographic options include life stages such as college students or new parents. Users may also be targeted based on interests via past search queries. Options are also available to serve ads to those who have been identified as being "in market" for several listed goods and services such as people interested in purchasing audio streaming services.

Performance metrics for the campaign, such as reach and frequency, are also available using Google Ads reporting. YouTube offers a Video Creative

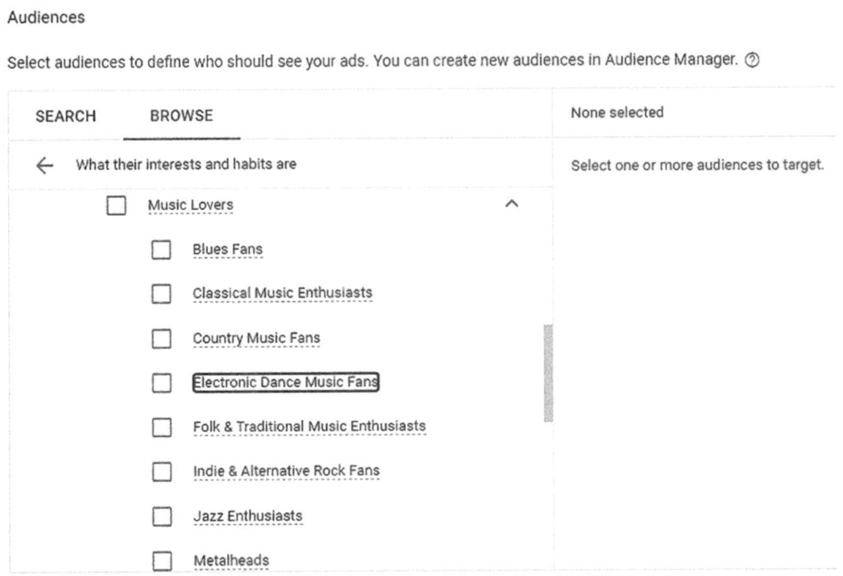

Audiences

Select audiences to define who should see your ads. You can create new audiences in Audience Manager. ⓘ

SEARCH	BROWSE		None selected
← What their interests and habits are			Select one or more audiences to target.

☐ Music Lovers ⌃

　☐ Blues Fans

　☐ Classical Music Enthusiasts

　☐ Country Music Fans

　☐ Electronic Dance Music Fans

　☐ Folk & Traditional Music Enthusiasts

　☐ Indie & Alternative Rock Fans

　☐ Jazz Enthusiasts

　☐ Metalheads

FIGURE 15.11   *YouTube Second-Level Targeting*

Analytics tool to help identify parts of the promotional video that are keeping viewers engaged and parts where viewership drops.

Snapchat

For marketers seeking to reach a younger audience, Snapchat is a top advertising platform. As with the other social media options, Snapchat lets advertisers choose a campaign goal first, such as building brand awareness and video views, web views and conversions, or engagement. Users engaging with Snap ads are encouraged to use the swipe up feature to see the ad. The units are three to ten seconds in length and may be used to drive fans to a website, view a video, or download an app. In addition to ads within Stories, the platform offers branded tiles that appear in the Discover section.

Of course, filters are very popular on Snapchat, and although some such as the cat-face filter are native to the app, brands may create their own. In addition, brands may create augmented reality lenses to encourage interactivity. Last, collection ads are used to showcase multiple products. These video units place thumbnail images at the bottom that scroll horizontally.

The targeting options are very similar to other platforms and include both the ability to create custom audiences as well as audience lookalikes; however, Snapchat does provide the source of the data provider so advertisers can see if the data is coming from a third-party provider of through internal data.

The platform offers several performance metrics to measure success, including being able to analyze swipe ups by age, gender, and device type.

Influencer Marketing

Influencer marketing has been one of the fastest-growing marketing channels over the last several years, with MediaKix reporting consistent year-over-year growth in budgets focused on this area. Influencer marketing in music typically involves paying a person of influence to post about new music to generate buzz around the project and to encourage their large social following to stream the music or buy the album. Although working with a social media influencer is an option on any platform, it is most popular currently on Instagram, specifically with Instagram Stories. In fact, Hootsuite reports that approximately one-third of influencer campaigns on the platform occur as Stories. According to Instagram, influencer campaigns have shown to be very effective in a recent study, with 34% of daily U.S. Instagram users reported in a recent survey that they had made a purchase due to the recommendation from an influencer or blogger.

While influencer campaigns are very popular across all product categories, efforts to promote music-related products have had mixed results. Perhaps the most famous example of a failed campaign is the one in support of the 2017 Fyre Festival. The campaign was launched with more than 60 influencers, who were reportedly paid $20,000 to $250,000 each to simultaneously post a graphic to their accounts. The campaign rapidly resulted in millions of impressions; however, those impressions failed to convert into ticket sales. The festival was billed as a luxury and exclusive event, but it ended up disappointing ticket holders, as the organizers ended up cutting corners and eliminating promised deliverables to compensate for a massive budget shortfall due to anemic ticket sales. The brands of many of the participating influencers were damaged as a result.

Of course, not all music influencer campaigns are so expensive. Influencers who agree to promote tracks may do so for $50–$10,000 per post, with the fee largely dependent upon reach, audience, and the number of impressions they generate on their Stories. While other product categories may focus on coupon redemptions or other sales metrics in stating an objective for a campaign, most music marketing efforts are intended to generate brand awareness and increase trial of the music. Although most big brands target influencers with millions of social followers, a growing trend among some marketers is to engage more niche audiences with **micro influencers**, who generally have social audiences of fewer than 10,000.

Key to implementing a successful influencer campaign, as it is with any type of partnership, is finding a good fit for the music artist being promoted. Authenticity is key. For music, this type of marketing typically involves finding an influencer who is already a fan, so followers believe the influencer when they say they love the music. If the promotion comes off as a paid promotion or one that looks staged, followers are sure to notice and reject it. When choosing the right influencer, the following questions may be helpful:

- What type of music does the influencer prefer, and what genres have they promoted in previous campaigns?
- How familiar is the influencer with the artist?
- What are the interests of the influencers' followers?

One tactic used often is to ask the influencer to use link stickers on Instagram (a minimum of 10,000 followers is required to utilize this feature) to suggest followers listen to the artist's music, but London-based music marketing agency Burstimo suggests flipping campaigns by asking the influencer to be involved in a music video rather than asking them to push the music through a standard social media post. New Zealand group Drax Project secured YouTube star Liza Koshy to be featured in their video "Woke Up Late." Koshy's fan base watched the video in droves, exposing many new fans to Dax Project in the process. One of the disadvantages to influencer campaigns is the short lifespan, but this type of effort offers longevity, which is typically an advantage afforded to traditional product placement in music videos, film, and television.

Once the campaign is completed, marketers will want to make sure they are provided with analytics, including the number of views the program received and the number of people who took action from the post in order to access if the partnership accomplished its intended goals.

Streaming Platforms

Spotify

Although other streaming platforms such as Pandora and Soundcloud also offer advertising options, Spotify is one most frequently included in paid marketing efforts. There are two main ways to secure Spotify advertising: (1) go through Ad Studio, Spotify's self-service ad platform, or (2) work directly with an agency or with the Spotify ad team.

Spotify's Ad Studio offers the ability to place two ad types on the platform with a minimum spend of $250:

1. Audio ads – these ads play between songs.

2. Video Takeovers – these ads appear during a listening session when the user is browsing the catalogue. They include a companion banner with a call-to-action button meant to drive engagement.

○ Interests ⑦
Target your audience based on their interests, as indicated by recent podcast and
playlist listening, as well as streaming platform.

☐ Books		☐ Business
☐ Comedy		☐ Commuting
☐ Cooking		☐ Culture & Society
☐ DIY Hobbies & Crafts		☐ Education
☐ Fitness		☐ Gaming
☐ Health & Lifestyle		☐ History
☐ In-Car Listening		☐ Love & Dating
☐ News		☐ Parenting
☐ Partying		☐ Podcasts ⑦
☐ Running		☐ Science & Medicine
☐ Sports & Recreation		☐ Studying or Focusing
☐ TV & Film		☐ Tech
☐ Theater		☐ Travel

FIGURE 15.12   *Spotify Interests List. Additional types of ads may be accessed through an agency or in working directly with the Spotify advertising team:*

Advertisers start the process by first identifying what is being promoted, then creating an ad set that includes choosing an ad format and platform (audio, horizontal video, and vertical video) as well as platforms (iOS, Android, and desktop). Once the desired schedule and budget is set, the targeting fun begins. Targeting options in Spotify include geographic regions, age, and gender. Additional targeting options:

- Fan base – targets the fan base of artists you are promoting or other artists or accounts. For example, an artist may target their own listeners or the listeners of a similar artist. They may also target genre-focused accounts such as Hip-Hop Beat Nation.
- Genre – targets listeners based on the genre of the song they heard immediately prior to the ad.
- Interest targeting is based on a listener's recent podcast and playlist activity.
- Real-time context – targets listeners based on the type of playlist they are listening to at that moment such as dance, party, or holidays.

Once a target is identified, the system walks the advertiser through payment options; then the ad is reviewed for accuracy and confirmed.

- Podcast ads may feature copy read by talent or a pre-produced ad if it fits naturally into the content.
- Sponsored sessions provide the user with a 30-minute ad-free listening session after the user watches the brand's video ad. Once the video ad is finished playing, a clickable display ad appears.
- Overlay ads feature call-to-action buttons such as "Buy Now" or "Visit Profile" and appear when a user opens the app. The clickable display ad drives traffic to the brand's website or app.
- Homepage Takeover display ads showcase a brand's message on the front of the desktop homepage for 24 hours. The units are clickable and support rich media content.
- Leaderboard visual ads are clickable units that are only shown for 30 seconds when Spotify is in view.

Amazon Music

Although Amazon offers brand marketers audio, video, and display advertising options across its many platforms, one of the most popular options for paid ads supporting new music releases is placing audio ads to run as consumers listen to Amazon Music's ad-supported free tier. These audio ads run on desktop, tablet, mobile, and even on Echo devices and include companion banners that appear on mobile, desktop, Echo Show, and FireTV. As is the case on other platforms, listening to the occasional ad unlocks a block of uninterrupted music.

Pandora/SoundCloud/SiriusXM

Sirius XM has built an extensive and impressive advertising platform with the 2019 acquisition of Pandora and its 2020 investment of $75 million in SoundCloud.

Although many music marketers would love to strategically place ads for their music on Sirius XM's many genre-specific channels, these music-focused channels, unfortunately, do not accept advertising. However, paid placement such as live reads, segment sponsorships, and targeted brand spots are available on approximately 50 talk, comedy, sports, and news channels.

Collectively, SoundCloud and Pandora enable advertisers to reach an audience of more than 100 million unique listeners, making it one of the largest marketplaces for digital audio advertising. SoundCloud offers artists and labels the ability to promote monetizing content directly through their self-service tool Promote on SoundCloud, but larger campaigns might be wise to utilize the advertising options through Pandora that allow for advertising placement on both platforms.

Targeting options through Pandora include:

- Demographics such as age, gender, income, family status, and culture
- Genre, artists, lyrical content
- Geolocation
- Activity while listening
- Language, device, carrier, platform
- Lifestyle characteristics such as travel and activities

Perhaps one of the most recognized advertising options is related to sponsored content that allows for the specific branding of stations or the unlocking of uninterrupted listening after a user watches an ad. While standard audio and display ads are offered across the spectrum of products, Pandora offers several options that utilize their vast technology to pinpoint a desired target market:

- Dynamic audio ads use intelligent technology to insert customized messaging based on triggers such as location, weather, and time of day.
- Sequential audio uses exposure or clicks to target customers as they move through certain stages of the process (the first ad might build overall awareness of the product and the second may drive a desired action).
- Creative targeting allows different ad creatives or landing pages to be presented to different targets by age, gender, or geographic area.
- Amplified display ads utilize rich media to present ad options such as a countdown clock to an album release.

RADIO

Although terrestrial radio might seem to be less relevant than it was in years past, it is still a critical component in the marketing plans for genres such as country and moderately important in others. Radio advertising may be implemented on the local level through an area radio station, as well as regionally or nationally by going through radio groups or conglomerates such as iHeart or Cumulus. Advertising opportunities are also available through programs that are nationally syndicated such as Delilah, Bobby Bones, Country Countdown USA, and more. Often, paid radio campaigns are placed in conjunction with a broader promotional effort with the station that might include on-air promotions. Units may be placed in the terrestrial broadcast; the station's online stream; or on the station's website, within the station's app, or a sponsored post on socials. Sponsorship of special events is also available. Advantages for radio are that it is already music oriented and that makes product sampling easy. It also boasts a short lead time, high quality, and the ability to segment by musical taste or genre and by geographic area; however, radio spots are generally only 30 or 60 seconds long, they do not offer the opportunity to include visuals (on the over-the-air units), and there is a lot of clutter with other ads, so it is easy for a message to get lost.

ADDITIONAL PAID MEDIA OPPORTUNITIES

The previous media options are the ones regularly used in music marketing campaigns, but many other options are available depending upon need and budget. Savvy marketers know going against the grain is sometimes effective. Each campaign must be approached individually and with a fresh viewpoint to ensure the best mediums are utilized vs. automatically choosing what the label has traditionally placed. Following are additional considerations.

TELEVISION

Television reaches a broad audience, and now with cable, it is easier than ever before to target by interest: homeowners, gardeners, and DIYers might be reached through HGTV, and those who love and/or own animals can be reached through the Animal Channel. Placement within specific programming is also available at a premium price. For example, ads for new music placed by record companies often appear inside award show broadcasts like the Grammys or a music-oriented series like *The Voice*. TV also has the benefit of sight and sound, which can be attention grabbing and can create an emotional connection to the product. Disadvantages are the high cost for placement and production of the spots and the short life span – it is often difficult to say everything you need to convey in 30 seconds. Also, according to Nielsen, DVR penetration continues to grow, with more than 50% of all television households now having at least one DVR. And what is their favorite way to use those DVRs? You know it – to skip that great promo spot you just developed.

TELEVISION TERMINOLOGY

ROS or run-of-schedule refers to the scheduling of a promo spot at any time the station wants to run it within a specified timeframe. Typically, this is much less expensive than if the marketer specifically request a daypart or placement within a particular program. For example, an ROS spot may run at 10:45 a.m. on the station one day then 9:45 p.m. the next day.

Road blocking is another term that is referenced mostly in television and radio. Have you ever switched a channel to avoid an annoying commercial, only to see it running on the other major networks at the same time so that you really cannot escape it? That is called road blocking, and it is common in both television and radio advertising.

PRINT

Magazines fall under the print medium and have the advantage of being able to communicate a lot of information, and it has a longer life span, with some magazines passed along to others and some kept for years. Marketers may also target particular audiences through specialty publications like *Car and Driver* or *Seventeen*. Disadvantages include the long lead time, as magazine creative has to be turned in weeks prior to the printing. And unless you pay for premium positioning, your ad might be placed in a less-than-desirable location like next to another ad or next to unrelated editorial. Of course, most entertainment publications now have online versions that are even more popular than the printed piece.

Most media outlets offer a **media kit** that compiles important information for advertisers including rates, the audience profile, special opportunities, creative specifications for designers, closing dates, and production deadlines, as well as terms and conditions. Within the media kit is a **rate card** that provides information like the **rate base** (the minimum number of subscribers and over-the-counter purchasers projected for those specific issues), the rates for four-color ads and black-and-white ads, and pricing for preferred placements. A higher price is charged for guaranteed premium positions like the inside front cover, inside back cover, and the back of the magazine. Typically, on any type of rate card, the publisher will list the open rate, which is the most you would be charged; however, the rates are often negotiable. The rate card also provides an audience profile that details who reads the magazine so advertisers can know if they are hitting their desired target audience or not. Many publishers utilize top syndicated research service MRI-Simmons for this data.

For large, national publications like *People*, there is often an option to run a split ad, meaning that your ad gets placed in one half of the national circulation. This essentially means that every other copy contains the ad message. This is a great option for advertisers who desire a national push but do not have the budget available to cover the cost of full distribution.

Newspapers are another form of print advertising, although they are not used frequently in music marketing plans. Newspaper ads are great if you have short lead time and you can run an ad in the entertainment section, for instance, to reach people interested in those types of products. Disadvantages are the poor quality, short life span, and the inability to really do any type of product sampling, except if you were to print a digital code for a download or special content that they must go online to "redeem." Also, the printed version of newspapers skews older, so knowing your target demographic is important.

OUT-OF-HOME

Out-of-home advertising refers to advertising that attempts to target the consumer outside of his or her home while the consumer is "on the go" or place-based media such as in-theatre advertising, gas pump advertising, and event marketing.

Types of OOH Media:

- Outdoor: Billboards, posters, murals
- Aerial/inflatables: blimps, hot-air balloons, aerial banners
- Street furniture: Bus shelters, kiosks
- Transit: Ads inside buses, subways, and trains; ads in taxis; and airport advertising
- Mobile advertising: Mobile billboards on trucks
- Vehicle wraps: Display panels on sides, rear, front of trucks and buses
- Building wraps
- College campus: Display racks and college magazines
- Indoor: Billboards in restrooms, ads in health clubs, arenas, in-theatre advertising (on-screen and in the lobby)
- In-flight: In-flight programming and print publications
- In-store: Shopping carts and front-of-store displays
- Stadium/arena: Wall displays, scoreboards, and programs
- Events: Flags and other paid messaging throughout an event

Billboards

The advantages of billboards are that there are many people reached and a low CPM. It is also easy to target geographically using them. Disadvantages are that, hopefully, anyway, drivers are more concerned with watching the road than reading your billboard. And there really is not a way to target a particular demographic group, so there is the potential to have a lot of waste. One car may be driven by a 65-year-old male, while the next car has four teenage girls in it.

Billboards are best used to convey simple messages such as a new music release date, as the text used on billboards need to be very brief – generally no more than eight words. The creative should include a large, easy-to-read font and lowercase letters. Also, think about which font colors are easiest to read, and use those (in other words, steer away from harsh color combinations such as purple letters on a black background). Simple or eye-catching photos or phrases work well.

In-Game

In-game ads are those that appear within a video game or ads that appear inside mobile apps like Sudoku, Wordscape, or Solitaire. For example, a billboard within a video game might promote an artist. These are paid promotions that appear as part of the scenery. This practice is considered an innovative way to reach certain demographics with a promotional message, but it may be frowned upon by gamers due to the disruptive nature of it.

Direct Mail

Direct mail has the advantage of being highly targeted by zip code or even by a particular block or individual. However, this medium is associated with junk mail, so it would be easy for an ad to be overlooked by the recipient. Depending on the type of direct mail utilized, it can also be very expensive.

Mobile/Texting

Mobile is a form of digital advertising that is displayed only on mobile devices like smartphones and tablets. It is great for "just in time" promos, and it is very personal in that most people do read every text message they receive. You might be able to use it for geotargeting, but be careful with that – many users are taking their phone numbers with them when they move away from the city where the area code is associated. Disadvantages are that you have a very short space in which to place a message, and it is easy to annoy consumers with too many text messages or with content that they do not find useful, either of which can prompt them to unsubscribe.

TRADE ADVERTISING

Most of the information we have covered thus far has been advertising directed to the consumer; however, many marketing plans include paid tactics that are meant to reach industry gatekeepers such as label executives, managers, talent agents, venue operators, and more. These ads may tout the streaming success of a track to get the attention of the radio community or promote the inclusion of an artist's song in an upcoming film to pique the interest of a label executive in the hope of eventually securing a record deal. Most entertainment business outlets offer many advertising choices that include distributing paid ads across their social media outlets or in print publications. *Billboard* is one of the more popular choices for industry-focused campaigns, as they offer a wide variety of placement options, including "Tentpole Issues" such as the Grammy Preview edition, Women in Music, or Country Power Players. Advertisers may also tap into their sister publications such as *The Hollywood Reporter* and *Vibe* to reach those audiences. *Billboard* also has sponsorable podcasts, newsletters, and events targeting the industry, such as the *Billboard* Live Stream Summit.

To build general industry awareness, *Billboard*, *Rolling Stone*, *Spin*, and *Variety* might be appropriate; however, if the marketer's goal is to focus on a particular geographic area like Nashville or if the artist they are promoting is in the country genre, a presence on *MusicRow*'s website might be more appropriate. Industry options are vast, so knowing your campaign objective and the specific audience you are trying to influence is key.

HOW ADVERTISING EFFECTIVENESS IS MEASURED

As reviewed in the radio chapter, the total number of unique views of an ad over a period of time is referred to as the **cume**. A household or person is counted only once even if the ad was viewed by that person multiple times. This is also known as net unduplicated audience or net reach. Cume, or **net reach**, may be expressed as an absolute number or as a percentage of the population.

Frequency is the number of times a consumer or population is exposed to the ad message.

Reach × Frequency = Gross Impressions

Gross impressions are the total number of times an ad was viewed. The figure counts duplicates or repeated viewings by the same person. For example, 1 million gross impressions could be 1 million people each exposed once, 10,000 people each exposed 100 times, or any other such combination of numbers. In other words, it includes duplicate views by the same person.

Another example:

Visitor A viewed: Ad1, Ad2, Ad3, Ad2
Visitor B viewed: Ad1, Ad4
To count: Ad1 has two unique (views) and two gross impressions
Ad2 has one unique (views) and two gross impressions
Ad3 and Ad4 have one unique and one gross impression each

A **rating point** is a value that is equal to 1% of the total population or households that are tuned in to a particular program or station at a specific time. For example, a six rating for women 18–49 means that 6% of all women 18–49 in a specified geographic region were viewing that station or program.

Gross Rating Points (or GRPs) quantify impressions as a percentage of the population reached rather than in absolute numbers.

GRPs measure the sum of all rating points during an advertising campaign without regard to duplication. Just like in our recent gross impressions example, a GRP of 100 could mean that you bought 100 spots with a 1% reach or that you bought 2 spots with a 50% reach.

GRPs are calculated by determining the reach of the spots (which is a function of the stations or programs purchased) times the frequency or number of spots.

GRP = Reach × Frequency

Each GRP represents 1%. For example, if 25% of all televisions are tuned to a show that contains a spot, that results in 25 rating points. If, the next time the show is on the air, 30% are tuned in, the total number is 25 + 30 or 55 rating points.

Targeted Rating Points or TRPs are similar to GRPs, but they express the reach times frequency of only the advertiser's most likely prospects. For example, if a brand were marketing makeup for women and knew that the audience for the program in which your message ran consisted of 50% women, your TRP would be half your GRP.

In other words, GRPs relate to the total audience exposure to the message, whereas TRPs relate to the target audience exposure.

Cost per Point is used by most media planners to determine the cost for reaching 1% of the target audience. The average cost per spot divided by the average rating point per spot would give you this number.

CPP = Avg. cost per spot/Avg. rating point per spot
or
CPP = Cost of schedule/Gross ratings points
Source: www.tvb.org

Cost per Thousand or CPM is the cost of reaching 1,000 homes or individuals with your message. By the way, the "M" in CPM represents the word "mille," which is Latin for "thousand." So, in this example, if Option A costs $50,000 and reaches 100,000 people and Option B costs $25,000 but only reaches 40,000 people, how do you determine which option gives you the biggest return? Figure out the cost of reaching 1,000 people, and you can see that Option A is the most efficient buy.

Example:

Option A costs $50,000 and reaches 100,000
  CPM = $50,000/100 = $500
Option B costs $25,000 and reaches 40,000
  CPM = $25,000/40 = $625

Advertising and its impact are not easily measured. Thankfully, digital advertising provides a plethora of assessment data, including click-through rate, video views, and engagement statistics. However, national campaigns that incorporate national and local radio, national and local television spots, online ads, publicity efforts, promotions, and other elements that all contribute to the success or failure of a project present a significant challenge in trying to determine which of those elements helped accomplish your objectives (and how much they each contributed).

Television and Radio Campaign: Here is the schedule of an actual ad buy for the CMA Awards show in Atlanta that included both TV and radio. You can see the rating points purchased; the cost ($6,000); and the cost per point, which is 800, or 6,000 divided by 7.5. One spot was purchased, and that generated 75,000 impressions among women 25–54, for a cost per thousand of 80.

This television advertising buy was bolstered by additional ad exposure at radio. These time buys were during the week leading up to the CMA Awards Show.

The combined advertising gross rating point for the Atlanta market was 82.5. Although $15,000 was spent on radio, the CPM was dramatically lower than at television. Additionally, many more women aged 25–54 were reached because of the focused advertising that radio and its specific targeted demographics can deliver. Generally, radio advertising does not include gross impressions of homes.

Table 15.2   Sample TV Advertising Buy

Market	Purchased			# Of Spots	Wm: 25–54		Homes	
Daypart Atlanta	Pnts	Dollars	CPP		IMPS	CFM	IMPS	CPM
PRI	7.5	6000	800	1	75	80	115	52.17

	Market:	Atlanta
PRI	Daypart:	Primetime
PNTS	GRP:	7.5
	Cost:	$6,000.00
CPP	Cost/Point:	$800.00
	# of Spots:	1
IMPS	# of Impressions of Target Market – Women 25–54:	75,000
CPM	Cost/Thousand on Target:	$80.00
IMPS	# of Homes:	115,000
CPM	Cost/Thousand of Homes:	$52.17

Table 15.3   Radio Buy

Market	Purchased			# Of Spots	Wm: 25–54		Homes	
Daypart Atlanta	Pnts	Dollars	CPP		IMPS	CFM	IMPS	CPM
AMD	40	8000	200	27	440	18.18		
PMD	35	7000	200	23	375	18.66		
TOTAL	75	15000	200	50	815	18.4		

	Market:	Atlanta
AMD	Daypart:	AM Drive
PMD		PM Drive
PNTS	GRP:	75
	Cost:	$15,000.00
CPP	Cost/Point:	$200.00
	# of Spots:	50
IMPS	# of Impressions of Target Market – Women 25–54:	815,000
CPM	Cost/Thousand	$18.40

With any budget, the greater the GRP, the better the advertising "bang for the buck." An actual advertising campaign would be comprehensive in attempting to reach the target market while maintaining a budget. To do so, buys at many outlets would be secured. The optimal equation to increase the GRP includes varying the combinations of media, dayparts, and number of spots. Depending on the agenda as well as the budget, all advertising should help increase visibility and sales of a specific artist.

Coordinating With the Other Departments

Any media campaign should be designed in conjunction with the publicity department. A coordinated media campaign is necessary because the vehicles targeted for advertising are the same vehicles targeted for publicity, and some synergy may occur. For example, placing an ad in a particular publication may increase the likelihood of getting some editorial coverage.

CONCLUSION

Advertisers have a myriad of choices when deciding how to alert consumers as to a new release from an artist, and as technologies emerge, these options will continue to increase. What is important is to keep the target market top of mind, the budget close to the vest, and an eye on the campaign objectives.

APPENDIX: PRINT TERMINOLOGY

Although print advertising has decreased substantially in recent years and is not generally included in paid marketing campaigns for new music releases unless it is a trade publication such as *Billboard*, there are occasionally partnership opportunities that might provide exposure in print. Knowing key print terminology will aid you when the occasion arises. The **center spread** is the ad unit in the center of a publication. If you dropped a magazine on a table, it might naturally open to this spread because of the way the magazine is bound. **Circulation**, or circ, is the average number of copies per issue. **Closing date** is the final deadline that you may reserve space in a publication.

The cover of a mag is referred to as **Cover 1**. When you open the cover and are looking at the page on the left, that is referred to as the inside front cover, or **Cover 2**. **Cover 3** is the inside of the back cover, and **Cover 4** is the back of the magazine. These are all referred to as "premium positioning," which means that they generally cost a lot more than just a run of print ad. Why? Because more people look at those than any other page in the publication.

A **double-page** spread refers to an ad that spreads across two facing pages. **FP4C** is shorthand for full page four color. **Guaranteed position** ensures a particular placement in a publication (as opposed to run of print, which means your ad could be placed anywhere in the publication).

If you want to place an ad in a magazine, even if it's an ad that's being offered as a promotional trade, you will be asked to fill out an **insertion order** specifying the date of the issue you want. The **primary reader** is the one who receives the subscription or pays for the issue at newsstand. The readers per copy considers the initial subscriber or purchaser, as well as any other people who may have read the issue. These people are called **secondary readers**. For instance, the primary reader might leave an old magazine at her doctor's office, where ten additional secondary readers may have thumbed through it as well.

GLOSSARY

Ads Manager—The advertising platform that allows businesses to develop, manage, and track the performance of Facebook and Instagram ads.

Banner advertising—A type of contextual advertising that includes a designed message that allows for a click-through interface to another site, engaging the consumer with a sell-through missive.

Carousel ads—An ad type that features two or more scrolling images.

Consumer advertising—Targeted advertising that speaks directly to consumers through mediums such a radio, television, and publications that are consumed by consumers.

Contextual advertising—Advertising on a website that is targeted to the specific individual who is visiting the website based on the subject matter and then featuring products that relate. A contextual ad system scans the text of a website for keywords and returns ads to the web page based on what the user is viewing, either through ads placed on the page or pop-up ads.

Cooperative advertising (co-op)—An advertisement where the cost of the advertising is funded by the manufacturer of the product and the retail outlet agrees to feature the product within the store.

Cost per click (CPC)—Also referred to as **pay per click (PPC)**, advertisers are charged for an online ad only when consumers click on it

Cost per point (CPP)—A way to determine the cost of reaching 1% of the targeted audience.

Cost per thousand (CPM)—A dollar comparison that shows the relative cost of various media or vehicles; the figure indicates the dollar cost of advertising exposure to a thousand households or individuals.

Cume—Total number of unique viewers of an advertisement over a period of time.

Custom audiences—A feature on several social media sites that allows an advertiser to upload a customer email list or other database, then retarget those consumers within the social platform's ecosystem.

Dayparts—Specific segments of the broadcast day; for example, midday, morning drive time, afternoon drive time, late night.

DIY or do-it-yourself—Refers to the independent artist who does not have a record deal and pursues a musical career using social media tools while creating the music that they like.

Facebook Pixel—A few lines of computer code that are placed within the header section of a website. The code provides detailed data on a consumer's interactions on a website after they have clicked on an ad.

Frequency—The number of times the target audience will be exposed to a message.

Geotargeting—A method of detecting a website visitor's location to serve location-based content or advertisements.

Google AdSense—The platform used by publishers to monetize their websites and apps by agreeing to let Google place ads there.

Google Ads—Google's main advertising platform that is used by marketers to place paid ads.

Google Display Network (GDN)—A network of more than 2 million websites and apps that have agreed to let Google place advertisements within their properties in exchange for compensation.

Gross impressions—The total number of advertising impressions made during a schedule of commercials. GIs are calculated by multiplying the average persons reached in a specific time period by the number of spots in that period of time.

Gross rating point (GRP)—In broadcasting/cable, it means the size of the audience during two or more dayparts. GRPs are determined by multiplying the specific rating by the number of spots in that time period.

Influencer marketing—A marketing tactic that provides payment to a person of influence to promote a product to their social media audience.

Keywords—Search terms or words consumers are likely to use in a search engine when looking for a particular product or type of website.

Lookalike audience—A way within social networks, such as Facebook, Instagram, and Twitter, to target an ad toward audiences with similar traits and interests as the identified account.

Media fragmentation—The division of mass media into niche vehicles through specialization of content and segmentation of audiences.

Media planning—Determining the proper use of advertising media to fulfill the marketing and promotional objectives for a specific product or advertiser.

Medium—A class of communication carriers such as television, newspapers, magazines, outdoor, and so on.

Out-of-home (OOH)—Advertising that targets consumers outside of their homes, with four main categories: billboards, street advertising, transit advertising, and alternative advertising such as building wraps.

Paid media—A form of marketing communication in which the message is created by a brand, then placed on platforms not controlled by the brand in exchange for compensation.

Pivot smartURL—A URL that directs consumers to a landing page that displays their preferred platforms for music consumption.

Product placement—A marketing tactic that refers to the incorporation of a brand or product into another creative work such as a music video, film, or television program.

Rating—In TV, the percentage of households in a market that are viewing a station divided by the total number of households with TV in that market. In radio, the total number of people who are listening to a station divided by the total number of people in the market.

Rating point (a rating of 1%)—1% of the potential audience; the sum of the ratings of multiple advertising insertions; for example, two advertisements with a rating of 10% each will total 20 rating points.

Reach—The total audience that a medium reaches; the size of the audience with which a vehicle communicates; the total number of people in an advertising media audience; the total percentage of the target group that is actually covered by an advertising campaign.

Rich media advertising—A type of contextual advertising that can include video and other interactive media with consumer.

ROI (return on investment)—Deals with the money you invest in the company and the return you realize on that money based on the net profit of the business.

ROS (run-of-schedule)—The scheduling of advertising such that the promotional spot will run any time within a specified timeframe, usually in radio or television. Typically, ROS is less expensive than if a specific daypart or placement within particular programming is requested.

Road blocking—When another channel in the same medium is running the same commercial in the same timeframe. This strategy is to catch channel surfers to expand the reach of the commercial to the same targeted market.

Search engine marketing (SEM)—Using paid strategies to improve search engine rankings.

Search engine optimization (SEO)—Organic efforts such as link structure and using strategic keywords on websites to improve a company's search rank.

Targeted rating points (TRP)—Like GRPs but express the reach × frequency of only the most likely prospects of your advertising.

Trade advertising—Advertising aimed specifically for retailers and media gatekeepers through trade publications and mediums.

Trade publication—A specialized publication for a specific profession, trade, or industry; another term for some business publications.

Vehicle—A particular carrier within the group, such as *Rolling Stone* magazine or the MTV network.

BIBLIOGRAPHY

Ahmad, Irfan. "Video Marketing Statistics for 2020 [Infographic]." *Social Media Today*, 30 Oct. 2019, www.socialmediatoday.com/news/video-marketing-statistics-for-2020-infographic/566099.

Bailis, Rochelle. "The State of Influencer Marketing: 10 Influencer Marketing Statistics to Inform Where You Invest." *The BigCommerce Blog*, 15 Jan. 2021, www.bigcommerce.com/blog/influencer-marketing-statistics/#what-is-influencer-marketing.

"Billboard Media Kit." *Billboard*, Billboard.com. Accessed 2021.

Bui, Brian, et al. "8 Types of Display Ads You Need to Know in 2021." *Directive*, 29 Jan. 2021, directiveconsulting.com/institute/blog/types-of-display-ads-you-need-to-know.

"Changing Channels: Americans View Just 17 Channels Despite Record Number to Choose From." *Nielsen*, www.nielsen.com/us/en/insights/article/2014/changing-channels-americans-view-just-17-channels-despite-record-number-to-choose-from.

Cooper, Paige. "The Complete Guide to YouTube Ads for Marketers." *Social Media Marketing & Management Dashboard*, 20 July 2020, blog.hootsuite.com/youtube-advertising.

Dopson, Elise. "Videos vs. Images: Which Drives More Engagement in Facebook Ads? | Databox Blog." *Databox*, 5 Mar. 2021, databox.com/videos-vs-images-in-facebook-ads.

"Drax Project – Woke Up Late Ft. Hailee Steinfeld (Official Music Video) Starring Liza Koshy." *YouTube*, uploaded by Unknown, 9 Apr. 2019, www.youtube.com/watch?v=Mc-v9NPveU4.

Geyser, Werner. "No, Fyre Festival Wasn't an Influencer Marketing Success (and Other Lessons From a Disaster)." *Influencer Marketing Hub*, 31 Jan. 2019, influencermarketinghub.com/no-fyre-festival-wasnt-an-influencer-marketing-success-and-other-lessons-from-a-disaster.

"Google Ads." *Google*, ads.google.com. Accessed 25 Mar. 2021.

Johnson, Sarah. "Case Study: Product Placement In 5 Music Videos [Infographic]." *Hollywood Branded*, blog.hollywoodbranded.com/case-study-product-placement-in-5-music-videos-infographic. Accessed 25 Mar. 2021.

"The Journal of Roots Music." *No Depression*, 4 Feb. 2021, www.nodepression.com.

Lister, Mary. "The Ridiculously Useful Guide to Snapchat Ads." *WordStream*, 18 Mar. 2021, www.wordstream.com/blog/ws/2019/01/09/snapchat-ads.

mediakix. "Types of Influencer Marketing Campaigns Marketers Should Know." *Mediakix*, 7 Jan. 2021, mediakix.com/blog/influencer-marketing-campaign-types.

"Musicians, STOP Wasting Your Money on Influencer Marketing." *YouTube*, uploaded by Burstimo, 29 Feb. 2020, www.youtube.com/watch?v=LHnJNBozlK4.

Newberry, Christina. "How to Advertise on Instagram: A 5-Step Guide to Using Instagram Ads." *Social Media Marketing & Management Dashboard*, 24 July 2020, blog.hootsuite.com/instagram-ads-guide.

Newberry, Christina. "Influencer Marketing in 2019: How to Work With Social Media Influencers." *Social Media Marketing & Management Dashboard*, 23 July 2020, blog.hootsuite.com/influencer-marketing.

"Pandora Advertising | Pandora for Brands." *Pandora Advertising*, www.pandoraforbrands.com. Accessed 25 Mar. 2021.

Schneider, Marc. "SiriusXM Acquires Minority Stake in SoundCloud With $75 Million Investment." *Billboard*, 11 Feb. 2020, www.billboard.com/articles/business/8550690/siriusxm-invests-75-million-into-soundcloud.

Statista. "Digital Market Outlook: Digital Advertising Spending in the U.S. 2017–2024, by Format." *Statista*, 18 Mar. 2021, www.statista.com/statistics/455840/digital-advertising-revenue-format-digital-market-outlook-usa.

"TikTok for Business." *Tiktok.com*, www.tiktok.com/business/en-US. Accessed 2021.

Tran, Tony. "The Ultimate Breakdown of Social Video Metrics for Every Platform." *Social Media Marketing & Management Dashboard*, 23 July 2020, blog.hootsuite.com/social-video-metrics.

Vrountas, By Ted. "Contextual Advertising 101: How It Works, Benefits & Why It's Necessary for Relevant Ads." *Instapage*, 18 Mar. 2020 instapage.com/blog/contextual-advertising.

CHAPTER 16

Artist Support and Tour Sponsorship

Artists and their "handlers" recognize the need to monetize all assets that include the sound recording; licensing agreements; synchronization of their music in movies, television, and games; cross-merchandising with related products; sponsorships of commercial items; and, importantly, live performances and touring. Record labels help to broker many of these deals with artists and their management teams, assisting with sound recording placement, tour support, and ongoing branding of the artist in the marketplace.

Other than the few record stores still in existence, virtually little to no retail space is being dedicated to traditional music sales. Additionally, labels are finding it more difficult to break through the social media clutter and are having to throw the spotlight on their artist on their own. By using the power of partnerships and using the deeper pockets of brand sponsors, artists, in tandem with their labels, are able to leverage these strong allies and position artists and their new music in front of demographically specific audiences that would consume new releases.

ARTIST INCOME

At every level of performance, artists have looked intensely at their personal income streams, recognizing that revenue made primarily through "sales" of recorded music is no longer a "given" in today's marketplace. Artists must create a variety of sources of income to maintain a stable income that could include a combination of songwriting and publishing, touring, product endorsements, and merchandising, as well as starting up their own

DOI: 10.4324/9781003153511-16

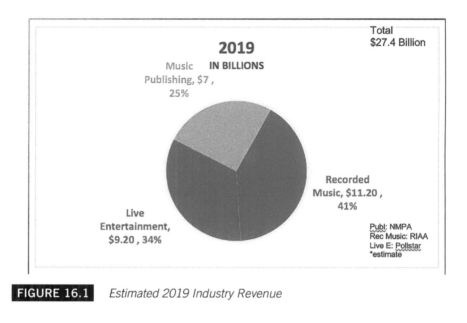

FIGURE 16.1   *Estimated 2019 Industry Revenue*

Total Top 100 Tours-North America

Year	Gross	Average Gross Per Show	Tickets Sold	Average Tickets Sold Per Show	Average Ticket Price
2015	$2,758,860,189	$692,311	37,354,278	9,374	$73.86
2016	$3,037,136,732	$787,639	39,212,642	10,169	$77.45
2017	$3,323,570,162	$849,366	42,208,278	10,787	$78.74
2018	$3,568,581,049	$982,809	39,766,234	10,952	$89.74
2019	$3,719,039,394	$958,726	39,219,086	10,111	$94.83

FIGURE 16.2   *Concert Industry Ticket Sales and Revenue*

(Source: Pollstar)

companies in support of other talented artists and songwriters. When looking at the 2019 data, the U.S. live entertainment revenue nearly equals that of recorded music, an achievement for touring which has been on the increase for over a decade, with the emergence of large festivals and specialty events.

In addition to touring, "the recording," along with the artist who produced it, has become marketable in so many more ways, via social media, licensing agreements, cross-merchandising, and using the artist's other talents to magnify and exploit all money-making opportunities. Artists have needed to become the "real deal" by having the talent to perform live. Some

Labels' typical investment in a major new signing	
ADVANCES	US$50,000 – 350,000
RECORDING COSTS	US$150,000 – 500,000
VIDEO PRODUCTION	US$25,000 – 300,000
TOUR SUPPORT	US$50,000 – 150,000
MARKETING AND PROMOTION	US$200,000 – 700,000
TOTAL	US $475,000 – 2,000,000

FIGURE 16.3   *Artist Label Investment*

(Source: IFPI)

artists in the past have been "created" in the studio, therefore handicapping themselves from being able to replicate their music live. By being a strong performer, artists and their business partners have the chance to realize a broader spectrum of financial gain, but getting started can be a challenge.

TOUR SUPPORT

In the traditional artist/label contract, the risk of artist development falls on the financial shoulders of the record label. The label acts as a bank and "advances" the artist monies to jump start the artist's career, with the idea that the label/bank will get paid back once revenue begins to be generated. As awareness of the artist begins to increase through the various marketing activities of the label such as internet marketing, social media, and radio airplay, the artist may have the opportunity to become the opening act for a larger tour that would give the developing act great exposure to a big audience. Oftentimes, the developing artist can "piggyback" on a tour of a much larger act that is on the same label, keeping it "in the family" and growing relationships internally as well as the potential revenue. Payment for the opening act slot may not cover the expenses of the developing artist, so the label often covers these costs associated with going on the road as "tour support," which is recoupable in most record deals. Once revenue begins to be flow, the label will pay itself back prior to paying royalties to the artist.

As noted by recent research of the International Federation of the Phonographic Industry, investment in an artist's career can be sizable, with no

FIGURE 16.4   *Coachella's Antarctic Dome 2019*

guarantee of success. And if an artist does not have a hit, no tour, no merchandise – and the relationship and contract dissolves, the artist owes no money to the label and walks away from the debt. Research and development is extensive in the music industry. Think about how many artists do not become household names, which is why many music companies are struggling to maintain the current business model.

TOUR SPONSORSHIP

To aid in offsetting the expenses of touring, many artists look for help from tour sponsors. This relationship tends to be a win-win between artist and sponsor, who is looking for a business relationship to help create a branding opportunity between a company's product and a target market to which the artist performs. In exchange for access to would-be consumers, the artist receives funds to help pay for touring expenses – everyone wins.

A variety of sponsorships are crafted where artists are paired with products or services in exchange for exposure. One example would be the Foo Fighters' relationship with the Capital One credit card. Starting in 2017, unique opportunities for Capital One card members included pre-sale ticket purchases and other special access experiences that allowed Foo Fighters fans early purchase of tickets to concerts. Capital One signage was also visible at live concert events. This relationship continued into 2021 as Foo Fighters co-headlined with Billie Eilish for the virtual concert iHeartRadio ALTerEGO Streaming Event on January 28, 2021, a presentation constructed in a virtual environment because of the COVID-19 pandemic. This "show" also featured Beck, The Black Keys, Mumford & Sons, and twenty one pilots and was in conjunction with the radio conglomerate iHeart Media. This ongoing sponsorship was built on a music platform Capital One worked to build over several years by developing customer loyalty for their card. What is unclear is the value of the relationship. How much did Capital One pay to become a sponsor of the Foo Fighters? Did card membership go up in order for members to gain access to exclusive packages? That information is not published. But, no doubt, there is a "cool" factor that is "priceless" when it comes to access to an artist.

FESTIVAL SPONSORSHIP

A different kind of sponsorship is the tour or festival scene, where multiple sponsors are involved with a multi-artist/multi-day event. In 2017, Coachella became the largest music festival worldwide with over 700,000 attendees in the two 3-day weekend event. Although artists are the draw, the entire affair is an "experience," which is anchored by sponsors creating encounters with their products. HP was the technical support for the entire

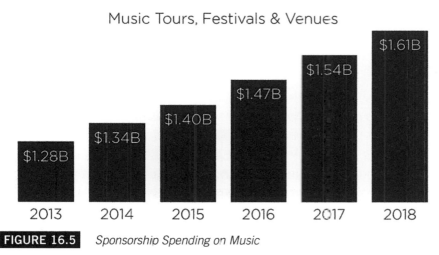

FIGURE 16.5   *Sponsorship Spending on Music*

(Source: IEG)

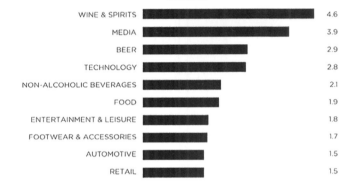

WINE AND SPIRITS BRANDS ARE 4.6 TIMES MORE LIKELY TO SPONSOR MUSIC THAN THE AVERAGE OF ALL SPONSORS

MOST ACTIVE CATEGORIES SPONSORING MUSIC FESTIVALS
Global

WINE & SPIRITS	4.6
MEDIA	3.9
BEER	2.9
TECHNOLOGY	2.8
NON-ALCOHOLIC BEVERAGES	2.1
FOOD	1.9
ENTERTAINMENT & LEISURE	1.8
FOOTWEAR & ACCESSORIES	1.7
AUTOMOTIVE	1.5
RETAIL	1.5

FIGURE 16.6 *Sponsorship Spending on Music by Product Category*

(Source: IEG)

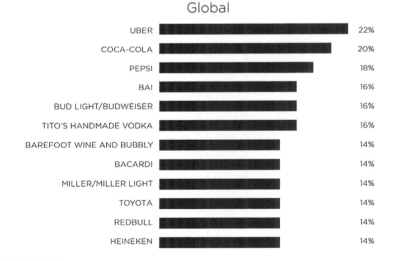

MOST ACTIVE SPONSORS OF MUSIC FESTIVALS
Global

UBER	22%
COCA-COLA	20%
PEPSI	18%
BAI	16%
BUD LIGHT/BUDWEISER	16%
TITO'S HANDMADE VODKA	16%
BAREFOOT WINE AND BUBBLY	14%
BACARDI	14%
MILLER/MILLER LIGHT	14%
TOYOTA	14%
REDBULL	14%
HEINEKEN	14%

FIGURE 16.7 *Most Active Sponsorship Spending on Music Festivals by Brand*

FIGURE 16.8   *2018's Largest Music Festivals by Attendance*

(Source: IEG)

festival but featured consumer interactions, including the HP lounge and the HP 360 Antarctic Projection Dome, giving concert-goers a needed break from the outdoors and festivities (Mediakix).

Alcohol is a significant presence at music events, and both Heineken and Absolut had hang-outs for patrons to sample product. Clothing and cosmetics were also present, with Calvin Klein, Revolve, Sephora, and NYX sporting their wares. BMW continued their tradition of "Road to Coachella" and showcased Khalid at the wheel of their specially designed BMWi en route to the gig (Pometsey).

The trend in brand sponsorship continues to grow, according to IEG Sponsorship Consultants. Sponsorship in music venues, festivals, and tours for 2018 topped $1.61 billion, an increase of 4.8% from 2017, according to IEG, LLC. Top brand categories include wine and spirits, media, beer (its own category,) technology, and non-alcoholic drinks. The most active sponsors of festivals include Uber, Coke, Pepsi, Bai, and Bud Light/Budweiser. Looking at foot traffic and thinking about target markets and demographics, it is easy to correlate sponsorship dollars and activation of consumers.

PRODUCTS IN THE MUSIC

In 2019, MasterCard launched a new "sonic branding" tone to assure users that their transaction had been authenticated by MasterCard. This "sonic tone" is to be used in musical scores, sound logos, ring tones, whole music, and as the POS acceptance signal once a payment has been completed. The company integrated the tone into 7.6 million POS systems worldwide, banking on the research that "found that 77% of people felt hearing the sonic sound made the transaction more trustworthy." To continue the story, MasterCard collaborated with Swedish artist Randle and songwriter/producer Niclas Molinder to "organically and authentically integrate" its sonic identity into a single music release. Debuted at the Consumer Electronics Show in Las Vegas with iHeart Radio, "Merry Go Round" was released in January 2020 (Fast Company) (The Drum).

Branded Music

There was a time in the not-so-distant past when music companies had departments known as "Special Products and Special Markets." The efforts of these staffs were to look for opportunities and projects that were not attached to traditional sales and whose profits were considered "gravy" to the label and not projected funds to the bottom line. But with traditional record sales all but gone, labels look for these products and markets to make up for revenue that has since been lost to streaming.

There are many companies that create a "branded" music experience – where customers hear a playlist that reflects a brand's personality, image, culture, and/or corporate principle. In the case of Tuned Global, this innovative company has created a pliable platform that can be tailored to brand-specific ideas to create a one-of-kind trademark experience for consumers using the interface. Companies such as Coca-Cola, Samsung, Universal Music Group, and McDonald's have all created distinctive applications of Tuned Global's offerings. Pizza Hut created the "singing pizza" where the pizza box became "phygital." Using vibrant box designs and an app printed on this box for connectivity, Pizza Hut joined forces with Universal Music Malaysia and

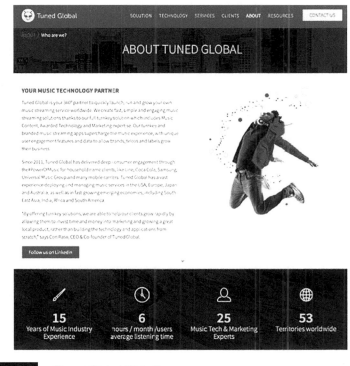

FIGURE 16.9   *Tuned Global Website*

PIZZA HUT

Turnkey - Brand Music Campaign

The Singing Pizza turned a regular pizza box into a music box! With a box that plays m
when combined with the app, and a range of vibrant designs, Pizza Hut collaborated v
Universal Music Malaysia and Tuned Global to create a truly memorable music foo
experience for customers. The app allowed a great, first of its kind, "phygital" executi
Pizza and Music - it just works, right?

FIGURE 16.10   *Tuned Global Pizza Hut Promotion*

Tuned Global to create a music-food experience (www.tunedglobal.com/tuned-global-white-label-music-solutions-case-studies).

THE BENEFITS

Why, then, are artists and music companies aligning themselves with brand partners? What is the benefit to them and their existing record label partners? The initial response would be money. The licensing of existing recordings from labels can be very lucrative. Because labels never know the value of their recordings long term, record companies have found it hard to calculate the value of licensed product. Historically, the licensing of music has been considered "gravy" – monies generated beyond the "meat and potatoes" of the business. But to address the evolving market as well as sagging record sales, segments of marketing departments are devoted to mining the catalog of their holdings and finding new, innovative homes beyond traditional music channels.

SOUNDTRACKS AND COMPILATIONS

Since moving pictures became "talkies," music has been an integral element to the success of the movie business. Early in the evolution of movies, the industry would hire actual singers to play a part, integrating music and story on-screen. Often, the music from these movies would become nationwide hits, and the popularity of the singing stars would explode, making many of them classic voices of the genre. Movie soundtracks have continued to evolve, with the 1960s delivering the Beatles *Hard Day's Night*, and *The Graduate* with its hits "Mrs. Robinson" and "Scarborough Fair" by Simon and Garfunkel. The 70s included *Urban Cowboy*, *Grease*, and *Saturday Night Fever*, all containing a common thread of one very hot actor, John Travolta. The 80s brought *Footloose*, *Dirty Dancing*, and Roy Orbison-famed *Pretty Woman*. The 90s elevated Whitney Houston to a new level of super-star-dom with her film debut in *The Bodyguard*. And Celine Dion reached new heights with her mega-hit "My Heart Will Go On" from the epic *Titanic*. All of these soundtracks contained songs which would eventually become huge pop hits, selling millions of units and igniting a new kind of consumer to purchase music. And in some cases, a single song has helped to launch a career and create a mega-star.

Soundtracks have learned the lessons of the past while keeping up with the purchasing practices of today's music consumers. Many music purchasers have become their own disc jockeys, curating their "personal soundtracks" containing the individual songs that they want to hear. By compiling songs that capture this spirit while representing the theme of a movie, soundtrack managers have fashioned a "genre" of music that has produced sales.

The holiday season of 2013 launched a game-changer for Disney, and what started as a flurry became an avalanche. *Frozen* was introduced to young audiences but was embraced by movie goers of all ages. And as only Disney can generate, all the accoutrements that adorn their movie flicks include memorable characters singing songs that have moviegoers humming the tunes as they exit the theater. The *Frozen* soundtrack topped the Top 200 *Billboard* Chart for most of 2014, generating sales of over 4.3 million, which is just the tip of the iceberg. *Frozen* spawned several spin-offs for the entertainment giant, including a Broadway version of the show, a television series, a *Frozen 2* movie, oodles of character-themed products from stuffed animals to pajamas, and entire sections of the Disney theme parks retrofitted to resemble Arendelle – the land of *Frozen*.

But Disney and sister label Hollywood Records do not stop there, and their soundtrack domination continued with the *Guardians of the Galaxy 1* and *2*, *Bohemian Rhapsody*, *Tron*, *Moana*, *Soul*, and more. Soundtracks are not restricted to the movie theaters. With the COVID-19 pandemic, many homebound consumers found themselves watching television series fed by various streaming services such as Netflix, HBOMax, Hulu, and Amazon Prime. These sources featured series such as *The Mandalorian* and *Stranger Things*, both with soundtracks positioned high on *Billboard*'s Soundtrack chart.

PRODUCT EXTENSIONS AND RETAIL EXCLUSIVES

McDonald's served up its first celebrity meal in nearly 30 years with the founder of Cactus Jack Records, Travis Scott. The $6 dinner included the Quarter Pounder with cheese, bacon, and lettuce; a medium order of fries with BBQ sauce; and a Sprite. But this collaboration found even more inspiration. Through Cactus Jack Records, Travis and his team created a line of McDonald's themed merchandise and wearables that could be purchased by both fans of the restaurant and devotees of Scott. These t-shirts and hoodies were being worn by McDonald's employees and embodied the marquee look of a Travis Scott design.

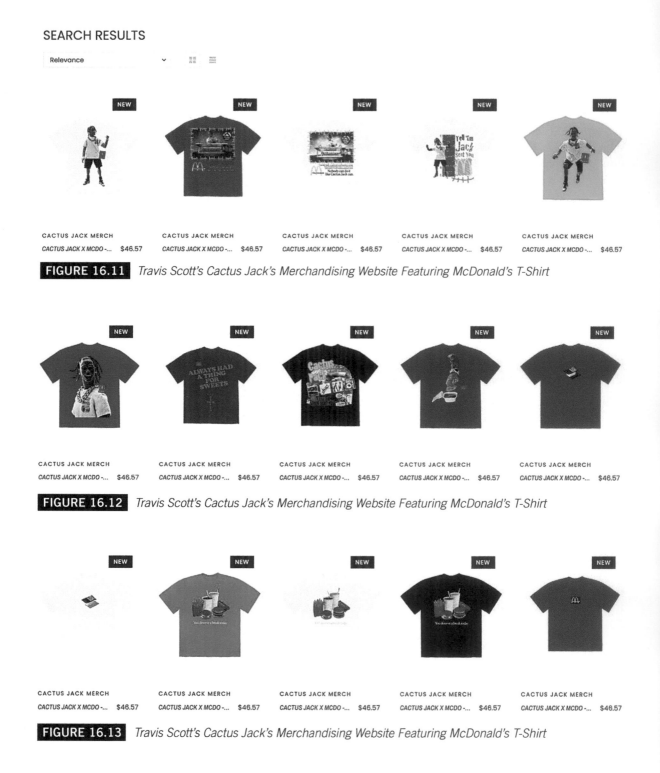

SEARCH RESULTS

FIGURE 16.11 *Travis Scott's Cactus Jack's Merchandising Website Featuring McDonald's T-Shirt*

FIGURE 16.12 *Travis Scott's Cactus Jack's Merchandising Website Featuring McDonald's T-Shirt*

FIGURE 16.13 *Travis Scott's Cactus Jack's Merchandising Website Featuring McDonald's T-Shirt*

CACTUS JACK MERCH	CACTUS JACK MERCH	CACTUS JACK MERCH	CACTUS JACK MERCH	MAGIC CUSTOM
CACTUS JACK X MCDO -... $46.57	CACTUS JACK X MCDO -... $46.57	CACTUS JACK X MCDO~... $46.57	CACTUS JACK X MCDO -... $46.57	Sweat Col Rond McWeed ... ~~$58.10~~ −$11.64 $46.45

FIGURE 16.14 *Travis Scott's Cactus Jack's Merchandising Website Featuring McDonald's T-Shirt*

FIGURE 16.15 *Travis Scott McDonald's Promotion*

Advertising for this meal promotion included all mediums with a special meme of Scott created to personify the artist and theme of the meal deal. Considered a success, McDonald's is sure to consider more exclusives with other artists – but a partnership with Travis Scott was a winner.

ARTISTS IN THE VIRTUAL WORLD

Travis Scott has been busy. His April 2020 Fortnite concert was a landmark event, featuring a ten-minute show while embedded in the Fortnite world, featuring his mega-hits "Sicko Mode," "Goosebumps," and "Highest in the Room" to an audience of 45.8 million viewers over five shows with 27.7 million unique attendees. A new collaboration with Kid Cudi called "The Scotts" was also debuted. Leading up to this event, Travis Scott songs increased in streaming activity based on all the hype surrounding this never-before event, with "Stagazing" realizing a jump in activity by 50%. And post-concert, viewers received in their email inboxes related concert merchandising products, including a $65 Nerf gun and $75 action figure based on the night's events (www.rollingstone.com/music/music-features/travis-scott-fortnite-concert-989209/).

Roblox, a more kid-friendly gaming interface, highlighted Lil Nas X in a motion-captured incarnation of himself styled to fit in the game and matched to perform each of his four featured songs. Over 33 million viewers watched the "show" over a four-program weekend. Ava Max launched her fall 2020 release on the Roblox platform with the artist on a huge screen in front of a dance floor. Max talked about the inspiration behind the album and then performed a few songs for fans. Flashing lights, fireworks, and a platform that descends into a virtual hell were all a part of the show that celebrated her new release titled *Heaven & Hell* (www.theverge.com/2020/11/16/21570454/lil-nas-x-roblox-concert-33-million-views; www.digitalmusicnews.com/2020/10/06/ava-max-roblox-music-release-party/).

Whether it be a corporate sponsorship for a tour, product placement written into a song, movie and television soundtracks featuring new releases, streaming playlist access through innovative portals, crafted partnerships featuring artists and specific product alignments, or new media alignments that highlight artists and their originality – marketing music has never been more inventive and groundbreaking, with the future being a landscape of endless possibilities.

GLOSSARY

Branded CDs—A CD sponsored by and sporting the brand logo or mention of a company not normally associated with the release of recorded music. Examples include product designed for and sold at Pottery Barn, Pier 1 Imports, and Victoria's Secret.

Brick-and-mortar stores—Businesses that have physical (rather than virtual or online) presences, in other words, stores (built of physical material such as bricks and mortar) that you can drive to and enter physically to see, touch, and purchase merchandise.

CIMS—The Coalition of Independent Music Stores.

Compilations—A collection of previously released songs sold as a one-album unit, or a collection of new material, either by single or multiple performers, sold as a collaborative effort on one musical recording.

Cross-promotion—Using one product to sell another product or to reach the market of the other product.

Exclusives—Retail-exclusive marketing programs that are not offered to other retailers but arranged specifically through one retail chain.

Private label—A label unique to a specific retailer.

Superior products—A value-added version of a product that is sold in the "regular" version elsewhere.

Tour sponsorship—A brand or company "sponsors" a concert tour by providing some of the tour expenses in exchange for product exposure at the events.

Value-adds—To lure consumers into a specific store, music retailers offer exclusive product with added or "bonus" content or material for their customers.

REFERENCES

"12 Best Coachella 2019 Brand Activations & Influencer Sponsorships." *Mediakix*, 8 Oct. 2019, mediakix.com/blog/coachella-brand-event-activations-influencers/.

Global, Tuned. "Tuned Global White Label Music Solutions Case Studies." www.tunedglobal.com/tuned-global-white-label-music-solutions-case-studies.

Holmes, Charles. "I've Never Played Fortnite, but Was Forced to Attend Travis Scott's Fortnite Concert." *Rolling Stone*, 15 Mar. 2021, www.rollingstone.com/music/music-features/travis-scott-fortnite-concert-989209/.

IEG Sponsorship and Services sponsorship.com IEG Year End 2018 Report.

Kastrenakes, Jacob. "Lil Nas X's Roblox Concert Was Attended 33 Million Times." *The Verge*, 16 Nov. 2020, www.theverge.com/2020/11/16/21570454/lil-nas-x-roblox-concert-33-million-views.

King, Ashley. "Ava Max's Roblox Music Release Party Clocks 1.2 Million Visitors." *Digital Music News*, 3 Dec. 2020, www.digitalmusicnews.com/2020/10/06/ava-max-roblox-music-release-party/.

King, Ashley. "Mastercard Decides It's an Artist – And Releases Its First Single." *Digital Music News*, 13 Jan. 2020, www.digitalmusicnews.com/2020/01/10/mastercard-audio-branding/.

Pometsey, Olive, et al. "How Sponsorship Swallowed Coachella." *British GQ*, 11 Apr. 2019, www.gq-magazine.co.uk/article/coachella-2019-influencers-business.

Stewart, Rebecca. "Mastercard Drops First Single to Expand 'Sonic Brand Architecture'." *The Drum*, 8 Jan. 2020, www.thedrum.com/news/2020/01/07/mastercard-drops-first-single-expand-sonic-brand-architecture.

Wilson, Mark. "Mastercard Just Launched a Sonic Logo. Here's What It Sounds Like." *Fast Company*, 13 Feb. 2019, www.fastcompany.com/90305949/mastercard-just-launched-a-sonic-logo-heres-what-it-sounds-like.

www.ifpi.org International Federation of the Phonographic Industry.

www.nmpa.org National Music Publishers Association.

www.pollstar.com Pollstar.

www.riaa.com Recording Industry Academy of America.

CHAPTER 17

Merchandise

When an artist tours, merchandise sales can make up to 50%–60% of the artist's earnings some nights. Every artist's income sources will differ, depending on their strengths and which revenues streams they develop and cultivate over time. The prolific songwriter who earns their revenue primarily in publishing and from other artists recording their songs might not need or wish to tour non-stop. The charismatic band that lives to play live, has built audiences across many markets, and tours every chance they get will have very different income and cash flow statements than that songwriter. But in general, the industry standard is that merchandise sales should generally make up 5%–10% of overall revenue for an artist, be it a developing act, an established artist, or a superstar. The key to profitability in merchandising is **inventory management**.

WHO, WHAT, AND WHEN MERCH SELLS

While there is no central industrywide tracking system for merchandise sales, like the charts and research that exists for music sales and consumption, it is meaningful to consider directional data that can provide a snapshot of consumer behavior with respect to merchandise.

Venue software management company atVenu has been tracking data from hundreds of thousands of shows they have supported. Here are some highlighted findings atVenu reported:

- With average show attendance of 1,548, an average of 297 items were purchased per show.

DOI: 10.4324/9781003153511-17

CONTENTS

- Artists offered 17 items for sale on average, and 4 items usually drove 75% of sales.
- Average revenue per fan has increased every year (2019 – $5.54, 2018 – $4.85, 2017 – $4.37).
- Country tours had the most shows, but K-pop, pop, hip-hop and alternative fans bought the most merchandise, in that order. K-pop fans outspent all other genres, $7.95 per fan. K-pop generated the most revenue per show as well, averaging $30,000 gross sales per event.
- What sells most? T-shirts! 50% of all revenue generated (59% of t-shirt sold were black).
- When looking at all items sold, generally four items generate 75% of all merchandise sales per night – which varies from artist to artist, show to show.
- Sixty-five percent of sales were paid for using credit cards, an increase over the previous year.
- Most merchandise was sold before the show, between 4–6 p.m.

atVenu's findings reported in 2020 were based on 2019 tour merch sales data from over 125,000 concerts. That data may seem old already at the time of publication of this book, but it is invaluable to marketers seeking to evaluate

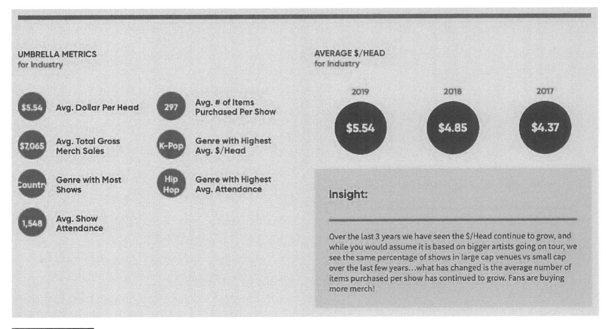

FIGURE 17.1   *atVenu Insights 2019*

a high-level snapshot of the merch industry landscape. 2020 was an exceptional year. Data analysts would call it **anomalous**. Touring was halted due to the pandemic, which significantly impacted how merchandise was sold to fans. To look at 2020 artist tour sales data would not be representative of what can be expected in the future, and it would also not be representative of what artists actually sold in merch that year. 2019 tour data is the most recent full active year of touring when business operated as usual, and we can learn a lot from the dynamics of how merch sold with touring at full throttle. In 2020, merch was primarily sold online, direct to consumer, and that change has led to exciting advancements in eCommerce tech and music company online sales practices. Looking forward, with the return of concerts and touring, smart marketers will take advantage of everything we learned in 2019 about merchandising on tour *and* everything we learned in 2020–2021 about merchandising online (www.atVenue.com).

CONTENTS

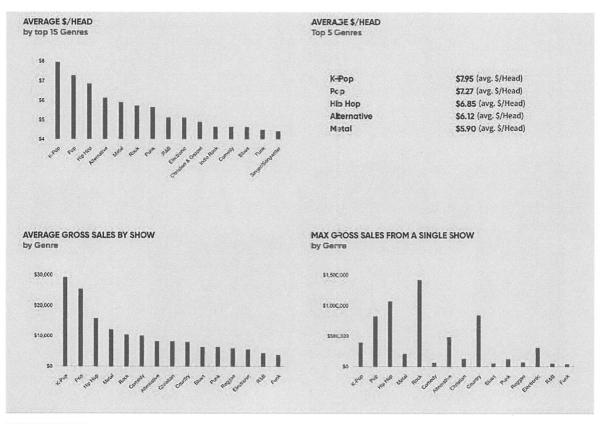

FIGURE 17.2   *atVenu Insights 2019*

WHY DOES TOUR MERCHANDISE SELL?

Many concert-goers like a memento or souvenir of the event, reminding themselves of great night of music. Others wear the gear like a "badge of honor" – they endured the event or festival and shared that experience with others who were there. Some merchandise is considered fashionable among certain circles of peers. Some merchandise is collectible and increases in value over time. In the moment, all reasons are justifiable when purchasing various items to take home. When looking at actual items that sell, 70% of merchandise sold is considered "soft goods" or wearables. The other 30% are table items or trinkets such as koozies; pop sockets; jewelry; posters; and a long list of various other fun, inspiring, and unusual things artists and their teams dream up for fans.

Opportunities for artists to monetize their art, brand, and style continue to grow, with new technologies, platforms, and consumer trends emerging daily.

- **Music products:** CDs, vinyl records, cassette tapes, sheet music
- **Soft goods:** t-shirts, sweatshirts, bandanas, tote bags, and similar
- **Digital products:** MP3s, MP4/MOV, on-demand streams
- **Services:** fan club/exclusive content access, livestream concerts, private meet and greets

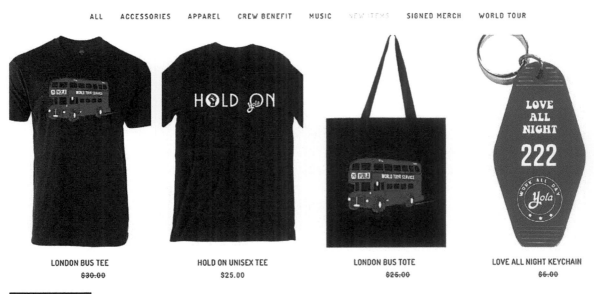

ALL   ACCESSORIES   APPAREL   CREW BENEFIT   MUSIC   NEW ITEMS   SIGNED MERCH   WORLD TOUR

LONDON BUS TEE
$30.00

HOLD ON UNISEX TEE
$25.00

LONDON BUS TOTE
$25.00

LOVE ALL NIGHT KEYCHAIN
$5.00

FIGURE 17.3   *Sample of Concert Merchandise*

Selling merchandise generates significant revenue for artists. It also empowers fans to spread the news about an artist and evangelize their brand, hopefully leading to greater awareness, loyalty, engagement, listening, and more sales. In this way, merchandising is both marketing and sales. It generates revenue for artists and engages audiences to ignite their passion for the artist, celebrating the fan's connection and affinity with the artist and their music.

To **"merchandise"** (verb) means to promote and sell goods. In music, when we talk about merchandise (noun), we most often are talking about goods sold in person at concerts and festivals, in retail stores, and online. Goods sold online are done so directly to consumers on an artist's website or via third-party sellers who are authorized to conduct eCommerce on behalf of the artist.

To sell merchandise effectively, and artist's team needs to take into consideration design factors that both impact a product's likelihood to sell, and also the costs associated with taking orders, manufacturing, and distributing the products to consumers.

DESIGNING MERCH THAT SELLS

There are many great t-shirt ideas, but an idea is just the first thing to consider when developing new merchandise ideas.

> *Aesthetics is the branch of philosophy that is concerned with the nature of beauty and taste.*
>
> (Tate www.tate.org.uk/art/art-terms/a/aesthetics)

Often people think of aesthetics in terms of its intrinsic value, but modern marketers know well that aesthetics have marketing value as well. What does that mean in music? Fans are willing to pay for beautiful things, or quality goods.

The modern music merchandiser is always choosing between beauty, quality, and profit because most often goods that are quality are more expensive to manufacture. In merchandising, the artists earnings are entirely based on profit. When an artist sells a t-shirt at a show, it might feel to the fan like they are paying the artist $25, but more often than not, this is more likely the scenario:

- $7.00 manufacturing
- $5.00 to venue (if 20% merch deal)
- $3.75 to manager (if 15% of gross income deal – closer to $1.75 if net deal)

$7 MANUFACTURING

$6.50 ARTIST

$5.00 VENUE

$3.75 MANAGER

$1.00 MERCH SELLER

$1.00 SHIPPING

$0.75 CREDIT CARD FEE

FIGURE 17.4   *Breakdown of T-Shirt Sales, T-Shirt Sold at $25*

- $1.00 shipping (from mfgr to warehouse, from warehouse to tour)
- $1.00 to merch seller (pro-rated, usually paid hourly or lump sum)
- $0.75 credit card transaction fee (3% of transaction amount)

= Artist earns $6.50 on the shirt = Approximately 25%

Ever wonder why t-shirts are so expensive at concerts? That is why!

If manufacturing and shipping costs were held constant, and the artist sells shirts for a lower price, it is the difference between them making money, just breaking even, or losing money. From this example, you can also see how important it is to manage manufacturing costs. In all of these scenarios, if the shirt cost less to make and ship, the profitability would improve significantly.

Which brings us back to aesthetics and balancing decisions about beauty, variety, quality, and costs.

WHAT FANS WANT . . . THAT WE CAN MAKE

If you ask fans what they want, they will tell you they want a multitude of options – every size, fit, color, format, and feature – with very little regard for what it costs to offer the things they want. That is okay. It's not the fan's job to think in practical terms or worry about what the artist and their team go through to make things. The more enthusiastic they are about the artist, the less practical they may become in terms of merchandise. This is to be encouraged! The marketer knows to cultivate both excitement and practical ability to execute and sell. Thankfully, manufacturers, embroiderers, embossers, engravers, and printers have all experienced advancements in technology in recent years that have enabled them to make it easier for music marketers to provide more options to customers at lower costs while still maintaining consumer-acceptable quality products.

MERCHANDISE DECISIONS THAT IMPACT SALES

For every product an artist sells, someone on the artist's team makes five to ten decisions about design, colors, fit, materials utilized, vendor, and process to manufacture and ship that product. Marketing professionals are often the individuals making at least some of these decisions. Many elements chosen when developing merchandise directly influence the **sales conversion rate** of that product, so it is important that marketers build knowledge and expertise about what sells.

Sales conversion is a measure of the rate a product sells when fans are given an opportunity to purchase it. Another way to phrase it is the percentage of customers presented with an opportunity to buy who make a purchase.

Sales Conversion Rate % = # of People Who Came to the Show ÷ # of Sales

While each product will have an overall sales conversion rate – accounting for sales across channels, across methods, across time – usually marketers will also have a sense of sales conversion for the variables and versions of that product. For example, a unisex extra-large size might have a higher or lower conversion rate than a medium size of the same t-shirt – or the black version of a sweatshirt might have a higher conversion rate than the blue one.

Why does this matter? Knowing how likely certain products are to sell will inform how many units of each item you manufacture; how many an artist carries on the road touring; and in some cases, whether a product remains available or is discontinued.

Seasonality can also be identified in conversion data – so before discontinuing an item due to declining conversion rate, be sure to consider if season is a factor and if an item should just be temporarily retired and brought back in future months.

COLORS, FIT, AND SIZING

The reality remains that black and dark gray t-shirts continue to be the most popular items on merch tables at concerts. Manufacturers have revolutionized their businesses and become more nimble than ever in their ability to deliver variety in different stock colors, fits, sizes, and colors, and yet, the black t-shirt still reigns.

Most manufacturers of stock used for clothing merch promote their ability to:

- Provide fashionable styles and variety of colors
- Maintain inventory of sizes in all colors for easy re-ordering
- Use premium fabric and quality stitch work
- Promise their products are ethically made – often referred to as "sweatshop free"
- Offer environmentally friendly process, dyes and ink

Fit in clothing goods represents style and shape and the core look of the item for purchase. Most clothing manufacturers offer style collections by gender – usually men's, women's and unisex. While men's and unisex stock is sold primarily on the basis of intended use, in categories like loungewear and streetwear, manufacturers use language like flowy, slouchy, relaxed, and slim fit for women's styles.

Sizes offered by manufacturers are standard – usually small, medium, large, XL, 2XL. Sometimes 2XL shirts cost more per unit than other sizes, since they utilize more fabric than other shirts – but marketers do not usually pass that cost thru to fans or charge more for those items. Some manufacturers do not offer 2XL in all women's styles – and suggest unisex XL and 2XL are more likely to be sold to customers in that body size range. Fans have varying opinions of this and are often very vocal expressing their wishes at merch tables and to customer service.

 FIGURE 17.5   *Sample of Merchandising Shirt*

FIGURE 17.6   *Sample of Merchandising Shirt*

FIGURE 17.7   *Sample of Merchandising Shirt*

Music marketers must listen; consider the artist's audience; and learn about their preferred styles, their range in body types and sizes, and messages (intended and unintended) that can be sent from the artist based on the choices they make in merchandise fits and sizes. To a fan, every decision made maps back to the artist – even merchandise! For example, if an artist does not make extra large or small sizes available to their fans, or if an artist

 Sample of Merchandising Shirt

chooses a style that runs very small such that fans have to order a size or two larger than they believe they are, it can relay messages about body size and body image the artist would never want to send.

Colors vary greatly from item to item, but almost all items come in a black or dark gray option given the popularity of sales of merch in those colors. Colors options differ based on the type of fabric used to create the product. The most common terms marketers need to know are:

- **Material** – what is used to make the fabric that is used to make the product, usually expressed in terms of percentages (example 100% cotton or 60% cotton/40% poly; new alternatives some people are using include bamboo, hemp, and viscose)
- **Triblend** – fabric made of three materials (usually cotton, polyester, rayon)
- **Fleece** – not always consistently used, this term is sometimes used to refer to standard sweatshirt material and sometimes to refer to its fuzzier polar fleece cousin
- **Heather** – a color created by mixing multiple colors or shades of the same color, creating a textured effect that can be overt or subtle depending on the colors mixed
- Other factors marketers consider when making merchandise recommendations to artists include:
- **Reorder speed and availability** – how easily products can be manufactured and inventory procured based on demand

- **Costs and shippability** – merch is shipped multiple times, from manufacturer to warehouse and warehouse to tour, festivals, retail and other sales locations
- **Roadworthiness** – ability to carry the item on the road easily, integrate with other products being sold, costs related to storage, space, and methods
- **Breakage** – frequency rates and procedure of what to do when a product is flawed (on arrival at warehouse and/or when received by a fan)

DESIGNING GREAT MERCH

Great merchandise combines form and function. Great merch designers have a sense of trends, what fans are buying, and also a deep sense of the artist's brand, personality, and what it is that fans identify with about the artist so much so that they would want to wear it on their bodies, drink out of it, carry it around town, or skateboard on it.

Marketers who work effectively with designers have clear parameters for the project. They set deadlines and share any "must have" text or images that need to appear on the merchandise, specifications if already known, color palette and themes in the campaign, and known design likes and dislikes of the artist or whoever is making final decisions/approving art. A marketer must clear licenses to use illustrations, photography, lyrics, and the artist or other people's likenesses being used. For this reason, many illustration fees for use on merchandise are "all in" or works made for hire. Designers will usually present multiple concepts for consideration, and the marketer will home in on which ones to pursue, providing feedback or new direction. Often the color of shirts and inks are the last elements to be chosen, unless color is inherent in design concepts.

Graphic designers' estimates for merchandise are sometimes different from everyday promotional design, because they know the images they are creating will be monetized directly. As a result, graphic designers often charge more for merchandise design than for promotional graphics, posters, or other work. Similarly, album cover art and album-related packaging usually cost more, too.

WHO MAKES MERCH

An artist's merchandise is usually organized and approved by their management, with involvement of the label and many other professionals involved in their specific role creating merch. Who specifically is involved depends on the artist's career size, the scale of their tour operation, and the artist's preference for how involved they wish to be in the extended details of their operations.

For major acts on large tours, merchandise companies are employed to handle merchandise design, manufacturing, logistics, sales, and settlement – beginning to end. Those companies are focused on ability to scale, reliability amid the complexity of other logistics going on behind the scenes on a tour, integration with retail partners, forming unique partnerships, and brand extension deals, as well as basic coverage of online selling and design.

Crown Merchandise in Nashville, Tennessee, handles merchandise for mainstream, internationally recognized country artists including Johnny Cash, Reba McEntire, Sturgill Simpson, Little Big Town, Kelsea Ballerini, Dan + Shay, and more. Consider how Crown Merchandise describes their services:

Mid-size acts may also hire companies to handle their merchandise needs, but often those companies are more focused on flexibility around shipping product to artists on the road directly, willingness to do short runs of product incrementally, and some of the added-value services artists might need that larger management companies and labels provide.

WHAT WE DO

On Tour

Crown provides full scale touring and production services, complimenting our artists' performances and allowing the fan the opportunity to connect through branded merchandise.

In Stores

We utilize the full force of our retail relationships and sales team for our artists through mass market chains to mom and pop stores. We will get your brand into the right stores and provide another channel for you to connect with your fans.

Online

Our e-commerce services compliment all our other merchandising channels. Crown's team is consistently ahead of the latest innovations in online merchandising, social media commerce, print on demand technology and everything direct to consumer.

Strategic Partnerships

Complimenting the work of all other merchandising channels, our team of experts will work to find partnerships and collaborations that fit your brand.

Product Development

Excellent creative is at the heart of the merchandising business. Crown's creative services team will work with your vision to create on trend world class design.

 FIGURE 17.9   *Crown Merchandise Service*

Hello in Phoenix, Arizona, handles merch for independent artists across genres, including Black Pumas, Edie Brickell & the New Bohemians, Garbage, Joan Osborne, the Wood Brothers, and more. Now look at how Hello describes their services, and consider the similarities and differences of how their services compare with what Crown does:

There are also merch companies that focus on specific music scenes and sub-genres. For example, Kung Fu Nation in Raleigh, North Carolina, handles merchandise for roots, Americana, string, and jam-band artists, including Beck, Mandolin Orange, Mavis Staples, Sara Watkins, Steep Canyon Rangers, Steve Martin, Tame Impala, the Avett Brothers, the White Stripes, Vampire Weekend, Wilco, Yo La Tengo, and more.

When artists are developing but not yet selling a high enough volume of merch to justify hiring a merchandise company, they can work with the merchandise or direct-to-consumer department at their record label, or their management company might handle their merchandise entirely. Before artists have management, they often work directly with graphic

FIGURE 17.10   *Hello Merchandise Service*

FIGURE 17.11 *Kung Fu Nation Logo*

designers and manufacturers to create merch. They might sell and ship orders to fans directly, as well. As a result, it is always good for marketers to know the ins and outs of how to manage merchandise projects, as inevitably at some point every marketer will handle merch, if they are not handling it all of the time ongoing.

Independents who are not working with a **merchandise company**, **label services**, or management will need to research manufacturers to find the right partners for each merchandise item they wish to create. Like in printing, where they have print brokers who handle sales on behalf of printing presses, you will sometimes run into manufacturing **resellers** who form relationships with **aggregators**, who work with many **raw goods** manufacturers to enable customers to order more than one item from them at a time. For example, if you want to manufacture t-shirts, sweatshirts, tote bags, bandanas, and posters, those can all be produced by one screen printing company. But if you want embroidered hats, mugs, flasks, custom socks, and stickers, you might need to go to other companies for those items. By working with a reseller, they handle the sourcing of different vendors to create the products, and you deal with one person. The core benefits of that approach are convenient ordering and reordering, and you pay one company for everything. The primary drawback of that approach is you have less control over the quality of goods and what other companies are involved with your merchandise. It can be more difficult to do something custom.

Often marketers decide to work with a particular merchandise reseller out of loyalty or to support that business – either because it's local, it has a commitment to the environment, or other causes you and the artist care about. Some newer merch companies have social impact business models in which they not only ensure safe **no-sweatshop** manufacturing conditions and have options to use **eco-friendly** materials, they might also contribute a percentage of sales to philanthropic causes and/or commit to **offsetting estimated emissions** related to operating their business.

EVENT MERCHANDISE OPERATIONS

Separate from artists' merchandise teams, often music festivals have their own **festival merchandise companies** – given the complexity of managing merchandise on that scale, dealing with many artists arriving and performing and departing at different times throughout a festival.

In the industry, we often get caught up with people we are used to working with, focusing on companies that specialize in music goods, but that doesn't mean those are the only companies that independent artists and their teams can work with – especially at the start of their careers when they are still learning and working with local businesses. At any career stage, if an artist is committed to sourcing from local vendors to support a specific regional economy, they might be surprised to find some merch services could be handled in their own backyard! In almost every city and most towns, one can find a local screen printing and engraving companies. Many short-run local screen printers and engravers have long served youth and school sports teams, providing uniforms and awards and can source materials needed to generate merchandise. The difference working with these companies will depend on what volume of that work they are doing. Cost efficiency, after all, is a matter of scale. The more t-shirts a company sells, the less expensive t-shirts become for them and everyone they sell to, on a per-unit basis. Since these companies would be more accustomed to using fonts and designs for athletics, the artist working with a company like that would also need to provide more creative direction than they might with a music merchandise company. The benefit, however, is that these companies are used to making short runs quickly, and some are able to screen print, embroider, and engrave, all on the premises.

DECISIONS THAT IMPACT PROFIT AND SUSTAINABILITY

Once a marketer identifies merchandise item(s) to make and sell, and they identify potential partners they may work with to manufacture their merch, they request a quote.

A quote request should include:

- Your name and contact information
- Your deadline/when you need the finished merch by
- Shipping address(es) for completed job (or if you are picking it up in person)
- Clothes: brand, fit, colors, quantity of each size, print locations, # of colors
- Posters: quantity, size, paper stock/weight, # of colors
- Artwork you wish to use
- Any questions you have

A manufacturing quote will include:

- A mockup of the item using your design on a stock image of the product
- Item description, color, size, quantity
- Unit price and extended subtotal
- Set up costs (fees charged for initial run)
- Up-charges for additional ink colors or other features
- Tax + estimated shipping

INVENTORY MANAGEMENT

Additional expenses related to merchandise that are often overlooked at the time of ordering but should be considered when deciding how many units of a product to manufacture include storage, supplies, and warehousing. Often marketers focus on the costs associated with manufacturing each item on a per-unit basis, but they forget that inventory management has its costs.

Will your merchandise be shipping to a single warehouse or multiple? Will you be selling merchandise in multiple countries? How will the merch get to the first location on the tour? Where on the tour will more merch be sent ahead to meet the artist on the road? Inevitably, the merchandising manager will face decisions about which products to make more of and which to allow to sell out of inventory.

Many artist teams develop a sense of priority around product availability and the duration for which certain products will be sold. The following are elements that impact those decisions:

- **Essential goods** (top priority to keep available inventory on hand)
- **Non-essential goods** (allowed to sell out, reorder when convenient/affordable)
- **Premium goods** (keep available, but in short, exclusive supply volume)
- **Evergreen products** (always available, career-long relevance)
- **Seasonal products** (available at certain times of year)
- Release or tour-related products (available for campaign period)
- One-off or vanity products

When ordering merchandise from a record label, music products are often ordered from a separate department from soft goods and other items. Ordering physical music products like CDs and vinyl records is usually done in multiples of "box lot" quantities – the number of units packaged per box. CDs are often packaged in boxes that fit 30 or 40 CDs, which are then in turn packaged inside a larger box that often holds 100 or 120 CDs. That amount is called the carton or case lot quantity. Vinyl records are often packaged in boxes that fit 30 or 50 records, which are then in turn packaged inside larger boxes that hold 100 or 150 records. If a marketer wishes to maximize their shipping efficiently and reduce unnecessary chargebacks to the artist's account, they will order music merch products in increments of those amounts.

It is important to consider merch needs well in advance of when they need to be in hand – whether ordering merch from a record label system, a merch company, or an independent reseller or manufacturer. Most companies require at least two weeks advance notice; some require four weeks, especially for newly offered products (as opposed to re-orders). This is important not only to ensure the artist's merch lands where it needs to be in time – whether it's for pre-order shipments, concerts, festivals, or other critical merch selling moments – but also to reduce costs. Too often budget is wasted on merch that needs to be shipped last minute, using expedited services, and so at a premium rate. Those shipping costs add up and eat directly out of artist profit. When they are incurred on the label side, while labels may have favorable shipping rates given the volume of products they ship daily, they are not usually able to be as attentive to cost on a per-shipment basis to monitor the premiums paid on merchandise shipped, and that part of the artist's recoupment bill can climb if merch needs are not managed proactively.

FORECASTING DEMAND AND REVENUE

Merchandising managers rely on sales and shopping cart analytics data in order to manage supply to meet future anticipated demand. This is done through a process called **merchandise sales forecasting**, which can be carried out using advertising, website, and tour sales data. The merchandiser leverages past sales statistics to develop benchmark assumptions and, using those assumptions, builds a spreadsheet forecast model to determine how much merch they might sell on a daily or weekly basis.

Some of the assumptions commonly used in merch forecasting are:

- **Conversion Rate** – As mentioned earlier in this chapter, this is the percentage of people who had an opportunity to purchase and did. In the context of music, this usually is used for live concerts or eCommerce and is calculated as follows.
    - *Concert Conversion Rate %* = # of People Who Came to the Show ÷ # of Sales
    - *Website Conversion Rate %* = # of People Who Came to the Site ÷ # of Sales
    - *Advertising or Email Conversion Rate %* = # of People Who Clicked on an Ad or a Link in the Email ÷ # of Sales Directly Attributed to Those Clicks

■ **Revenue Per Head** is calculated by dividing your total revenue of a show by the number of people attending in the audience. This calculation is usually done inclusive of ticket sales earnings and merch, but some tour managers and managers like to see this figure just for merch revenue so they can forecast without the variables present in performance fee deal structures that differ from promoter to promoter. In broader marketing communities, you will more often hear this term referred to as average order size (AOS) or average revenue per user (ARPU), but revenue per head is the term most often used in music by tour managers, managers, and merchandising companies.

Once a tour manager or manager has identified trends and benchmarks for these figures, they can use them, alongside known given data points for events and campaigns coming up, to estimate future performance. This is done in a forecasting and actualization process that is ongoing. First the merchandiser uses what they know to estimate what might happen, then they actualize what happens to build on what they know, and then they use that new information to estimate what will happen. Naturally, the more data that is collected, the stronger the assumptions become and the more accurate the forecast is.

FORECAST AND OPTIMIZATION PROCESS

This process not only helps the merchandiser forecast how much money can be made selling merch to manage expectations of the label, artist, and other team members. It also helps the artist's team estimate what quantities of merchandise items they should order at the start of a campaign and the dates they may need to order more or source more from warehouse and ship to the tour. For eCommerce and retail, it can help those teams identify which warehouses across the world need to have more or less inventory on hand to efficiently satisfy consumer demand.

MERCHANDISE SALES ON TOUR

One of the major revenue drivers for a tour can be merchandise. Some artists are heavily dependent on merchandise sales for income – some see merch as more as a little extra money that can be channeled back into the operation and value it more for the extra opportunity for their team to engage with fans in person to cultivate enthusiasm for the artist and build loyalty.

When selling merch at shows, pricing is usually set in advance, but on some tours can be altered in real time based on locally relevant information

FIGURE 17.12   *Forecast and Optimization Process*

the tour manager may or may not know before arriving at the venue. In addition to the content earlier in this text about how prices are set, add this layer of nuance to how merchandise is priced at shows. All of the rules of cost-based and consumer willing to pay prices still apply, but merchandise managers also consider additional factors in the moment and make decisions from experience about what they think will maximize merchandise earnings for any given night.

Prior to a tour, a manager will set parameters or a range in price that is acceptable. If the price a tour manager thinks is necessary is outside that range, the tour manager might contact the manager for approval – but most often those decisions are being made on the fly.

Amidst the logistics of things that need to happen at a concert, there is a moment in the schedule when the merch area needs to be set up. Marketers put a lot of thought into the presentation of merchandise at a show because a merch area is essentially a pop-up store – and how things are laid

Table 17.1 Pricing Deals at Venues	
Known in Advance	**Sometimes Learned on Premises**
Venue percentage – the amount of money the venue or promoter is going to retain as a fee for the opportunity to sell at the show	**Prices other artists are charging** – this most commonly occurs at festivals, or if the artist is on a multi-act bill or as an opener for a headline act they don't know well
Fee paid to seller – sometimes there is a separate fee to be paid to a merch seller, whether provided by the venue or if the artist is hiring someone to sell	**Knowledgeable seller** – sometimes an artist or tour manager will decide to pay a merch seller extra, or a bonus, if they appear to be a knowledgeable seller who might be incentivized by earning a bit more
Prices across tour – artists will strive for price consistency, or they may have a strategy of selling merch at shows or online for a lower price, depending on what channel of sales they want to incentivize	**Prices fans are used to paying there** – while one would think this would be known in advance, sometimes it is not evident to a tour manager until they land on the ground and see evidence that fans are used to paying more or less for merch in that location

FIGURE 17.13 *Merchandising Table*

out, signage, and the way items are displayed makes a huge impact on what people buy.

In a merchandise area or table, the seller has limited space and a limited amount of time to make as many transactions as possible. Buyers are motivated but easily distracted and confused if any step in buying becomes unclear. It is important to make efficient use of space and develop a streamlined sales process. The merchandise seller is the store's welcoming committee, sales person, teller, inventory stock handler, and security guard. It's a big job! While every team has their own process, most commonly merchandise is handled in this fashion:

1. **Counting in** is the process a merch seller conducts in order to know what amount of inventory of each item they started with.

2. They **set up the merch area** with a limited supply of display samples so customers can view an array of products and home in on what they want to purchase before they get to the front of the line. During setup, the merch seller also organizes the inventory around their selling area behind the merch table in a way that works for them to easily access the sizes and options customers request at time of purchase. It is also important to make it clear to fans how the merch line will flow, and anticipate the volume of people waiting to ensure they will be able to purchase in the time allotted between doors and the show starting, during intermission, and after the show before curfew or closing of doors.

3. They **sell the merch**, utilizing whatever tracking systems are preferred. Some sellers use tablets, mobile devices, and specialized software to process transactions. Some even offer the ability for fans to opt between

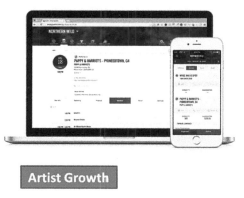

Welcome to modern tour and music business management in the cloud.

Forget installing computer-based software. As a cloud based platform, Artist Growth is everywhere your connected devices are. Plus, on our iOS and Android app your latest synced data is available from anywhere - on *and* offline.

Artist Growth

FIGURE 17.14 *Samples of App Interface of Merchandising*

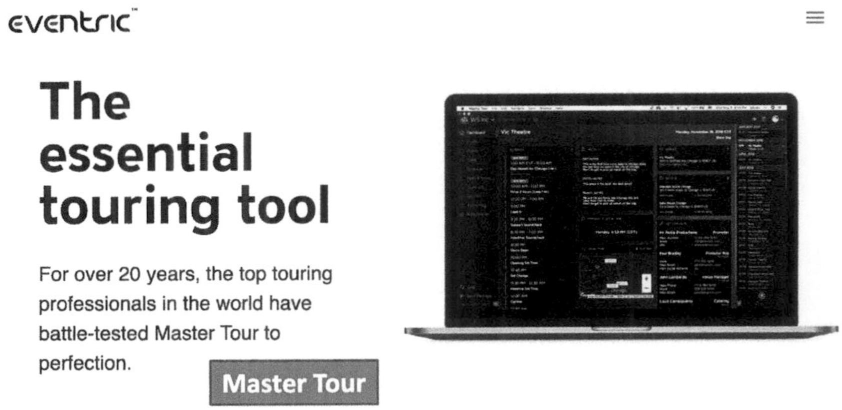

FIGURE 17.15 *Samples of App Interface of Merchandising*

FIGURE 17.16 *Samples of App Interface of Merchandising*

taking products with them at that moment or having them shipped so they do not have to carry merchandise around at the show.

4. They **settle the merch** after the show, doing a **count out** and calculating how much was comped, how much was sold, and what was earned in revenue for the night. Following is an example of a blank worksheet a merch seller might use to count their "register," account for inventory, compare that to make sure it matches, and calculate what is owed to the venue in way of the percentage of sales fee due to the house.

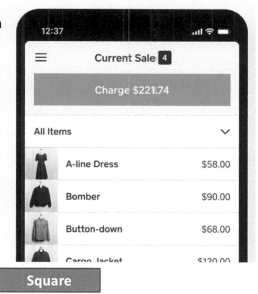

Powered by the free app that helps you run your whole business.

✓ Ring up customers quickly with an intuitive checkout experience

✓ Make smarter business decisions with real-time inventory and reporting.

✓ Easily add devices and team members as you grow, and keep sales synced to one account.

Learn more about Square Point of Sale ›

▶ See Square POS in action

 FIGURE 17.17 *Samples of App Interface of Merchandising*

5. A **settlement meeting** happens between the promoter and artist representative – usually the tour manager on mid-size to larger tours. Sometimes this is the artist themselves early in an artist's career and as they are developing. In that meeting, the promote reports final ticket sales to the tour manager, and the tour manager reports merchandise sales to the promoter (unless the promoter-controlled merch selling, in which case the promoter will report merch sales to the tour manager, too).

It is not unusual for there to be an extra round of negotiation about merch percentage in this moment, in light of how the show has gone. There is no obligation on the part of a promoter to alter anything they agreed to in writing in the show's contract, but at times they will make amendments in favor of the artist if they feel that is appropriate. For example, if there are still many tasks to be performed in settlement and they do not want to delay the settlement process by calculating every last sale before settlement can happen, a tour manager and promoter might agree on a per head merch buyout amount or a "buying out the house" amount. This potential occurrence re-emphasizes the importance of having a merchandise forecast in advance that the tour manager can rely on to make informed decisions like this on the spot.

Table 17.2 Sample Merchandise Settlement Sheet

Merch Settlement Sheet

Merch Sales	How Many?	How Much?
CASH	n/a	$
CREDIT CARDS		$
CHECK		$
GROSS EARNINGS	n/a	$
(-) VENUE	%	$
=NET EARNINGS	n/a	$

ITEM	UPC	PRICE (S)	START (#)	END (#)	COMPS (#)	SOLD (#)	SALES (S)	VENUE FEE ($)	ARTIST PAID ($)

GIG REPORT *(please complete this section after settling merch – so this information is not seen by public shoppers at the merch table)*

				ARTIST EXPENSES LOG
Date & Venue Name				
Address				Hotel:
Deal				Flights:
Capacity & Headcount				Car:
Ticket Price(s)				Meals:
#Tix Sold/# Comps				Other:
Earnings	Gross:	Artist:	Presenter:	

WHAT ELSE MAKES A DIFFERENCE IN TOUR SALES

At one time, the most impactful thing that could happen for sales at the merch table was the artist personally coming out to sign products. While that still happens, and it still has a huge impact on sales when it does, artist merchandisers and promoters realized that was in itself a monetizable event – so more often now artists are offering VIP fan packages and enhanced experiences. VIP meet and greet experiences also allow the artist's team to control the space and security around artist in-person interaction with fans. Whether an artist is willing to conduct these types of sales-boosting activities, it remains true that simply a mention from the stage that a merch area exists, and, even better, where it is, makes a huge impact on sales by making fans aware. If an artist is not comfortable selling their own wares in that way, a sponsor or person introducing the artist can sometimes make that announcement, depending on the show environment.

If fans feel the artist is personally connected to the merchandise in some way, it increases interest in the merch. Some artists have employed their friends, or they promote their relationship with the folks behind the table selling their merch, as a way to show a closer connection – and that type of communication on social media, if not in the show environment, can also increase traffic to the table, sales conversion rate, and average order size.

SELLING MERCHANDISE ONLINE

The other way artists sell merchandise is in online environments – primarily their website or a store connected with their official online presence. Merchandise can now also be sold through some of the music listening platforms via deals with merch companies.

The artist's website presents an opportunity to sell **exclusive** items that are not available anywhere else, as well as more variety than can be offered at a merch table, where there is limited space and restrictions around what amount of inventory can be carried. Some merchandise items don't travel well on a tour but can be offered online and promoted in person. That tie between the physical world and virtual world is so important, and sometimes it is **a lead generation opportunity** – a way to create interest offline and capture fan contact information online, which creates ability to follow up and re-market to fans at later dates, provided they opt in to communications.

Some artists entirely control their website, including the store there. Some artists control their website but not the store page. Usually when an artist does not control their web store, it is a "direct-to-consumer site" or "D2C" controlled by their record label or another partner they outsourced merch management and fulfillment to. A D2C store may be integrated seamlessly into an artist's website, or it could be an outbound link to a secondary site that fulfills for the artist. Either way, a transaction is occurring with critical information being exchanged. Most merchandising companies have a privacy policy in place that guarantees that customer information will not be sold to third parties. Additionally, most merchandising companies also agree that consumer information is proprietary to the artist, meaning that consumer data is captured for the artist's future use and will not be shared with other artists. The primary artist can mine the data to upsell for future events and merchandise to this active consumer. Anyone representing an artist in negotiations about direct-to-consumer deals should protect the artist's right to their fan data.

INCREASING AVERAGE ORDER SIZE

The fastest way to increase revenue from eCommerce is to increase average order size (AOS), because a strategy for that is simply based on selling more merchandise to the customers you are already reaching and communicating with. A common selling technique to increase AOS is to offer exclusive bundles – grouping products together and offering them all at a single packaged price. Sometimes a bundle price provides a discount compared to if the customer purchased the items individually, and sometimes it is the same amount of money but has emotional cache or convenience when offered together. Sometimes items are not available individually, and fans can only buy them as part of a bundle. Bundling not only increases average order size for online sales, but it also presents ability to tell a **marketing story** around products being promoted. Often bundles are named witty titles that match the artist's brand or the campaign or tour they are pushing, and promotional banners, social media graphics, and entire campaigns are waged around bundled packages.

INCREASING ORDER VOLUME

While increasing average order size may be as simple as altering prices and bundling items, increasing orders overall is the main objective for most marketers, and it is the goal merchandisers are constantly chasing and optimizing. Increasing order volume is influenced primarily by capturing **leads**, making contact with potential customers who are interested in the artist – and then generating **awareness** and **engagement** with those leads. Sometimes merchandisers leverage digital media, advertising and performance, and **affiliate marketing** to reach the artist's audience.

Another method to increase order volume is to constantly expand the variety and options of merchandise people can buy. Some artists update certain items within their merchandise seasonally, or even ongoing. Frequent new offerings and announcements about merchandise keep fans coming back to see what is available and keep the artist's merch top of mind. These merchandise selling strategies are planned in advance and **rolled out** tactically in key moments in campaigns.

For example, if an artist has an appearance at a large festival, they might make merch that will launch exclusively at that festival but then later be available on the website or through a retail partner. Some festivals are creating their own labels and merchandise lines in partnership with artists and their teams to keep selling their alignment over time after the festival happens.

LEVERAGING OFFERS

In online selling environments, marketers need to gather data to understand when their customers are price sensitive and when fans are willing to pay. Fans who are eager to support artists by purchasing merchandise directly are not usually incentivized by discount offers and deals. Other fans are motivated by a notion that they are getting a better deal right now than they can get tomorrow – or because they have access to something special in the moment. Savvy online marketers are constantly striving to identify potential customers and reach them in the right moment with a message that compels them to fulfill that potential and make a purchase.

The most common discount offers used in eCommerce are:

- **Dollar Off** (or pound, Euro, peso or yen . . .) – when an amount of money is deducted from the price or total order amount at checkout
- **% Off** – when a % of a price or total order amount is deducted
- **Free Shipping** – usually offered on orders over a specified total dollar amount, as an incentive for customers to increase average order size

The beauty of digital media is its trackability. Used wisely, offers can help you detect which media drives your sales. For example, by using uniquely trackable verbal codes when offers are made on the radio, you can track how many orders on your website came from radio announcements.

The only thing to be cautious about when employing offers to drive sales is not over-using them. There are many stories of companies who gave offers so frequently that customers learned to wait for discount codes to make purchases and did not purchase if there was no offer available. If you train your customers to believe offers will come soon, they might not buy unless there is an offer.

PRE-ORDER CAMPAIGNS

While pre-order campaigns were at one time a way independent artists funded their projects, professional companies like Kickstarter normalized pre-orders with music lovers and artist fans. In the early days of Kickstarter, major and mainstream artists would not offer pre-order campaigns because they felt it showed weakness, that they were not famous enough, or that they could not afford to do the things they wanted to do creatively. Consumer perception of the pre-order has changed a lot in recent years – and now it is simply seen as good business. Fans are used to pre-ordering music and/or merchandise and in fact see pre-ordering as an opportunity to express their loyalty to an artist. To honor this, artists often offer their fans exclusive products that are only available during a pre-order period or early access to a merch item.

VIP PACKAGES

An entire cottage industry exists within VIP packaging for entertainment. VIP packages offer fans opportunity to purchase tickets in an exclusive location at the venue, with backstage access; meet and greets with the artists; a concert pre-package that includes gear wear, concert items, and exclusive activities that range from high-end dining to exclusive travel arrangements

FIGURE 17.18 *CID Entertainment VIP Packaging Sample*

Join fellow John Legend fans at the LVE Rosé Garden Party for tasty bites perfectly paired with LVE collection of wines, and a conversation with the man himself about his life, new music and more. All guests will take home limited edition merchandise, and enjoy the show from a premium seat. Looking to sweeten your summer tour plans even more? Go all in on a Meet & Greet experience.

Packages are on sale now!

JOHN LEGEND MEET & GREET EXPERIENCE
JOHN LEGEND LVE ROSÉ GARDEN PARTY EXPERIENCE FOR 2
JOHN LEGEND LVE ROSÉ GARDEN PARTY EXPERIENCE
JOHN LEGEND PREMIUM MERCHANDISE BUNDLE   FAQ

FIGURE 17.19   *CID Entertainment VIP Package Featuring John Legend*

VIP JOHN LEGEND MEET & GREET EXPERIENCE

- One (1) premium reserved seat in the first (5) rows
- Meet & Greet / Photo Opportunity with John Legend
- Pre-show sparkling rosé toast with John Legend
- Invitation to John's LVE Rosé Garden Party, featuring:
  - An interview with John Legend about the LVE Collection, his new music and his life
  - Two (2) complimentary glasses of LVE rosé
  - Appetizers selected to perfectly pair with the LVE collection of wines
  - Cash bar featuring LVE varietals*
- Exclusive John Legend merchandise gift
- Limited-Edition LVE travel wine glass
- One (1) rideshare credit**
- Official meet & greet laminate
- Merchandise shopping before general doors
- On-site experience host

*Local liquor laws apply

*LVE wines for sale where available

** One (1) per order

FAQ

FIGURES 17.20, 17.21, 17.22   *CID Entertainment VIP Levels of Experience*

JOHN LEGEND LVE ROSÉ GARDEN PARTY EXPERIENCE FOR 2

- Two (2) premium reserved seats in the first (10) rows
- Invitation to John's LVE Rosé Garden Party, featuring:
  - An interview with John Legend about the LVE Collection, his new music and his life
  - Four (4) complimentary glasses of LVE rosé
  - Appetizers selected to perfectly pair with the LVE collection of wines
  - Cash bar featuring LVE varietals*
- (2) Exclusive John Legend merchandise gifts
- (2) Limited-Edition LVE travel wine glasses
- (2) Official VIP laminates
- Merchandise shopping before general doors
- On-site experience host

*Local liquor laws apply

*LVE wines for sale where available

FAQ

JOHN LEGEND PREMIUM MERCHANDISE BUNDLE

- One (1) premium reserved seat
- Exclusive John Legend merchandise gift
- Limited-Edition LVE travel wine glass
- Commemorative ticket
- Merchandise shopping before general doors
- On-site experience host

FAQ

FIGURES 17.20, 17.21, 17.22 *(Continued)*

and exotic accommodations. Depending on the arrangement, if the merchandising company owns the merch, the VIP company has to purchase the items from the company and *not* make new items elsewhere. CID Entertainment is one such company that offers a variety of getaways that include travel to festivals, sleeping in a yurt, and access to artists.

PROTECTING BRAND EQUITY: KNOCKOFFS AND UNOFFICIAL MERCHANDISE

One of the biggest challenge artist teams face as an artist's career develops and there is real marketplace demand for merchandise is unauthorized fan-made art and merchandise. It is a double-edged sword in that some would say the fan enthusiasm should not be stifled, but managers struggle with the amount of revenue others earn using the artist's name and likeness in unofficial merch. It has become easier for user-generated merch to be sold without any connection with artists or labels, through the popularity of sites like Etsy and others.

Here are some examples of the confusion in the marketplace around merch:

Ads · Shop mavis staples merch

Mavis Staples Classic T-shir...	Mavis Staples - t-shirt	TeePublic | Mavis Staple...	Mavis Staples Classic T-shir...	Mavis Staples Studio Pop A...
$19.90	**$16.48**	**$22.00**	**$19.90**	**$19.90**
Redbubble	Etsy	TeePublic	Redbubble	Redbubble

FIGURE 17.23   *Unauthorized Merchandise*

MAKING AND SELLING MUSIC MERCHANDISE

Merchandise is a critical revenue stream and fan engagement stimulator for artists. To successfully capitalize on merchandising opportunities, creativity and having a strong sense of the artist's brand and fans is important. In merchandising, marketers must develop and navigate many relationships, plan in advance, persistently analyze sales data, develop sales processes, and manage inventory.

GLOSSARY

Aesthetics—Philosophy pertaining to beauty and taste

Affiliate marketing—Performance-based advertising in which a seller offers partners a commission or bounty on sales or leads driven by their efforts on the seller's behalf

Average order size (AOS)—Total sales revenue divided by number of orders

Average revenue per user (ARPU)—Total sales income divided by number of unique users acquired

Awareness—Building consumer knowledge of an artist, product, tour, or event

Box lot—The number of units packaged per box (also sometimes called "box quantity")

Breakage—The rate at which products are damaged and either unsellable or in need of repairs before they can be sold (expressed as a percentage)

Carton lot—The number of boxes packaged per carton (also sometimes called "case lot" or "carton quantity")

Conversion rate—The rate a product sells when fans have an opportunity to purchase (expressed as a percentage)

Digital products—MP3s, MP4/MOV, on-demand streams

Direct-to-consumer (D2C)—The practice of artists selling products directly to fans without a retail store or distributor involved

Eco-friendly—Environmentally conscious

Embroider—When stitching is used to add text or images to soft goods (like hats, shirts, or jackets)

Engagement—The rate and frequency of fans responding to messages or advertising (often measured in clicks, shares, comments, time on site, pages per user, duration of video watched)

Engrave—When etching is used to add text or images to hard goods (like plaques, trophies, jewelry, flasks, travel mugs, metal cups)

Essential goods—Top priority to keep available inventory on hand

Evergreen products—Items artists always keep available that they sell year after year and have career-long relevance

Festival merchandisers—Companies that provide merch sales and logistics services to festivals

Fit—Style of a shirt, sweatshirt, or other soft goods product

Fleece—Not always consistently used, this term is sometimes used to refer to standard sweatshirt material and sometimes to refer to its fuzzier polar fleece cousin

Fulfillment—Processing, stuffing, handling, and shipping customer orders

Heather—A color created by mixing multiple colors or shades of the same color, creating a textured effect that can be overt or subtle depending on the colors mixed

Inventory—Quantity of products available for sale

Label services—Ad hoc services provided by record labels to artists and their teams, usually beyond the core scope of the artist/label relationship

Lead generation—Marketing focused on acquiring contact information of potential new customers, making contact with the hope of selling to them in the future

Manufacturer—Company that assembles raw and stock goods to make products for sale

Manufacturing quote—Estimate of costs related to manufacturing products

Material—What is used to make the fabric that is used to make products, usually expressed in terms of percentages (example 100% cotton or 60% cotton/40% poly, new alternatives some people are using include bamboo, hemp, and viscose)

Merchandise (noun)—Products that are sold

Merchandise (verb)—To promote and sell goods

Merchandise company—Organization that an artist, manager, or label hires to handle the coordination, project management, design, manufacturing, fulfillment, and inventory management of their merchandising activities

Merchandise sales forecasting—Spreadsheet or data model used to leverage learning from past sales to predict future sales

Music products—CDs, vinyl records, cassette tapes, sheet music

Non-essential goods—Allowed to sell out, reorder when convenient/affordable

Offset emissions—An environmental term referring to the practice of reducing carbon dioxide and greenhouse gases or seeking alternative solutions to make up for energy consumption in the process of manufacturing goods

Optimization—Leveraging previous learnings to improve future performance while striving for best possible results

Premium goods—Keep available, but in short, exclusive supply volume

Profit—Income remaining after expenses are paid/deducted

Quote request—Form sent by seller to manufacturer to provide details of what they want to sell, so manufacturer can provide an estimate

Raw goods—Materials and unfinished goods to be used to make a product

Reorder—Purchase more of an item that has been sold before

Resellers—Companies that work with many different manufacturers on behalf of sellers, to make it possible for sellers to produce many different products but work with one vendor

Revenue per head—Total sales revenue divided by number of tickets sold (fans at show)

Roadworthiness—Resilience of a product in the face of the rigor of touring, both in terms of having low breakability while being easy to transport and sell

Screen print—A process of printing using metal plates, screens, and stencils to spread ink on cloth materials or paper, with each color being applied in separate steps

Seasonal products—Available at certain times of year

Settlement—Calculating the earnings at the end of a concert and meeting to form an agreement between promoter and artist representative (usually when the artist gets paid)

Soft goods—T-shirts, sweatshirts, bandanas, tote bags, and similar

Sustainability—Ability to continue to do something for a long period of time

Sweatshop-free—The company guarantees that garment workers were not forced, abused, or taken advantage of in the making of products, confirms that compensation to workers was fair

Triblend—Fabric made of three materials (usually cotton, polyester, rayon)

Vendor—Company hired by another company to provide a service

Warehouse—Location where inventory is stored before being sold

Wholesale—When goods are sold at lower than market cost with the expectation that they will be re-sold to the public at a higher price – in merchandising, this can refer to raw goods being sold to manufacturers, or it can refer to the price distributors pay before selling finished goods to retailers

REFERENCES

https://artistgrowth.com.

https://friendlyarcticprinting.com.

https://squareup.com/us/en/hardware/contactless-chip-reader.

www.atvenu.com.

www.atvenu.com/features/year-in-review-2019.

www.bellacanvas.com.

www.cidentertainment.com.

www.crownmerchandise.com.

www.eventric.com/master-tour-management-software.

www.google.com/search?q=mavis+staples+merch.

www.hellomerch.com.

www.iamyola.com/store.

www.kungfunation.com.

www.tate.org.uk/art/art-terms/a/aesthetics.

Index

Note: Page numbers in *italic* indicate a figure and page numbers in **bold** indicate a table on the corresponding page.